Arctic Animals

Arnoldus Schytte Blix

Arctic Animals

and their Adaptations to Life on the Edge

tapir academic press

ISBN 82-519-2050-7

Layout:	Laboremus Prepress AS
Cover Design:	Tapir uttrykk
Printing and binding:	PDC Tangen
Cover photo:	The author with polar bear cub at Point Barrow, Alaska (Photo: P. E. Reynolds).

Tapir Academic Press
N–7005 TRONDHEIM
Tel.: + 47 73 59 32 10
Fax: + 47 73 59 32 04
E-mail: forlag@tapir.no
www.tapirforlag.no

Preface

This book was written out of the frustration of having had to teach students in arctic biology for 25 years without a proper book to give them an introductory overview of the arctic region and its animals. – A book that could give them the answers to questions like: Where and how is the Arctic, what kind of animals live there, and how are they adapted to their environment? The opportunity to get started came with a sabbatical leave to the University of Cambridge, during which I had the great fortune to be accepted as a visiting fellow at St. Catharine's College, with additional dining privileges in St. John's College, while working at the Scott Polar Research Institute, with the financial support of the Norwegian Research Council. The result of all this is a book which is meant as an introduction to arctic biology for the undergraduate student at universities and elsewhere. However, it should also be of some value as a reference book to researchers with an interest in the Arctic. It should require little more background than an introductory course in biology at college level and should therefore also be of some interest to the educated arctic traveller.

The book is in many ways based on my own experiences and observations from more than 30 years of travel throughout the Arctic, and, on first hand and personal contact, either in the field, or in the laboratory, with almost all the true arctic mammals and birds. However, in writing a book which spends the entire animal kingdom, it has, of course, been necessary at times to lean, sometimes heavily, on published material outside my own special field of interest and expertice. This has been the case, in particular, with regard to geophysics, meteorology, oceanography and plant biology, which are necessary to be able to put the animals into the right environmental context, but also with regard to invertebrate biology, of which I am personally unfamiliar. Fortunately, several excellent books already exist in these specialized fields, and reference have been given to those I have found particularly useful, so that the reader can gain easy access to the vast scientific literature in these fields. In my own field of physiological adaptations, where few, if any up to date textbooks on Arctic animals are available, on the other hand, a number of references to original data are given.

In writing this book I have found it necessary to limit the text to the "true" arctic animals, which in my mind are those that have winter residency and their main area of distribution within the arctic region. This, unfortunately, excludes a great number of species that extend into the Arctic, while having their main area of distribution outside the region.

During the years I have spent working on this book, my companion, Turid Steen, has attended to most of my domestic affairs and thus allowed me to keep most of my attention on my work, for which I am grateful. I am also grateful to Sir Terence English and my old friends Richard Laws and (the late) Peter Jewell for opening the doors to college life to me, and to my new friends, John A. Crook and G. Clifford Evans for a series of excellent meals and most stimulating academic discussions, and to Robert K. Headland for a lot of fun, during my year in Cambridge.

During the preparation of the manuscript for this book I have benefited from advice and criticism from a number of colleagues, of which I am particularly thankful to : Terry V. Callaghan, Bjørn Folkow, Lars P. Folkow, Erling Nordøy, Egil Sakshaug, Karl-Arne Stokkan, Monica Sundset, John M. Terhune, Lars Walløe, David Walton and Karl Erik Zachariassen. Their effort has greatly improved the book, but they are not responsible for the errors which remain. I am also thankful to Rod Wolstenholme and Kjell Lund for help in the production of the many illustrations, and last, but not least, to my secretary, Elin Giæver, who has served for many years in the "war zone" of my office, where she has typed the many versions of the manuscript with endless patience, skill and charm.

Arnoldus Schytte Blix
Department of Arctic Biology,
University of Tromsø

Table of Contents

CHAPTER 1

The Arctic Region

The Arctic is the region under Arctos, the polar star, but aside from that, there is no general agreement on the limits of the region. In a purely geographical sense, the polar circles which are parallels of latitude 23° 28' from the poles, defined by the angle between the axis of rotation of the earth and the ecliptic, or plane of earth's orbit about the sun (Fig 1.1), may be useful limits for the polar regions. Stonehouse (1989) among others has, for instance, pointed out the basic facts that day length (Fig.1.2) and amount of solar radiation received (Fig.1.3) are ecologically important factors that vary with latitude. In fact, solar radiation is the main source of heat for the earth's surface, the main driving force behind winds and oceanic currents, and the ultimate source of energy for most living creatures. The areas within the circles are also colder than the rest of the world because they receive less solar radiation per year, but, there is no sharp transition at the circles themselves. The sunless period in winter and the period of continuous daylight in summer both increase from one day per year at the polar circle to 6 months per year at the pole (Fig. 1.4). However, there is no notable change in radiation levels at 23° 28' north, and the circles themselves have no ecological significance; neither plants nor animals respond to them. Air temperature, on the other hand, provide indication of thermal stress, and is the simplest of several distinguishable indicators of environmental conditions. The most useful definition of the Arctic, in an ecological sense, is therefore the area enclosed by the 10 °C isotherm for the warmest summer month, normally July, in the northern hemisphere (Fig. 1.5).

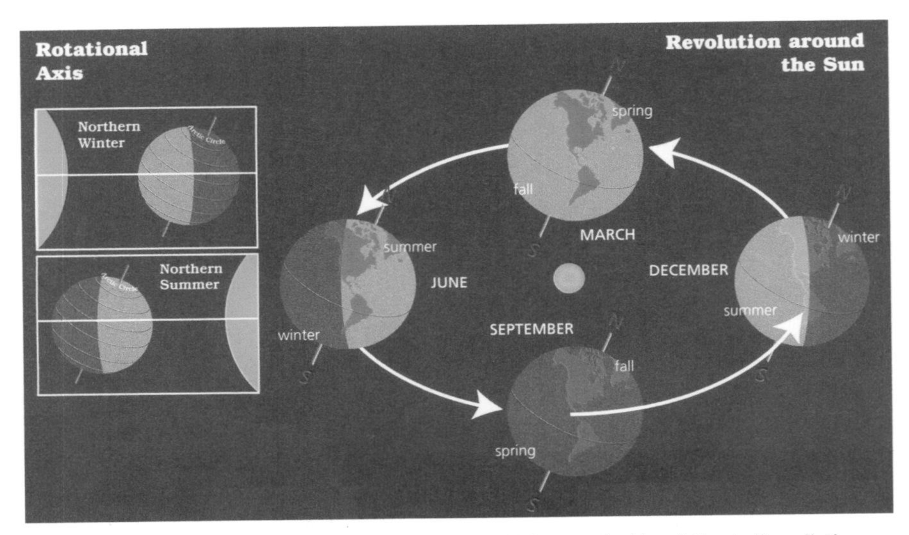

Figure 1.1 The earth's axis of rotation is tilted about 23° from vertical in relation to the ecliptic or plane of rotation of the earth around the sun. Resulting from the tilt, the earth receives differential illumination during its yearly circulation around the ecliptic. Regions poleward of the Arctic circle are fully exposed to solar radiation across midsummer and fully shaded across midwinter (Redrawn from Mac Quarrie, 1996).

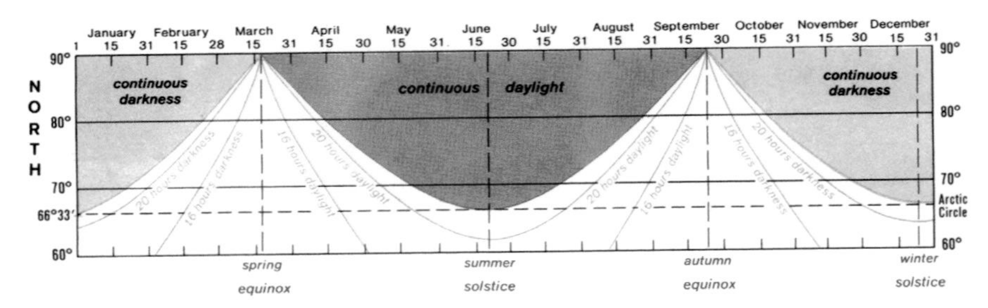

Figure 1.2 Duration of day and night in high northern latitudes (90°- 60°) (Anon., 1978).

In the Arctic this isotherm encircles the Arctic Ocean and includes Greenland, Svalbard, the northern 2/3 of Iceland and most of the northern coast and all off lying islands of Russia, Canada and Alaska. In the eastern Atlantic Ocean it is pushed anomalously northward by air masses associated with the North-Atlantic drift, or Gulf-Stream, to exclude most of Scandinavia. In the western Atlantic sector, on the other hand, the isotherm loops southward, influenced by cold sea and air currents from the polar basin to bring northeastern Labrador, northern Quebec and Hudson Bay into the Arctic. In the Northern Pacific

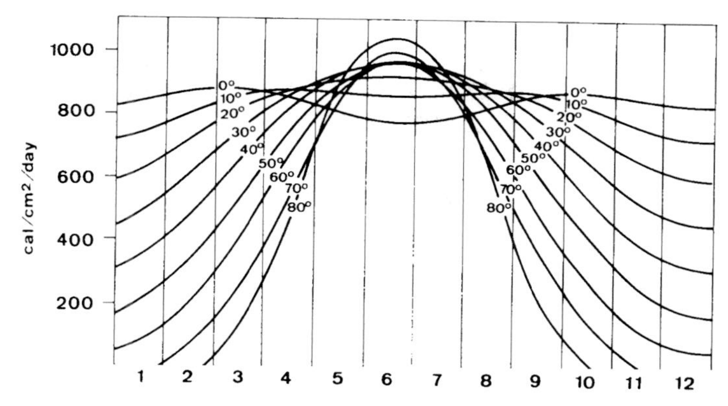

Figure 1.3 Insolation per day in the course of the year (abcissa) at different latitudes.

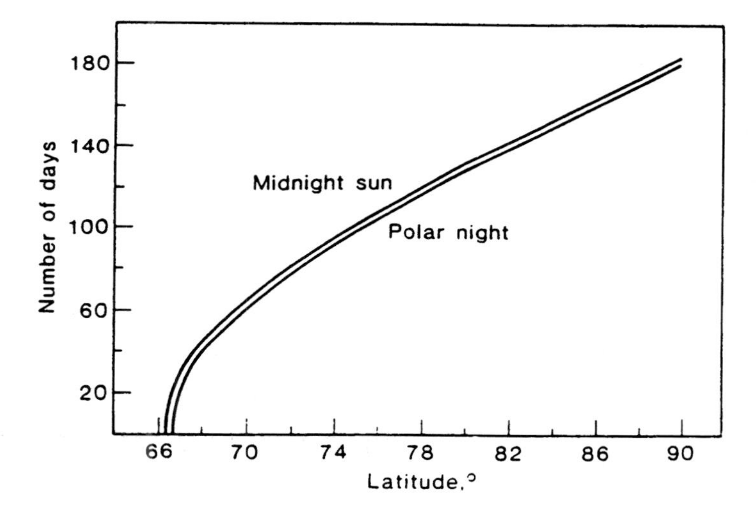

Figure 1.4 Duration of the periods of midnight sun and polar night at various latitudes.

similar currents push the isotherm southward to take in central Kamchatka and most of the Bering Sea (Fig. 1.5). The importance of the mean July temperature on the distribution of species is illustrated in Fig. 1.6.

The climatic elements that most concern living systems the world over are ground temperature, air temperature, wind, humidity and precipitation. All depend ultimately on the amount of solar radiation received. Polar regions are cold because they receive their radiation obliquely and reflect much of it away rather than absorbing it. A column of sunlight, shining on a tropical region,

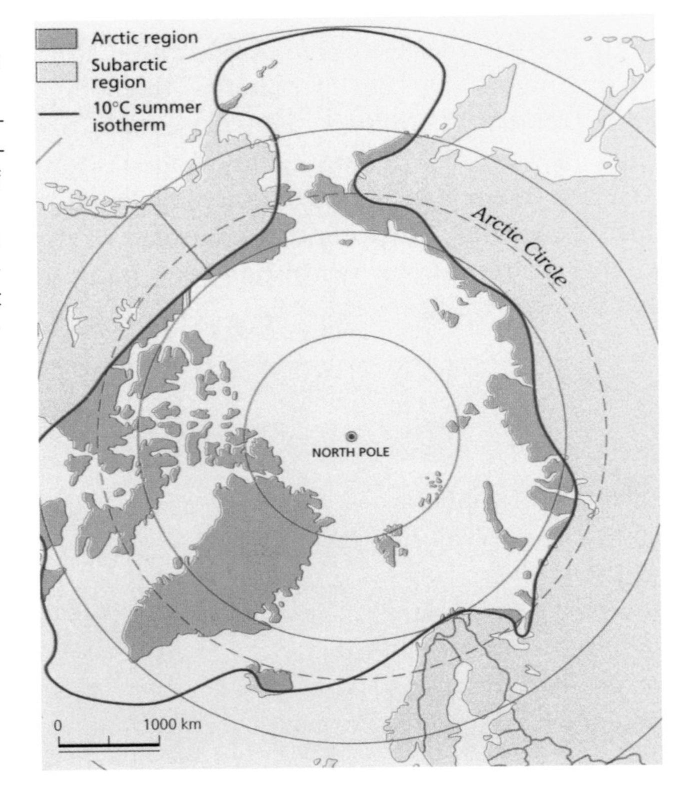

Figure 1.5 The 10 °C summer isotherm is a line through places that have an average daily temperature of 10 °C in the warmest month (usually July) of the year. The Arctic is defined as the area north of this isotherm, and arctic animals are those that have their main distribution inside that line in January (Redrawn from Mac Quarrie, 1996).

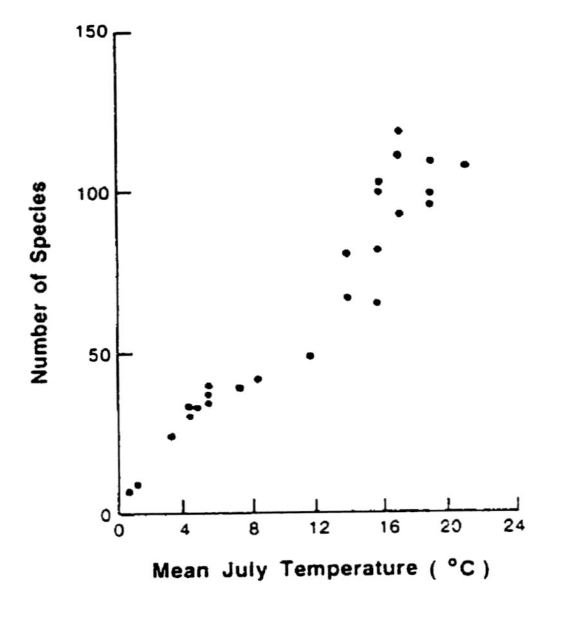

Figure 1.6 The relationship between number of nesting bird species and July mean temperature, western and middle Siberia (Chernov 1989).

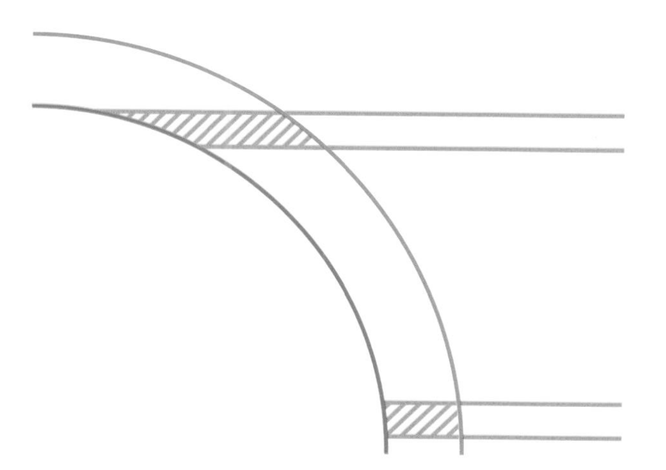

Figure 1.7 Solar rays that strike high latitudes pass at an angle, travel a longer distance through the atmosphere and spread over a wider area than those that strike at lower latitudes.

passes through a thin stratum of atmosphere and warms the ground beneath. A similar column shining on a polar region passes through a greater thickness of atmosphere (Fig.1.7), where more of the energy is reflected or absorbed, and spreads tangentially over a lager area; then it is mostly reflected away by snow and ice surfaces. Polar regions receive annually only about 40 % as much solar radiation as equatorial regions, and reflect away on average some 89-90 % of all they receive. The ratio of reflected short wave radiation to total short wave radiation received is the reflectivity or *albedo* (whiteness) of the surface (Fig. 1.8). Most of the energy it receives reflects and radiates back during the long summer days. Being warmer than space even in winter the polar regions continue to radiate energy through the sunless period when incomisng radiation is minimal, incurring a strong energy deficit. Spring thaw in the Arctic is initiated by surfaces warming from direct insolation, and the autumn freeze-up starts as daily insola-

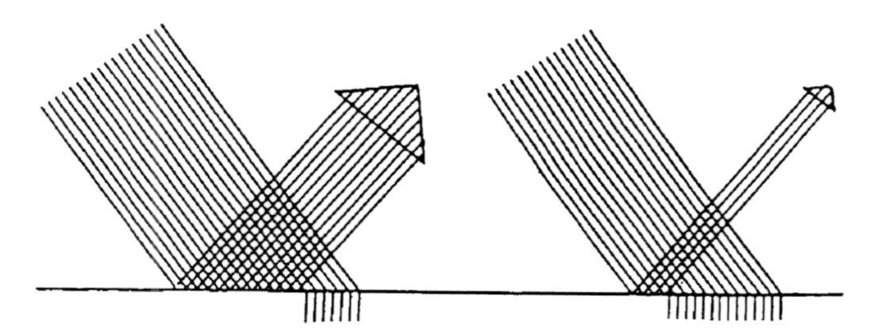

Figure 1.8 *Albedo.* With high albedo (*left*) much of the incident radiation is reflected away, and only a little remains to warm the ground. With low albedo (*right*) less is reflected and more absorbed; a lower total incidence can then result in higher ground temperatures (Stonehouse, 1989).

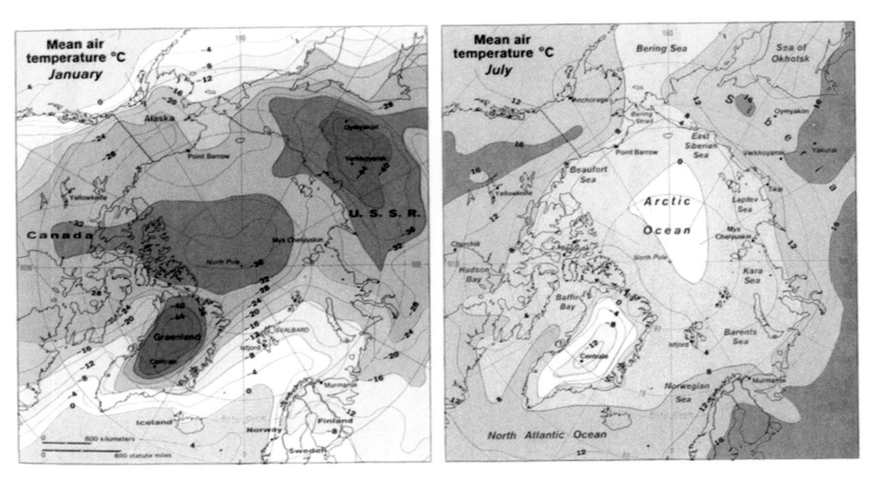

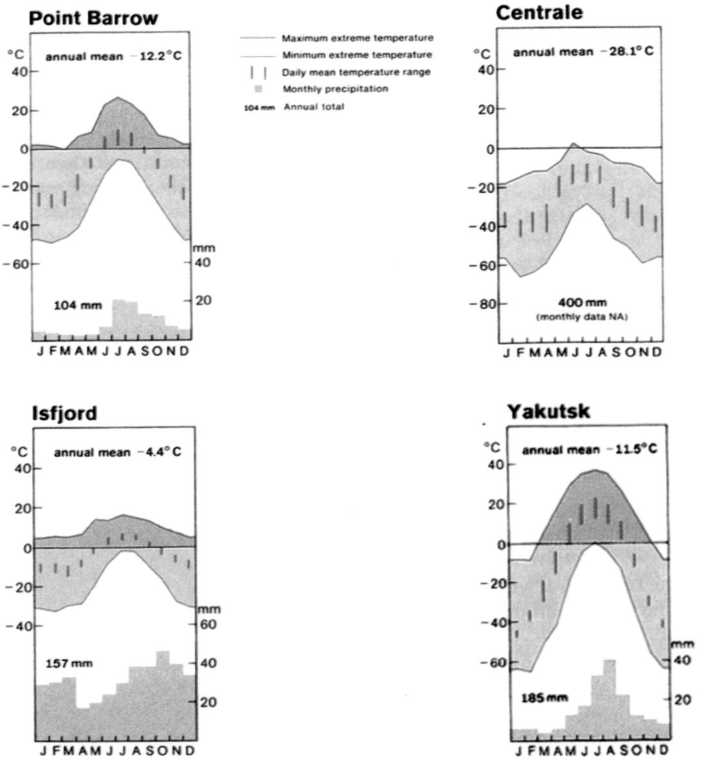

Figure 1.9 Mean air temperature in the Arctic in January (*top; left*) and July (*top; right*). *Bottom*: Climatic characteristics for Point Barrow, Alaska; Centrale, Greenland; Isfjord, Svalbard; and Yakutsk, Russia (Anon., 1978).

tion weakens. Because of their direct bearing on microclimates, the small scale climatic environments in which organisms actually live, insolation and radiation balance are particularly important factors in polar ecology.

Climatic records indicate that polar climate is typically cold, dry and windy, but the Arctic is large and its climate ranges widely from severe to subtemperate. These differences are due mostly to altitude, latitude and solar radiation, but distribution of land, open water and sea ice are important factors too (Fig. 1.9).

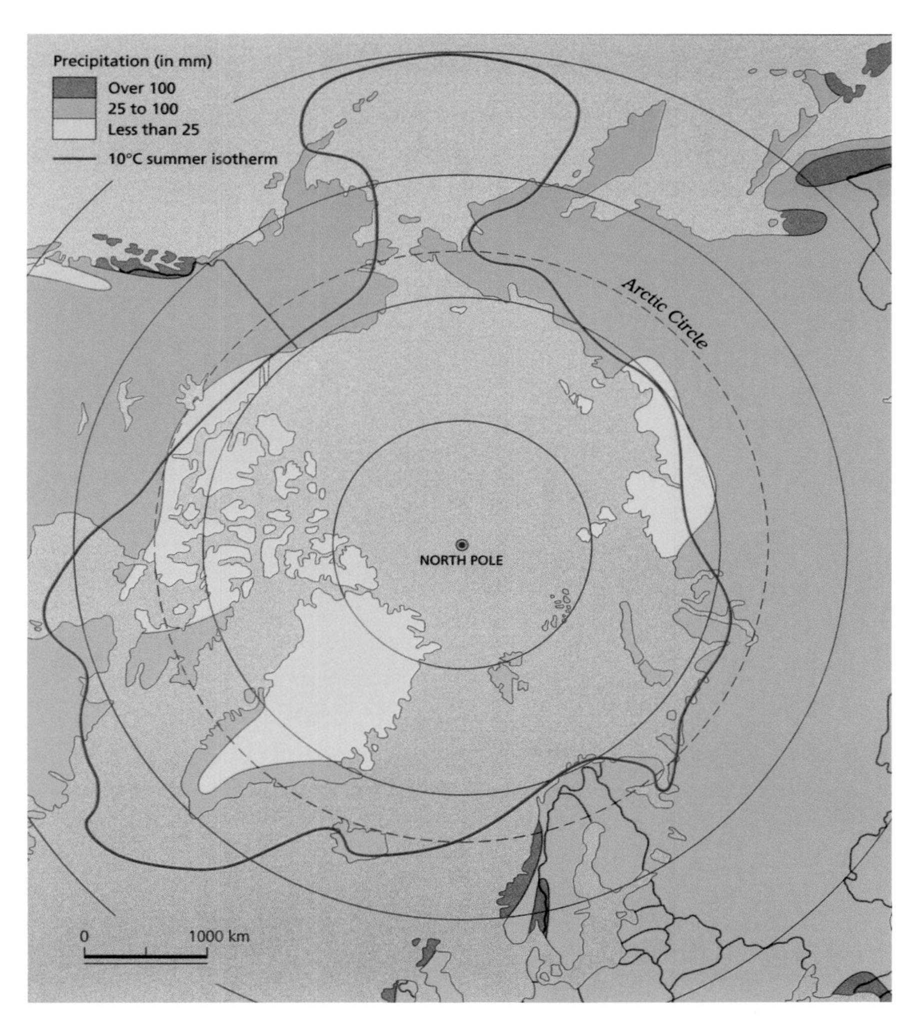

Figure 1.10 Precipitation in the Arctic (in July), is very low, mainly because water is covered by ice most of the year (Redrawn from MacQuarrie, 1996).

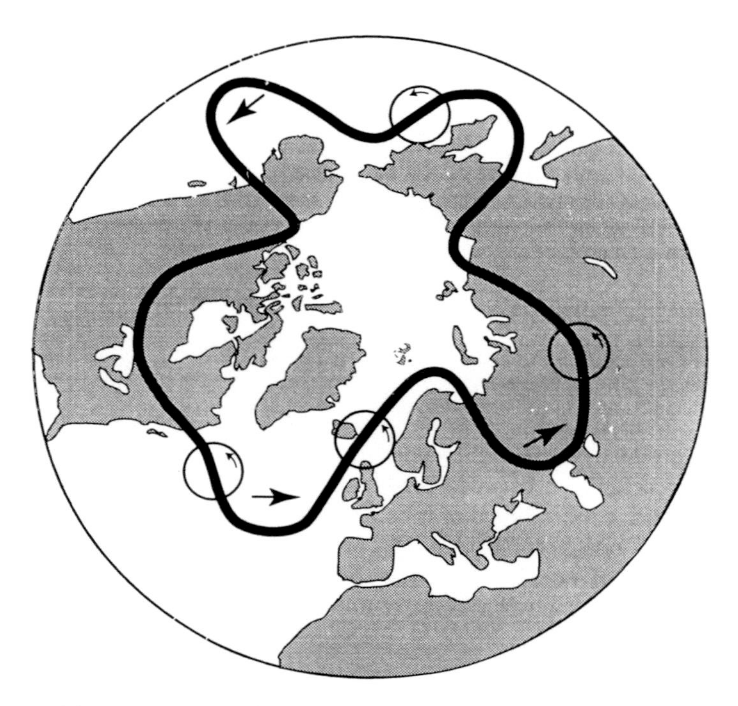

Figure 1.11 Schematic illustration of the jetstream (solid line) and atmospheric low pressures (circles) in the northern hemisphere. A low - pressure area may be more than 1000 km in diameter (Sakshaug, Rey & Slagstad, 1995).

Wind is caused by air pressure differences mostly due to differential heating of the atmosphere (systemic winds). Such winds include those prevailing in latitudinal bands – *westerlies* on fringes and *easterlies* closer to the pole.

The chronically cold air of polar regions contains very little moisture; though relative humidity may be high, absolute humidity is low, and the moisture available for living organisms is often meagre. Many polar areas are dry deserts; others are semi-arid for longs spells of the year, when much of the ground is frozen and neither rain nor snow are falling (Fig. 1.10). Throughout polar regions lack of available water may therefore be as limiting an ecological factor as extreme cold. Thus, precipitation is low in the high-arctic region of the polar easterlies, but significant at the fringes which are dominated by the westerlies that are associated with large variability and high storm frequency. These cyclones, or low pressure systems, occur along the atmospheric polar front or boundary between cold polar and warm air masses from lower latitudes (e.g. Rasmussen & Turner, 2003). In the Arctic they migrate eastward in circumpolar tracks at rates of 600 to

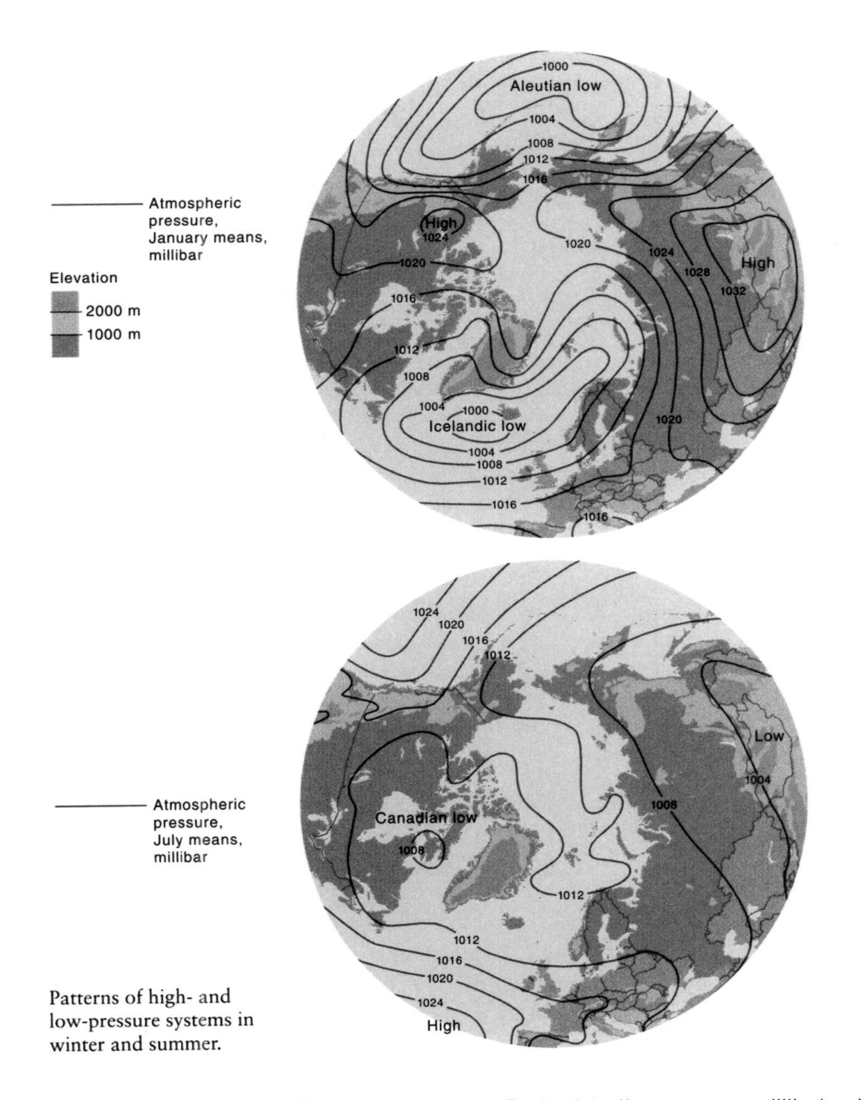

Figure 1.12 Patterns of high - and low - pressure systems. *Top*: in winter (January means, millibar) and *Bottom*: summer (July means, millibar) (Anon., 1997).

1,000 km per day (Fig. 1.11). Each depression brings a sequence of low cloud cover, rain or snow, and sharp wind shifts. These westerly winds have, as we shall see, an important impact on the mixing of the surface layer of the oceans, and hence on primary production. Their eastward movements are, however, diverted by other pressure systems. For example, persistent high-pressure areas keep them away from central Siberia and northern Canada in winter (Fig.1.12). The warm

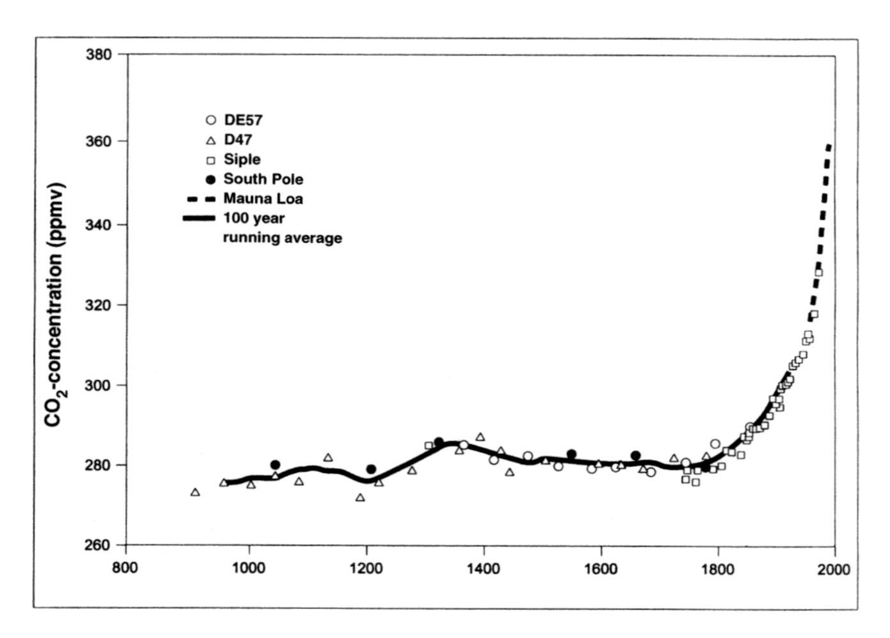

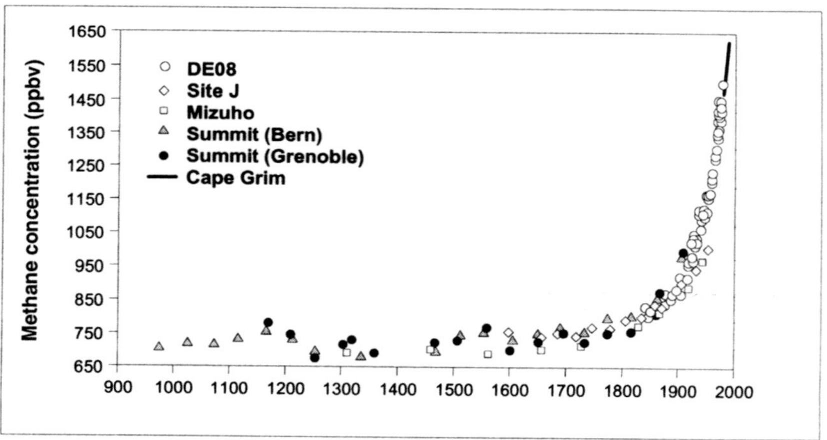

Figure 1.13 *Top*: CO_2-concentrations during the last 1000 years, as determined from ice-cores, and after 1958 in the air at Mauna Loa, Hawaii. *Bottom*: CH_4-concentrations during the last 1000 years, as determined from antarctic ice-cores (Houghton *et al.*, 1995).

air associated with the depressions brings mild winters and heavy snow fall that accumulates on the glaciers and ice caps. Anticyclones, or high pressure systems, form windows of clear air in the overcast, usually with a lifetime of two or three days, but occasionally lasting much longer. Their clarity allows free flow of radiant energy to and from the earth. In summer they therefore bring hot sunny weather, in winter clear days and intensely cold nights. The persistent winter anticyclones over Siberia and northwestern Canada (Fig. 1.12) make these the hemispheres' coldest and driest places. Thus, Verkhoyansk in Siberia holds the northern hemisphere record for extreme cold (-67.8 °C in January).

The atmosphere forms a protective greenhouse about the earth, letting in short wave energy but inhibiting the escape of the long wave energy derived from it. Without the atmosphere the earth's surface temperature would average 30-40 °C cooler, and vary far more diurnally and seasonally, like the surface of the moon. Several gases contribute to the greenhouse shield in the atmosphere, mainly oxygen, nitrogen and water vapour, but some that occur in low concentrations, such as CO_2 and methane, may also be important. Of these CO_2 is the most important of the gases that are released as a result of human activity, mainly as a result of combustion of fossil fuel. Concern is currently being expressed because the release, and hence, atmospheric concentrations, of CO_2 and methane have increased dramatically over the last 100 years (Fig.1.13), and the dangers of global warming as a result of it are much debated at the moment. Concern is currently also being expressed that the depletion of ozone over the Arctic (and Antarctic) is altering radiation balance by allowing more UV-radiation through to the earth's surface during polar summers. The depletion is due mainly to chlorofluorocarbons and other man made industrial gasses, which escape to the atmosphere and break down in the upper atmosphere to release radicals which destroy ozone (Fig. 1.14). These gasses are currently entering the atmosphere 5 times faster than natural processes are destroying them, with results which may well be environmentally harmful in the not so distant future.

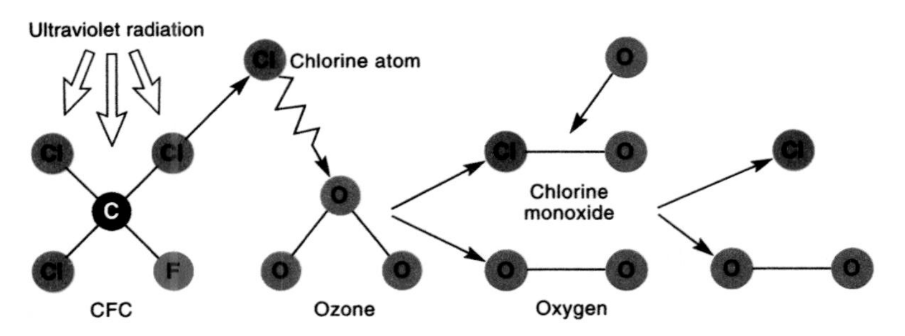

Figure 1.14 Atmospheric ozone destruction: When chlorine from stable man - made compounds, such as chlorofluorocarbons (CFC), reach the atmosphere the energy from ultraviolet radiation in sunlight splits off the chlorine atoms (Cl). The chlorine, which is very reactive, proceeds to attack ozone (O_3), splitting off one of its oxygens (O). The reaction leads to the formation of Chlorinmonoxide (ClO) and ordinary oxygen gas (O_2) (Anon. 1997).

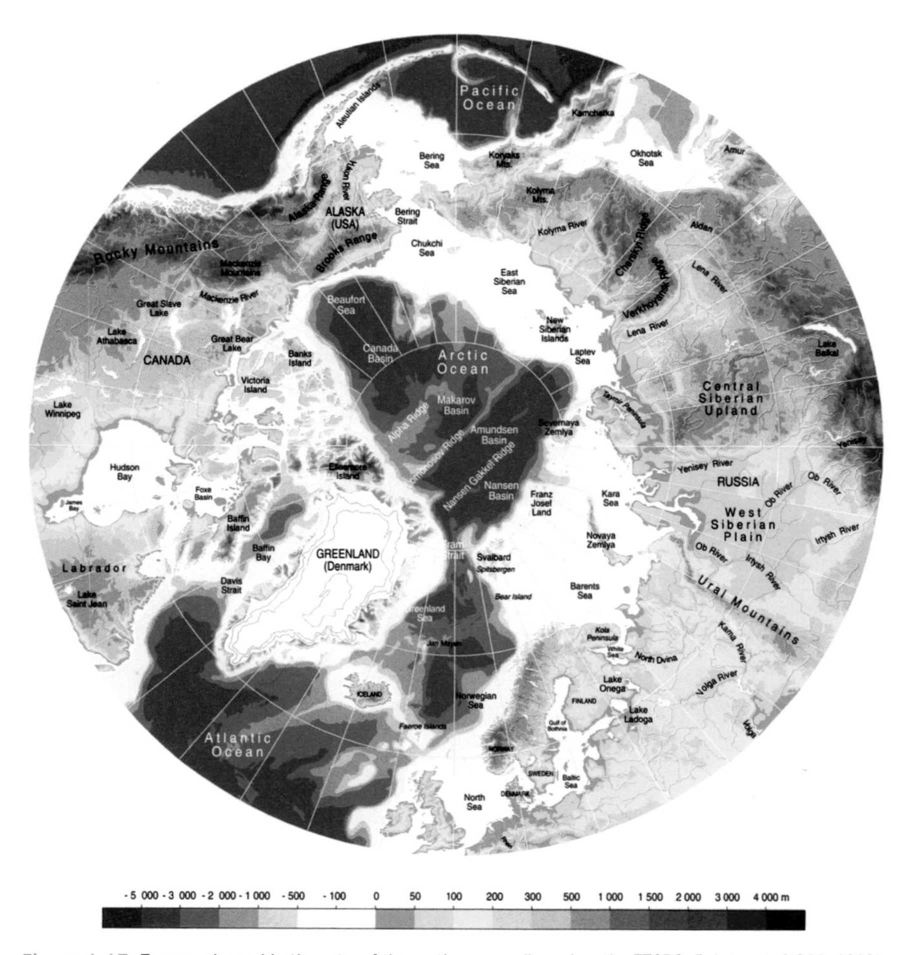

Figure 1.15 Topography and bathymetry of the arctic oceans (based on the ETOPO. 5 data set, NOAA, 1988).

The arctic oceans consist of the deep central Arctic Ocean around the North Pole and a number of rim shelf seas; the Barents-, the Kara-, the Laptev-, the East Siberian-, the Chukchi-, the Bering- and the Beaufort Seas. All these rim oceans are very wide and shallow shelf areas, with depths of only 20-60 m in the Chukchi-, and the East Siberian Sea, 10-40 m in the Laptev sea, about 100 m in the Kara Sea, and 100-350 m in the Barents Sea (Fig. 1.15).

The Arctic Ocean, with an area of about 10 million km^2 occupies a deep, complex basin. This basin is divided in two by a submerged mountain range, the Lomonosov Ridge, which runs between northern Greenland and the New Siberian Islands and raises from a seabed some 4,000 metres deep to within about

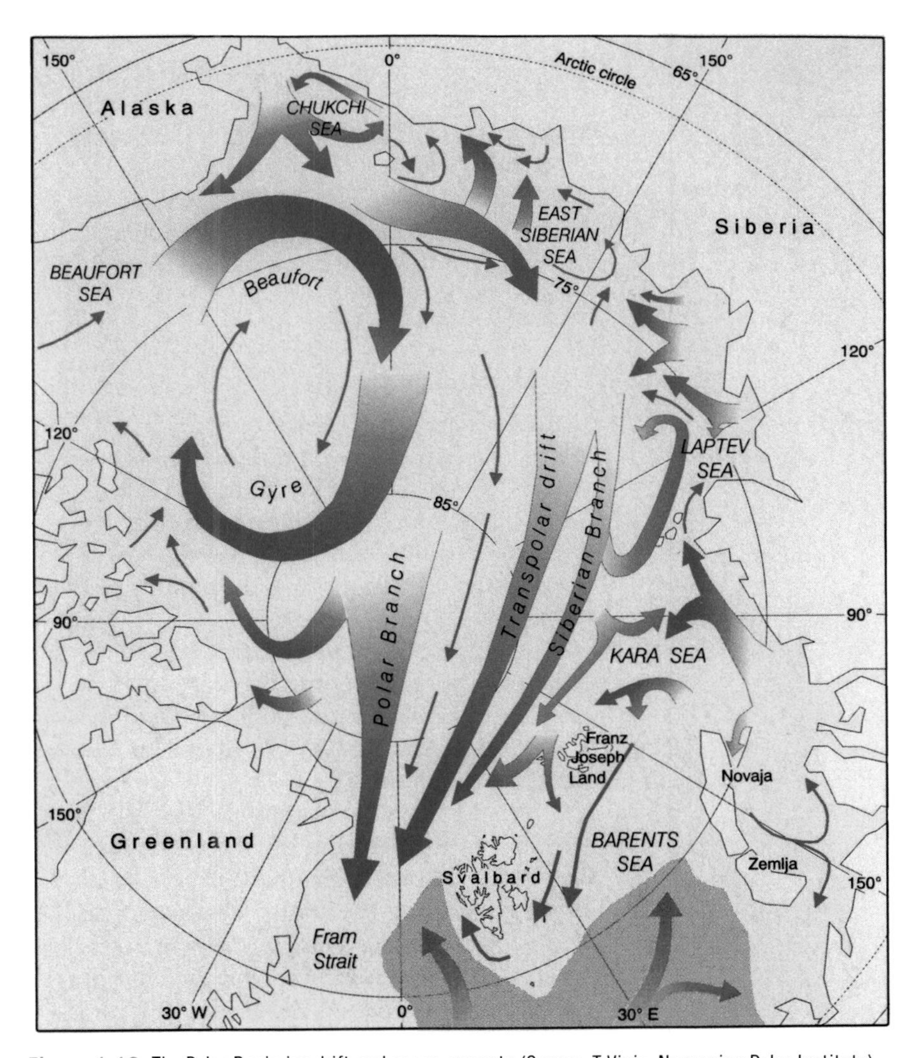

Figure 1.16 The Polar Basin ice drift and ocean currents (Source: T.Vinje, Norwegian Polar Institute).

1,000 metres of the surface. The Eurasian side is further divided into the Amundsen Basin and the Nansen Basin by the Nansen-Gakkel Ridge, while the Canandian side is divided into the Makarov Basin and the huge Canada Basin by the Mendeleyev- and the Alpha Ridges (Fig. 1.15).

Two main features characterize the circulation of the surface water in the Arctic Ocean; the Beaufort Gyre and the Transpolar Drift (Fig. 1.16). The Beaufort Gyre is a large clockwise gyre extending over the entire Canadian Basin. It

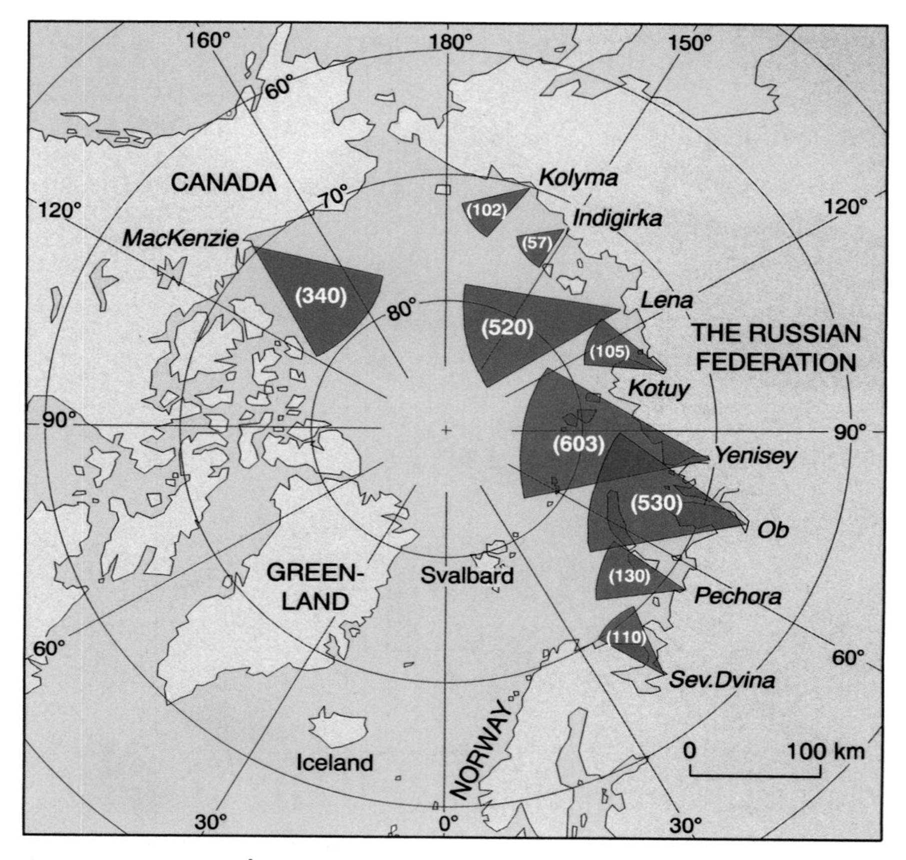

Figure 1.17 Volumes (km^3) of freshwater annually entering the Arctic Ocean from major rivers (Aagaard & Carmac, 1989).

circulates slowly between the pole and the Canadian Archipelago. In this way, water is exported both to Baffin Bay through the Canadian Archipelago and to the Transpolar Drift. The Transpolar Drift runs east to west across the Eurasian Basin from the Siberian coast out through the western Fram Strait. Our understanding of the surface circulation in the Arctic Ocean is derived from drift tracks of research stations on ice-islands and ships trapped in the ice either accidentally or deliberately. The most famous of these crossings of the polar seas was the drift of the research vessel "Fram", built for the Norwegian explorer Fridtjof Nansen for his 1893-1896 expedition (Nansen, 1897).

Tracer studies show that about 10 percent of the water in this current comes from the great northward flowing rivers of Siberia and the smaller ones of North America (Fig. 1.17), which bring year-round flows of water from the south. They

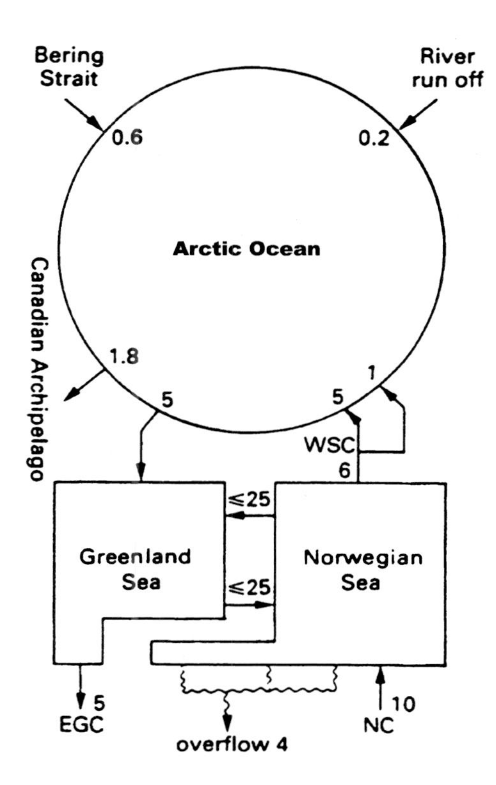

Figure 1.18 Mass budget for the Arctic Ocean. EGC: East Greenland Current, NC: Norwegian Current, WSC: West Spitsbergen Current. Volume transports are in Sv (10^6 m^3/sec.) (Tomczak & Godfrey, 1994).

bear thick ice in winter and their flow is strongly seasonal. Thus, 35 % of the annual flow of the Lena in Siberia is delivered during the peak month of June, less than 1 % in each of the three months of February, March and April. This flow of fresh water reduces the salinity of the shallow Siberian rim-oceans substantially and contributes to the almost permanent ice-cover of these seas.

A mass budget for the Arctic Ocean and the adjacent seas has been suggested by Tomczak and Godfrey (1994). The Arctic Ocean receives water from the Norwegian Current, and from the Bering Sea, through the Bering Strait, as well as from the above mentioned river run-off. On average, the same amount of water must leave the Arctic Ocean through the East Greenland Current, and through the channels in the Canadian Archipelago. The Norwegian Atlantic current transports about 10 Sv* of Atlantic Water northward to the Norwegian Sea, of which, on average, some 4 Sv goes back to the Atlantic Ocean, as outflow of Arctic bottom water across the Greenland-Iceland-Scotland Ridge. Of the remaining 5-6 Sv, the

* The unit, 1 Sverdrup (Sv), is named after the Norwegian oceanographer, H.U. Sverdrup and is defined as: 10^6 m^3/sec.

West Spitsbergen Current carries, on average, about 3-5 Sv into the Amundsen and Nansen Basins, while 1 Sv flows through the Barents Sea and enters the Arctic Ocean between Franz Josef Land and Novaya Semlya. Transport into the Arctic Ocean through the Bering Strait ranges between 0.6 Sv in winter and 1.1 Sv in summer, while river run-off is estimated to 0.2 Sv . Outflow in the East Greenland Current is estimated to, on average, about 3-5 Sv , which includes 0.1-0.2 Sv of meltwater and ice, while transport through the Canadian Archipelago into the

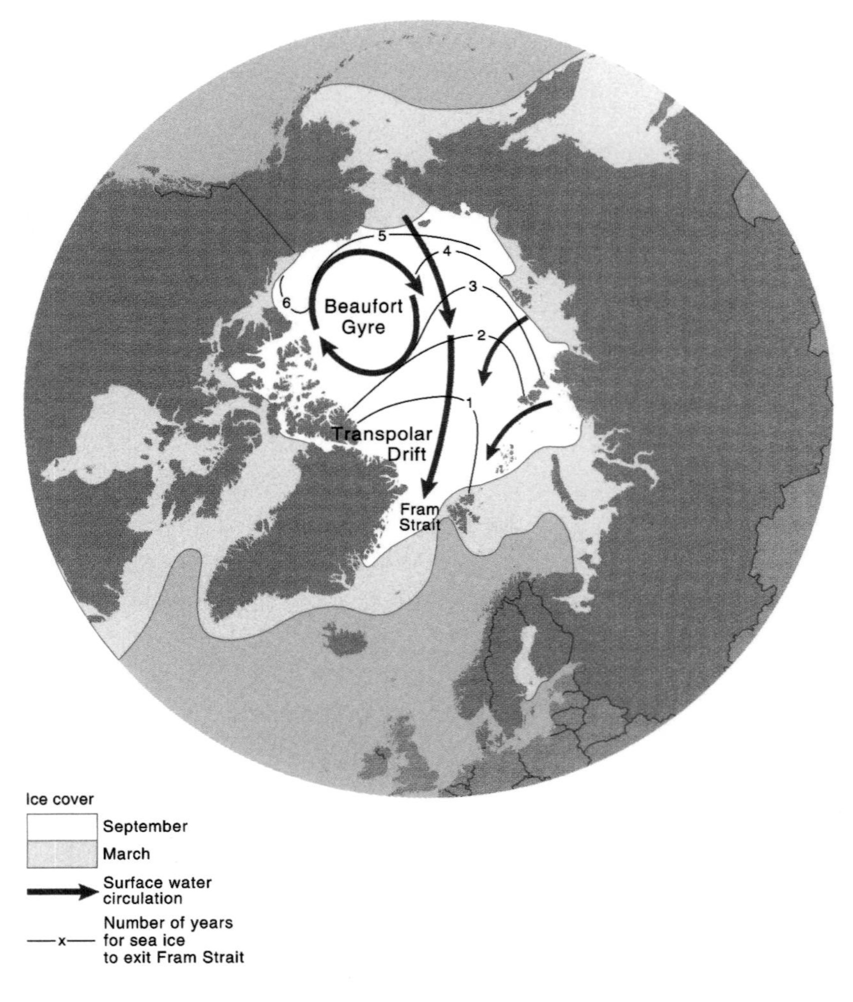

Figure 1.19 Sea ice cover in September and March and the major surface currents governing the transport of sea ice. The numbered lines show the expected time in years for the ice at that location to exit the Arctic Ocean through the Fram Strait (Anon., 1997).

Atlantic Ocean via Baffin Bay is estimated to in the order of 1-2 Sv. A summary of the mass budget is presented in Fig. 1.18.

The Arctic Ocean has at its centre a permanent cover of interlocking ice-floes slowly circulating with the ocean surface currents, most of them several years old and many 3 to 4 m thick. It is estimated that the trans-polar drift of ice takes about 5 years (Fig. 1.19).

Surrounding the core is a zone of seasonal pack ice (Fig. 1.20). The seasonal pack ice grows to about 1 m thick and disperses annually. Sea ice is important ecologically, providing a substrate for algae to grow and for seals to breed on. The ice distribution of the arctic oceans is remarkable in its extreme southward extension

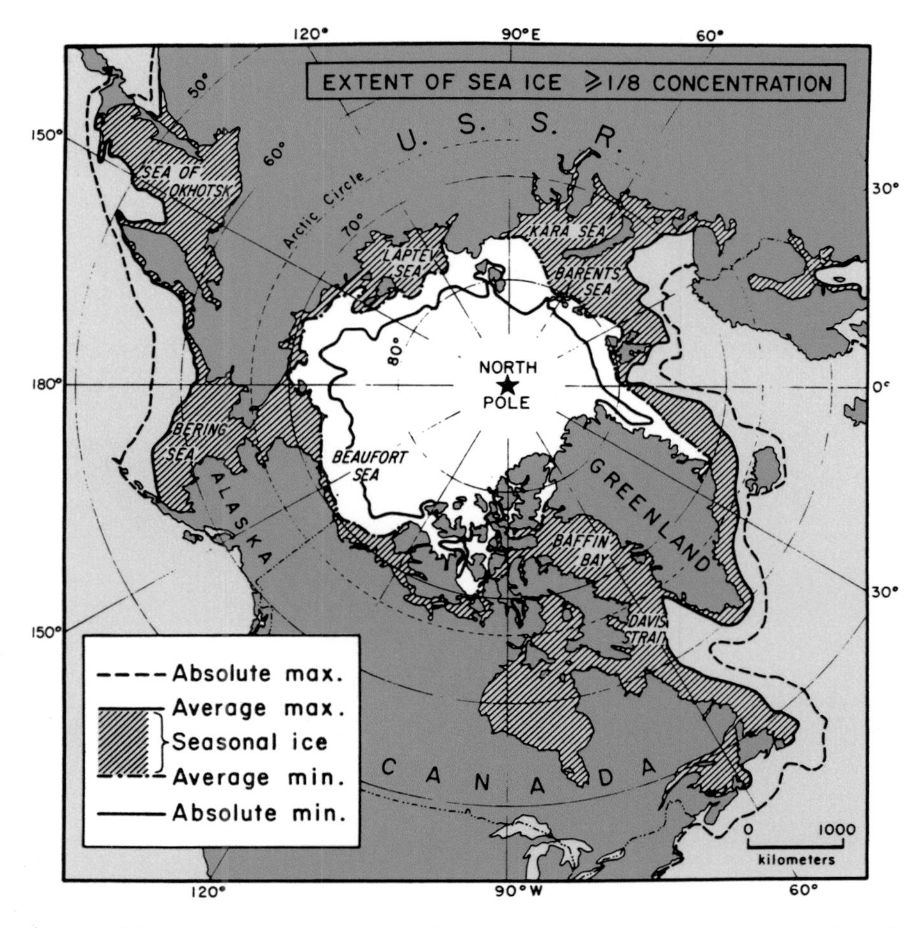

Figure 1.20 The average and extreme seasonal limits of arctic sea ice extent for ice concentrations of more than 1/8 (Anon., 1978).

along the American continent and the extreme northward extension of the region which is permanently ice-free along the Norwegian coast. Nowhere else in the world oceans can ports at 70°N be reached by sea during the entire year, such as is the case with the cities of northern Norway, and nowhere else do ice-bergs reach 40°N, as they do off the coasts of Newfoundland. This is of course the result of the temperature difference between the East Greenland and the Norwegian current. The average temperature of the Norwegian Currents is 6-8 °C, while the average temperature of the East Greenland Current, which exports several thousand ice-bergs southward annually, is below -1 °C. These climatic differences along the same latitude in the Atlantic Ocean are vividly illustrated in Fig. 1.21.

The arctic land area is topographically varied. Greenland, a mountainous island is almost completely hidden under ice. Its ice-cap rising to 3,230 m, and maintained by heavy snowfall, is the largest northern remnant of the last glacial period. Iceland by contrast is mostly icefree, with only small glaciers and is volcanically lively. Ice covers much of Svalbard and other islands north of Eurasia while much of the Siberian coast is low lying and snowfree in summer. Alaska's rugged mountain ranges carry permanent ice, but the costal plains are seasonally snowfree. The northwest Canadian Arctic is mostly rolling upland country with thin intermittent snowcover for much of the year (Fig. 1.22).

On land, the northern limit beyond which trees do not grow, where forests disappear and are replaced by tundra, is called the tree line. For much of its length

Figure 1.21 Illustration of the importance of ocean currents for the determination of local climate. The left picture shows water skiing off northern Norway, while the right picture shows a sealing vessel in heavy ice off the east coast of Greenland at about the same latitude (69°N) and time (June) (Photo: A.S. Blix)

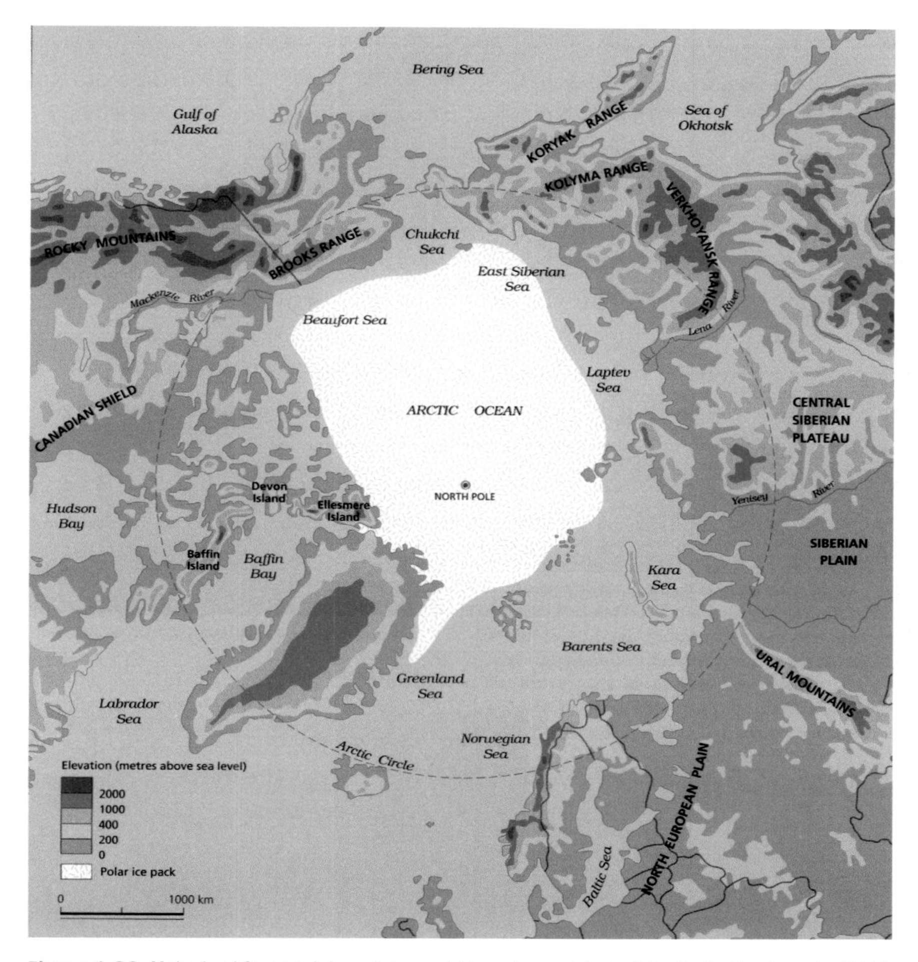

Figure 1.22 Major land forms (plains, plateaus, hills and mountains) of the Arctic (MacQuarrie, 1996).

the tree line keeps close to the 10 °C summer isotherm, reflecting the strong influence of summer temperatures on tree growth. However, departures arise because winds, soils, topography, availability of water and other local factors also effect tree growth.

Where the ground is frozen for most of the year, soils that contain water remain solid except for a brief spell each summer, when they thaw to a depth that varies with latitude and local conditions. The thaw does not usually exceed about 1 m on flat ground, but on south-facing slopes it may penetrate to 1.5 m. Most soil-forming processes are restricted to this *surface active layer.* Recent drilling for

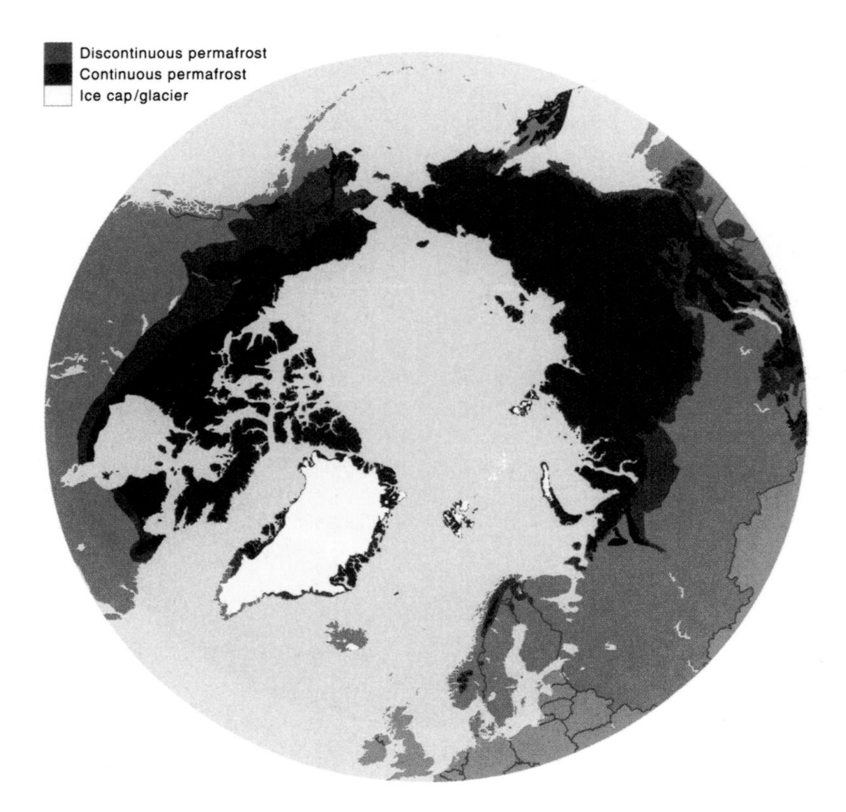

Figure 1.23 The distribution of ice caps, glaciers, continuous and discontinuous permafrost in the Arctic (Anon., 1997).

oil has shown that continuous permafrost underlies much of the Arctic, reaching thicknesses of 600 to 1,500 m in the coldest regions. The distribution of permafrost in the northern hemisphere is shown in Fig. 1.23.

With their limited capacity for obtaining water, bedrock and gravel are poor material for plant colonization. In completely dry inland regions where they predominate, there is therefore little vegetation, but where there is a hint of moisture, rocks that have stable surfaces and are sheltered from the wind support lichen, which is the characteristic vegetation of polar deserts. True organic soils require the presence of vascular plants with penetrating roots, and a lively population of soil flora and fauna, such as bacteria, algae, nematodes and other creatures that mediate decomposition and the incorporation of organic materials into the soil.

Bare or thinly covered polar soils often show characteristic patterns like circles, rectangles, polygons and stripes that may be marked out in shallow furrows

Figure 1.24 Patterned ground: A typical feature of permafrost areas which occur when the wet soil freezes and contracts (Photo: T. Kjærnet).

Figure 1.25 Polygon fields develop where stones and other surface rubble that have fallen into cracks, which occur when the soil freezes and contracts, are compressed into wedges when the soil expand again in spring. The cracks recur each year in the same places, the wedges thicken annually, and the ground eventually buckles into ridges with raised edges (Photo: R.Vik).

and ridges, or by lines of stones (Fig. 1.24). These sometimes spectacular patterns are formed as temperature falls with the approach of winter, and the soil freezes and contracts, forming cracks which open to admit snow and surface rubble. When the soil expands again in spring the contents of the cracks are compressed into wedges. The cracks recur each year in the same places, and the wedges thicken annually and the ground between buckles into irregular domes with raised edges forming irregular polygons (Fig. 1.25).

According to Walker (1995), the most recent estimates of the size of the Arctic flora include about 1,500 vascular plants, 750 bryophyte, and 1,200 lichen species. Approximately 60 % of the vascular flora is found throughout the Arctic, increasing to about 90 % in the most northerly areas. In fact, if one begins at some arbitrary point in the southern Arctic there will be a gradient of decreasing diversity to the north, with many species being lost and only a few new truly arctic species being gained. The gradient of decreasing diversity (Fig. 1.26), is related

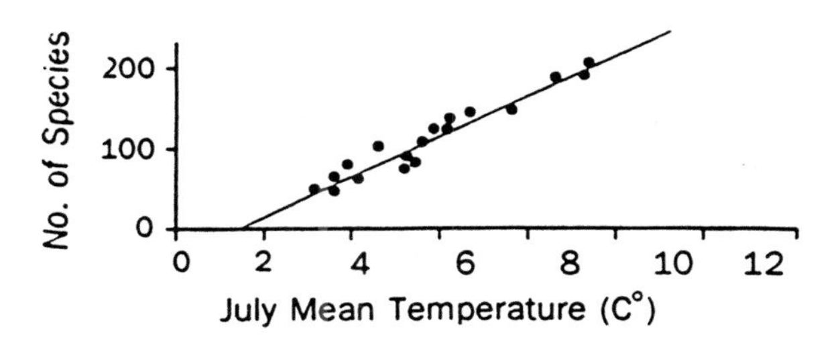

Figure 1.26 Correlation between July mean temperature and number of species for local floras of the Canadian Arctic Archipelago (Redrawn from Rannie, 1986).

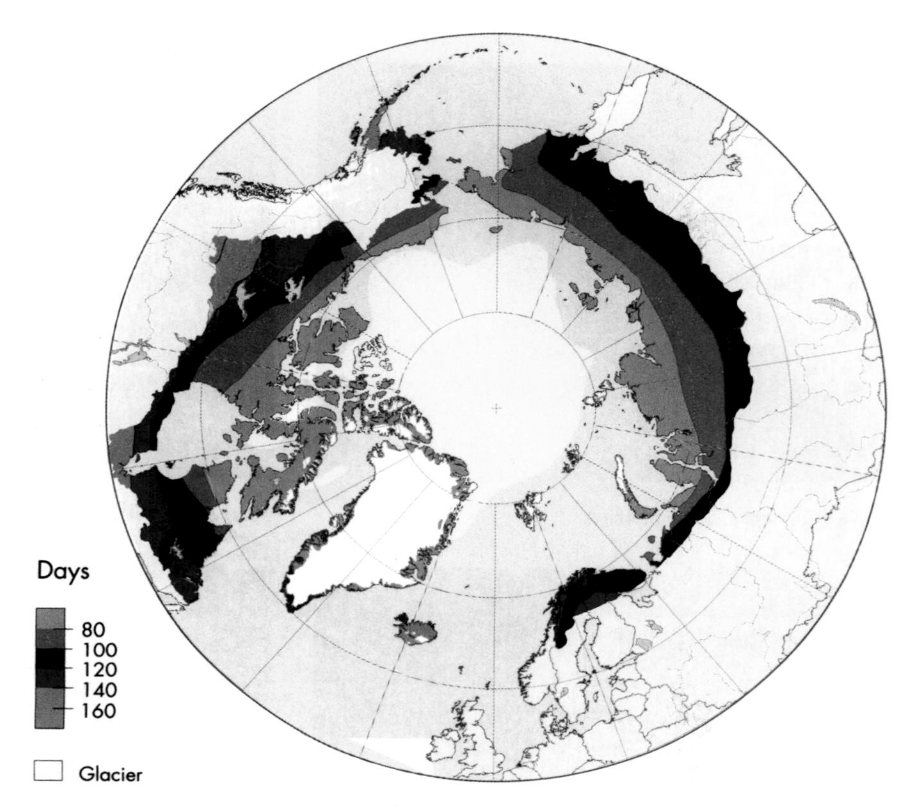

Figure 1.27 The average length of the growing season (days with an average temperature of + 5 °C or higher) across the Arctic (Redrawn from Anon., 2001).

primarily to changes in thermal energy, and the related length of the growing season (Fig. 1.27), as one goes north.

For many years western arctic botanists (e.g. Bliss & Matveyeva, 1992) have viewed the circumpolar Arctic as one floristic unit; but traditionally Russian scientists (e.g. Chernov, 1985) have recognized a larger number of vegetation zones in their north. Recently, an international effort has resulted in the agreed bioclimatic subdivisions which are presented in Fig. 1.28. These are not to be confused with vegetation zones, which would be too detailed for reproduction in this text. However, detailed arctic vegetation maps are available on the Circumpolar Arctic Vegetation Map projects WEB-page:http://www.geobotany.uaf.edu/arcticgeobot/cavmpage.html, while pictures of typical examples of the bioclimatic subdivisions in Fig. 1.28 are shown in Fig. 1.29.

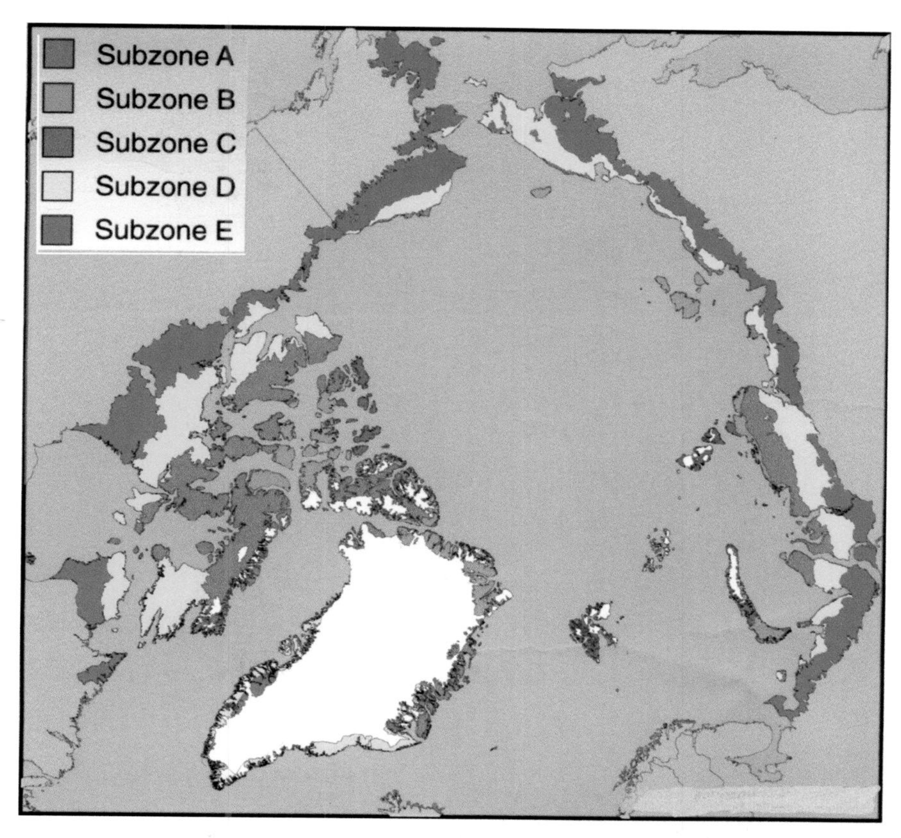

Figure 1.28 The bioclimatic subdivisions of the Arctic. A: Arctic polar desert; B: Northern Arctic tundra; C: Middle Arctic tundra; D: Southern Arctic tundra; E: Arctic shrub-tundra (Elvebakk, Elven & Razzhivin, 1999), with further information and vegetation maps on the Circumpolar Arctic Vegetation Map Program home page).

Now, then, is there anything special or different in evolutionary terms among arctic plants? The answer probably is: Not very much. Arctic plants photosynthesize, respire, absorb nutrient and grow as rapidly as temperate zone plants, so in physiological terms, it is the sum of adaptations conferring the ability to do this at low temperatures that distinguish these plants from others. Annual variation of seed production and seeding survival, for example, can be enormous. Therefore, it is the uncertainty of suitable environmental conditions from year to year that characterizes the Arctic.

Information on physiological adaptations in plants are available in a number of excellent textbooks.

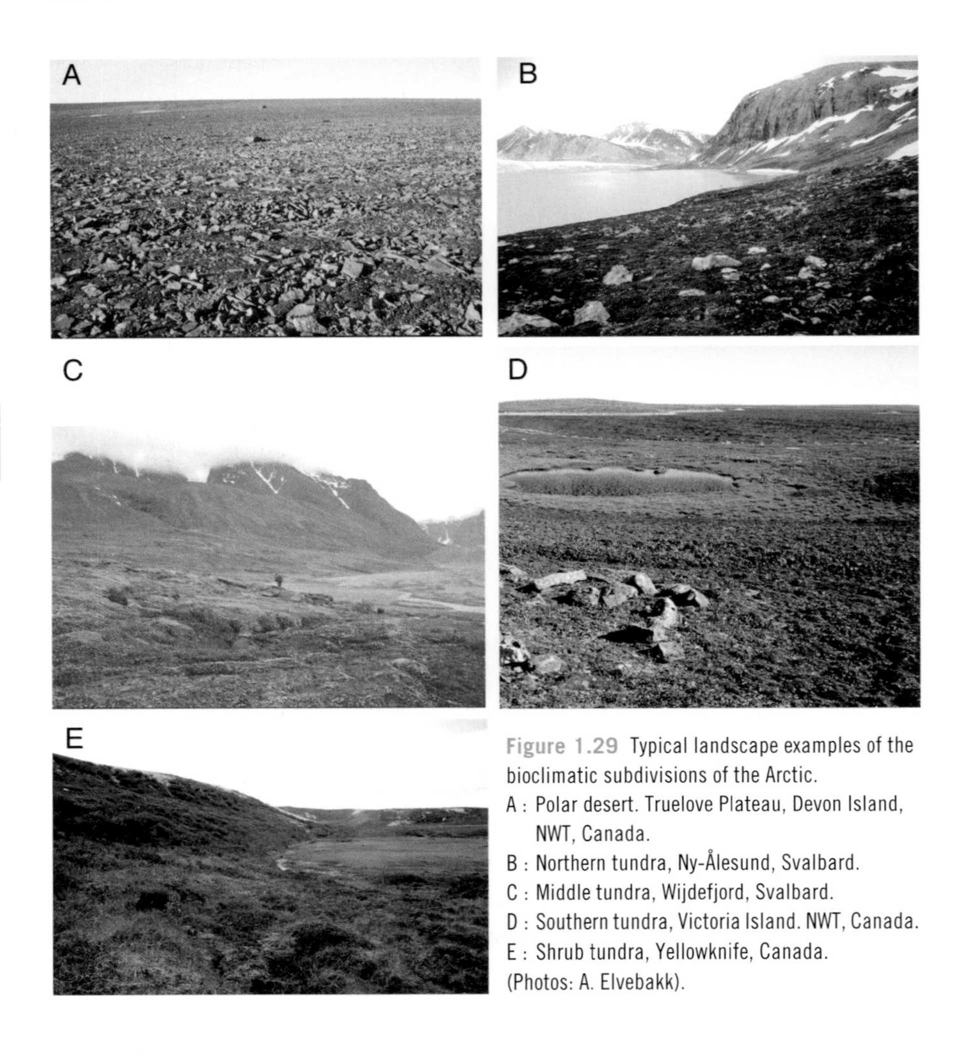

Figure 1.29 Typical landscape examples of the bioclimatic subdivisions of the Arctic.
A : Polar desert. Truelove Plateau, Devon Island, NWT, Canada.
B : Northern tundra, Ny-Ålesund, Svalbard.
C : Middle tundra, Wijdefjord, Svalbard.
D : Southern tundra, Victoria Island. NWT, Canada.
E : Shrub tundra, Yellowknife, Canada.
(Photos: A. Elvebakk).

CHAPTER 2

The Late Cenozoic Glaciations

Some 5-10 million years ago the climate of the Arctic changed from warm to cool-temperate, and between 3 and 5 million years ago it cooled further, and ice invested the uplands and spread to the plains. Thus, during the last periods of the Pleistocene glacial events (Wisconsin/Würm) tremendous masses of ice covered Greenland, Iceland, the Faroe Islands, Britain south to the Thames and spread over Scandinavia, the Netherlands, Germany, Poland and western Russia. Eastern Siberia had its own, relatively small, ice-caps, and North America was covered south to New York, St. Louis and Vancouver (Fig. 2.1). The ice-covered lands were lifeless and barred access from the south to the limited ice-free arctic regions in North America and Europe, and restricted communication between the Arctic and Southern Asia to a few ice-free corridors (Fig. 2.1). But, it is becoming increasingly evident that the withdrawal of water from the oceans into northern and southern ice-caps that took place during the Wisconsin epoch lowered the level of the sea by over 100 metres and exposed the shallow continental shelf between Alaska and Siberia to form dry land, as broad as Alaska, between the two continents (Fig. 2.2). This land bridge allowed many species of steppe mammals, like mammoths, bison and horses to pass from Eurasia to America. The land bridge region, which is called *Beringia*, remained dry land during many thousand years late in the last Wisconsin glaciation, only to be submerged with the sudden melting of the continental ice-caps that thereby flooded the continental shelfs some 10,000 years ago. When that happened, a residue of arctic life with asiatic origin became cut off from

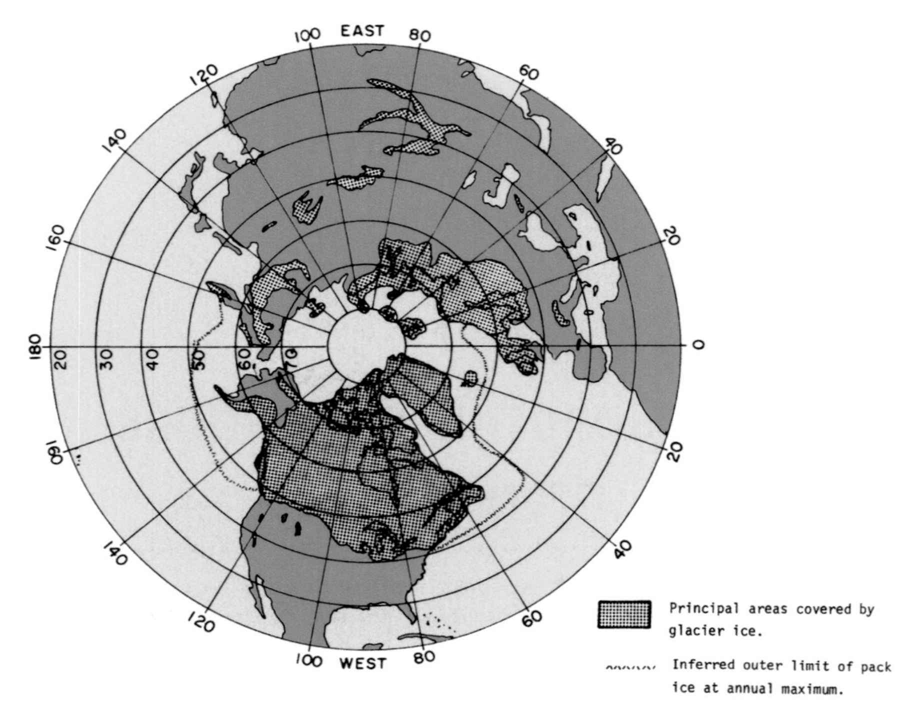

Figure 2.1 Extent of large glaciers and pack ice at the time of maximum glacier ice during the Wisconsin glaciation some 18,000 years ago (Flint, 1957).

Asia by submergence of the landbridge, but could now disperse south and resettle the great extent of northern North America as it became free from ice. A great area of North America is thus rather newly populated and is occupied by plants and animals derived from relicts in northern refuges, and by others from south of the glaciers in temperate America as they moved north. Thus, most polar species are currently still invading and adapting from neighbouring temperate regions; the paucity of polar species, in comparison with temperate or tropical species, is therefore due partly to lack of time.

It has hitherto been assumed that the extinction of the Pleistocene megafauna (mammoths and others) shortly after the termination of the last glaciation, was caused by climatic change, but Zimov *et al.* (1995) have recently put forward an alternative hypothesis: The populations of large grazing mammals were decimated by hunting at the end of Pleistocene, and the reduction of grazing pressure converted the previously highly productive grassland to a moss-dominated tundra, which eventually led to the demise of the large mammalian grazers.

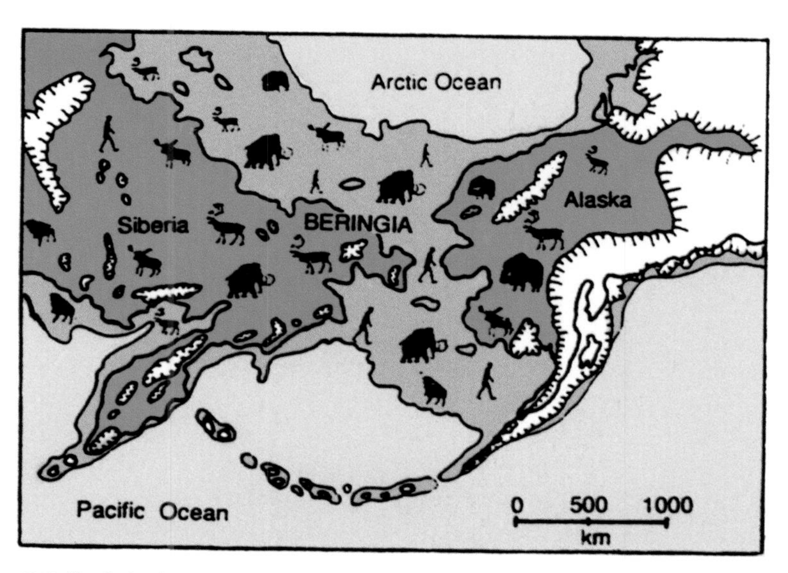

Figure 2.2 The Bering Land Bridge was an extensive emerged land area connecting Asia and America during the last ice ages. The map shows the approximate maximum extent of Beringia during the last glaciation, and the Alaskan glaciers are at their maximum extent. Note that the glaciers at this stage (about 18,000 years ago) blocked the passage from Beringia to the southern parts of North America (Redrawn from Hopkins, 1967).

CHAPTER 3

Marine Invertebrates and Fish

Before we start reviewing the invertebrate and fish fauna of the arctic oceans we have to agree on a few general concepts and definitions:

Populations which are defined on the basis of who eats whom in the food chain are called *trophic levels.* The lowest trophic level, the phytoplankton, are the primary producers (Fig. 3.1). Primary production is limited to the upper illuminated part of the water column, and the energy and solid matter of almost all food originates from photosynthesis, either on land, or in the ocean.

Organisms excrete waste products, for example in the form of urine and faeces, which contain nutrients, like ammonium and phosphate and/or uric acid (in birds), urea, amino-acids or others, which can be used as food by the primary producers.

A certain proportion of the pelagic organisms, that being algae, zooplankton and others, are not eaten and will eventually sink out of the illuminated layer of the water column together with excreta and other dead material. This sedimentation is of interest both climatologically and biologically, because it represents a transport of carbon away from the surface-layer (and therefore away from the atmosphere) down to the bottom. This process is called the "biological carbon pump" (Fig. 3.2). This process provides nutrients to benthic organisms and to animals living in the deeper part of the water column, and is the origin of much of the organic sediments on the sea floor.

Now, seawater contains most known natural elements, of which sodium and chlorine ions are the most abundant. The total amount of dissolved material in

Figure 3.1 Simplified arctic marine food web (Anon., 2001).

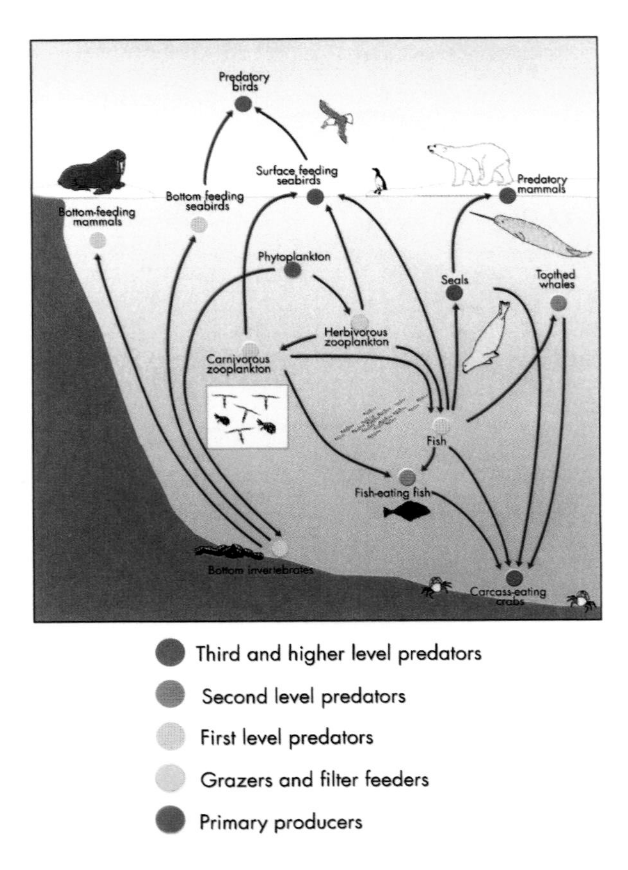

sea-water is termed the *salinity* and the average salinity of sea water is about 35 grams of salts per kilogram, measured as the electrical *conductivity* (C). Salinity and *temperature* (T) of sea-water are particularly important parameters because, together with pressure, they determine the *density* (D) of sea-water, which is measured at various depths by use of so called CTD-sampling equipment.

The density of sea water does not increase uniformly with depth. In equatorial and tropical regions there is usually a shallow upper layer of nearly uniform density, then a layer where the density increases rapidly with depth, called the *pycnocline*, and below this the deep zone where the density increases slowly with depth. The rate of change of density with depth determines the static stability of the water, or its "unwillingness" to be moved vertically. Where the stability is high, vertical movement and vertical mixing are minimized. Thus, the water in the pycnocline is very stable, which is to say that it takes much more energy to displace a particle of water up or down in the pycnocline than in a region of lesser stability. The pycnocline then, although it offers no barrier to the sinking of bod-

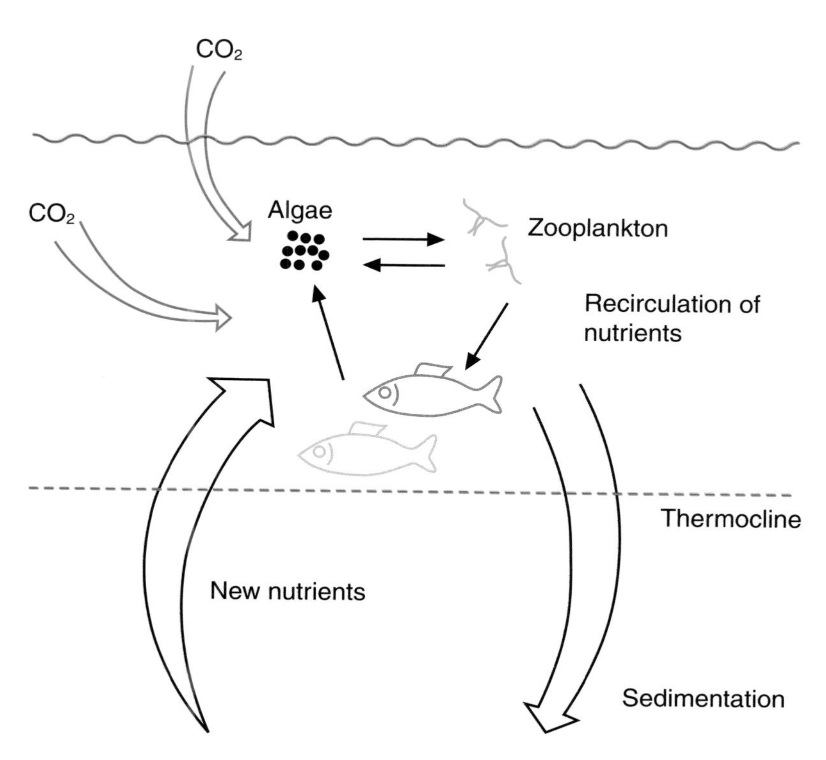

Figure 3.2 Regenerative and new production: In systems where, for example, a pycnocline (sharp density gradient) separates the upper illuminated water column from the deep ocean, primary production will depend on re-cycling of nutrients (*regenerative production*). The huge nutrient reserves in the deep water masses can only be utilized by the algae when the pycnocline is broken down by hydrographic mixing by wind, currents, etc., so that deep water ascend to the illuminated part of the water column to give *new production* (Redrawn from Sakshaug *et al.*, 1994).

ies which are much denser than water, offers a real barrier to the passage of water and its constituents, both ways, in the vertical direction. Moreover, there is also often a layer in the water column where the temperature and the salinity increase rapidly with depth; these are called the *thermocline* and *halocline*, respectively.

We now have to define the two important concepts of *new* and *regenerative production*: Imagine first a system, say the ecosystem in the upper illuminated part of the water column. Assume that it is limited in the lower part by a pycnocline, which excludes exchange between the layers above and those below. In other words, we have a closed system, in which the regenerative production is equal to the primary production, which is based on the nutrients that are excreted by the organisms which live in the upper water column. It follows, that regenerative production is based on re-cycling of nutrients, while new production, on the other hand, is based on nutrients which originate from outside the system. That is to say the nutrients below the pycnocline, where the deep water masses represent an enormous reservoir of nutrients. But, for new nutrients to emerge to the upper illuminated layers in any appreciable amounts, the pycnocline has to be broken down. This implies that our closed system has to be opened, and that, in turn, depends on hydrographic mixing processes caused by wind, currents and bathymetry (Fig. 3.2).

The most important ocean systems with primarily regenerative production are the central warm ocean basins, in which a distinct thermocline occurs at 80-150 meters of depth all year around, year after year, so that the contribution of nutrients from the deep water to the upper illuminated layers is minimal. Such ecosystems have high biodiversity, but extremely low productivity. The ecosystems which are characterized by high new production are mostly found at high latitudes (Fig. 3.3) and/or along continental shelf-breaks, where the physical processes that break down the pycnoclines are prominent. Since such systems are open it is not necessary to replenish what is taken out. Areas characterized by high new production usually have relatively low biodiversity, but very high productivity. It is therefore only the new production in open systems which provide the basis for sustainable harvest. This implies that from a management point of view

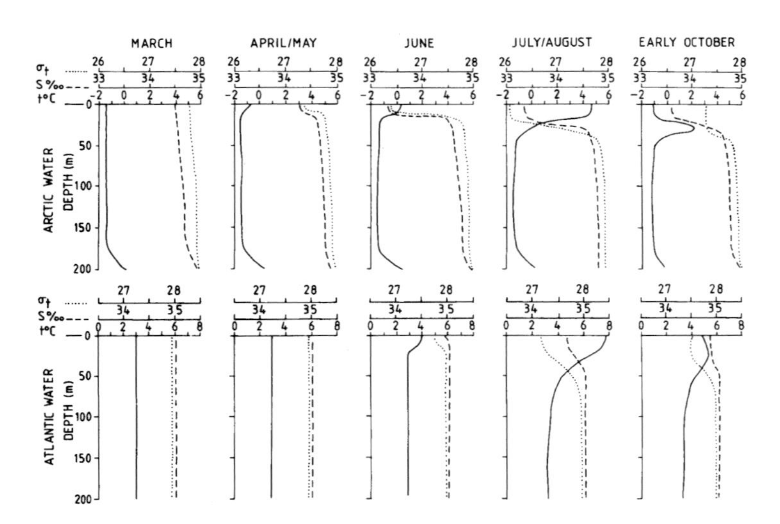

Figure 3.3 Illustration of the vertical distribution of temperature, salinity and density (sigma-t) from the surface to 200 m in arctic and atlantic water masses. The illustration is based on several years of observations (Loeng, 1990).

it is much more important to know how large the new primary production is, than to know the size of the total production. This may be illustrated by the situation in tropical rain forests, which may have 6 times the production of the Barents Sea per unit area, but the tropical rain forests are closed systems and almost 100 % of the production is regenerative. Therefore, any harvest will unfortunately soon turn them into deserts.

Further information on basic concepts in oceanography and marine biology are available in several excellent textbooks.

The Arctic Ocean

The Arctic Ocean is probably the least studied of the world oceans, yet, according to Aagaard and Carmack (1994), among others, it plays a global role in the Earth's surface heat balance and thermohaline circulation.

Because of its inaccessibility, the marine ecology of the Arctic Ocean is poorly known, and the notion of a barren central Arctic basin has been accepted until several expeditions to the central Arctic Ocean in the early 1990ies. The year-round presence of ice, a short photosynthetic season and low temperatures were thought to severely limit biological production, although the shortage of data has often been noted. However, Wheeler *et al.* (1996) discovered that algal production in the ice was about twice that in the water column in the Makarov and Amundsen basins. In the central Arctic Ocean the mean ice algal production was similar to the values for Arctic first year ice at lower latitudes. Meso-zooplankton in the upper 100 metres consisted mostly (87-98 %) of copepods, with *Calanus hyperboreus* being the most important in numbers and biomass. Increased zooplankton biomass was observed in the Makarov basin in the northward direction, with a peak at about 87°N. Meso-zooplankton biomass were 2-10 times greater than previous estimates for the central Arctic Ocean, and similar to biomass reported for the Greenland Sea.

Given the novel information presented above, it is rather intriguing that the central Arctic Ocean still seems to hold very few pelagic fish, seabirds and marine mammals, and clearly much more is to be learnt about the biology of the Arctic Ocean.

The arctic shelf seas

We will now turn to the arctic shelf seas, all of which with enormously different biological diversity and production. We will therefore deal with them one by one, and first examine the Barents Sea in some detail and use it to illuminate some general principles.

The Barents Sea

The Barents Sea is bounded on the north by the archipelagos of Svalbard and Frans Josef Land, to the East by Novaya Zemlya, and to the west by a slope towards the deep Norwegian Sea. The average depth of the Barents Sea is 230 m. The powerful warm and salt Atlantic Current flows in from the south and divides into the Norwegian coastal current, which skirts the North Cape and enters the Barents Sea through the broad passage between the North Cape and Bear Island, and the West-Spitsbergen current, which continues north along the west coast of Svalbard. A part of this branch even penetrates into the Barents Sea from the north through the straits off the eastern and western coasts of Frans Josef Land (Fig. 1.16). The inflow of cold water, on the other hand, occurs between Svalbard and Frans Josef Land and to an even greater extent through the strait between Frans Josef Land and Novaya Zemlya (Fig. 3.14).

The area of contact between the Atlantic water and the water from the Arctic Ocean is called the Polar Front. This front is remarkably distinct and stable in the southwest at about 76°N, and conveniently divides the Barents Sea into a Polar (northern) Domain and an Atlantic (southern) Domain of about equal areas. Thus, in the Barents Sea, the southwestern part is permanently ice-free, and in the summer only a relatively narrow rim of ice usually remains in the northernmost parts.

The vertical mixing which is the result of the "meeting" between polar and atlantic waters is of paramount importance for the transport of nutrients from the deep water to the surface layer, since it is only there that there is enough light for phytoplankton growth. In the winter, the atlantic water is homogenous from the surface and down to 200-300 m, primarily because the North Atlantic resides in the belt of northern atmospheric lows (Fig. 1.11).

In spring the surface water is warmed, and the water masses in the upper part of the water column become more stable and the layering more pronounced, in the southwestern part of the Barents Sea even down to 50-60 m of depth (Fig. 3.3), while at the ice-edge water stability is achieved by the melting of the ice.

In the autumn, cooling of the surface layer starts, and eventually the new ice starts to form. When the sea ice is formed salt is released into the water which then becomes denser and starts to sink. This greatly contributes to the vertical mixing and destabilisation of the water column. In the course of winter the water column becomes completely mixed down to 200-300 m, which in the Barents Sea in most places is to say down to the bottom. The great import of relatively warm atlantic water also explains the relatively warm winters along the west coast of Norway all the way to West-Spitsbergen, where the ocean is ice-free in mid-winter, while on the western side of the Atlantic the sea ice extends to the Gulf of St. Lawrence (Fig. 1.21), which is on the same latitude as Paris.

Unlike the central Arctic basin where the ice has an average lifespan of about 6 years, the ice in the Barents Sea is mostly annual. The seasonal changes in the extent of the Barents Sea ice-cover is mainly determined by the influx of warm Atlantic waters, and seems pretty well to follow the Polar Front. The extent of ice-cover is therefore very variable from year to year. The ice-drift in the Barents Sea is primarily wind-driven in winter, currents playing a minor role, while the summer distribution is determined mainly by the intensity of the solar radiation.

Phytoplankton

The distribution of primary producers in the Barents Sea is determined by the location of the Polar Front, the run-off from the Siberian rivers (Fig. 1.17) and the extent of sea ice cover. Arctic water, north of the Polar Front, and atlantic water to the south each contribute different groups of plankton. Silicified diatoms *(Bacillariophyceae)* are the most important group of primary producers during the spring bloom in the Barents Sea, where they sometimes occur in concentrations of 100 million cells /m^3 . In addition to the diatoms, other groups of algae, like the prymnesiophyte flagellate, *Phaeocystis pouchetii*, are also important and occur in concentrations of several billion cells per m^3.

In the Barents Sea there is enough light for the algal bloom to start in late February. The algal bloom is concentrated to a 20-50 km wide belt following the retreating ice-edge. North of the Polar Front the ice-edge bloom is responsible for much of the primary production, but this is not to say that the annual production in the ice-covered areas is particularly high. It is rather that a belt of high productivity sweeps over the area and provides a time of plenty for those that can follow, for 2-3 months, and a short fiesta for benthic organisms that can not.

Zooplankton

The most important zooplankton in the Barents Sea are the copepods, *Calanus finmarchius* south of the Polar Front, *Calanus glacialis*, north of the Polar front, and krill, mostly *Thysanoessa inermis*, *T. raschii* and *T. longicaudata.* The copepods and *T. inermis* and *T. raschii* are primarily herbivorous, while *T. longicaudata* primarily feeds on zooplankton. Adult krill live near the sea floor during the day

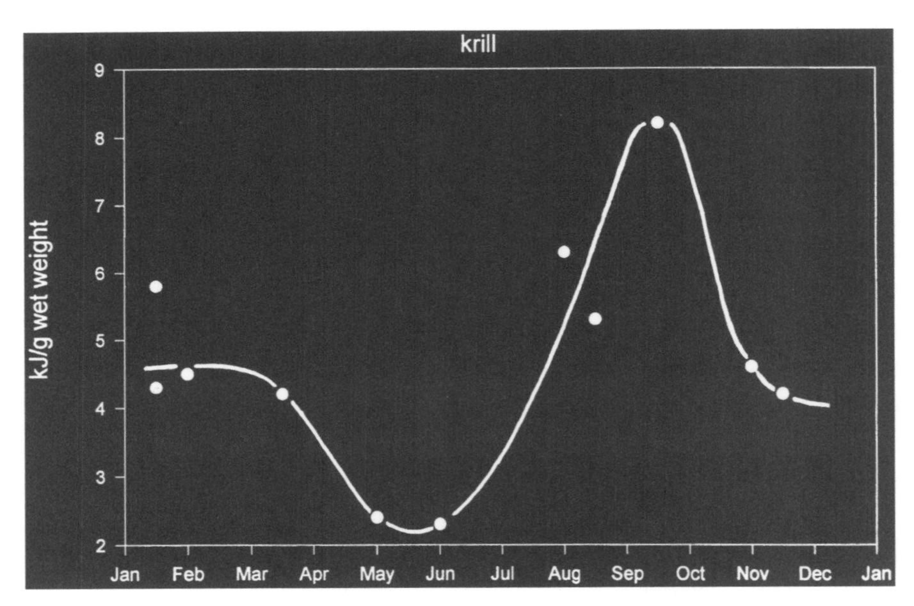

Figure 3.4 Changes in energy density (kJ/g wet weigth) in krill (*Thysanoessa sp.*) from the north-east Atlantic throughout the year (Mårtensson, Gotaas, Nordøy & Blix, 1996).

but make daily migrations to the surface during the night. This daily vertical migration of 100-300 m takes place in 2-3 hours. The vertical migrations are most pronounced in spring and autumn when the difference between day and night is most apparent. During winter when darkness prevails the krill are usually found deeper than 100 m around the clock. Krill is important as food for immature cod, seabirds and whales, but the energy density of krill changes markedly throughout the year (Fig. 3.4). In addition to copepods and krill, arrow worms (*Chaetognatha*), like *Sagitta elegans arctica*, and comb jellies (*Ctenophora*), like *Mertensia ovum*, are common, the latter particularly north of the Polar Front.

Life in the ice

Sea ice harbours a society of organisms from bacteria and algae through crustaceans and fish to sea-birds, seals, polar foxes and polar bears. The age of the ice is to a great extent determining the composition of organisms which are present. In this respect the main distinction is between new and multi-year ice. In areas characterized by new ice, the ice must be colonized every year, by either pelagic or benthic organisms which swim to the surface or from multi-year ice in the vicinity. Multi-year ice is, as we have already heard, typical of the Arctic Ocean, and some of this old ice is transported into the Barents Sea through the strait between

Svalbard and Frans Josef Land. This import of ice is important for the colonization of the sea ice in the Barents Sea.

Ice algae

Ice-algae are microscopic unicellulars which at some stage of their life-cycle are attached to sea ice. The ice-algae communities are best developed in multi-year ice. In the Barents Sea, where the annual ice dominates, such communities are therefore less developed than in the Arctic Ocean. The primary production in ice-algae communities is much less than for phytoplankton in the open oceans, primarily due to lack of light. Ice-algae occur as surface assemblages, interior assemblages and bottom assemblages.

There are two kinds of surface assemblages. One is found in summer and autumn in melt-pools on the top of the ice, and one called infiltration assemblages occurs when the accumulation of snow is heavy enough to press the surface of the ice under the surface of the sea, and create a habitat for the algae in the snow. Such infiltration assemblages are common in Antarctica, but rather rare in the Arctic.

Also the interior assemblages are of two different kinds. One is found in the brine channels which are created when seawater freezes, and the other is band assemblages, which are seen as thick brown bands, usually in the interface between two and one-year old ice. The latter consists of algae which originally grew under the one-year ice the previous season.

For good measure, the bottom assemblages also are of two kinds. The interstitial assemblages grow in the spaces between the platelets under the ice in layers a

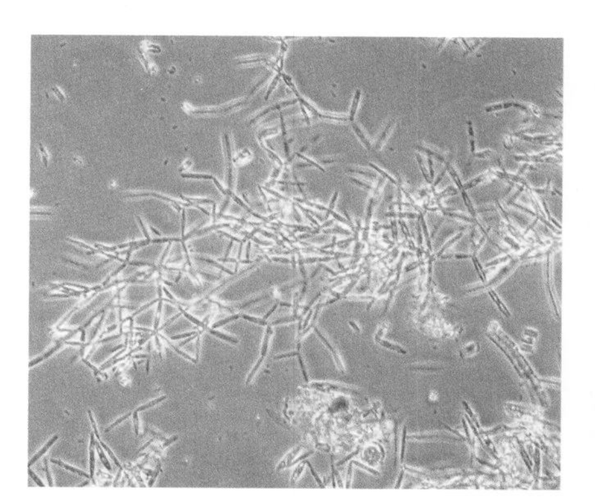

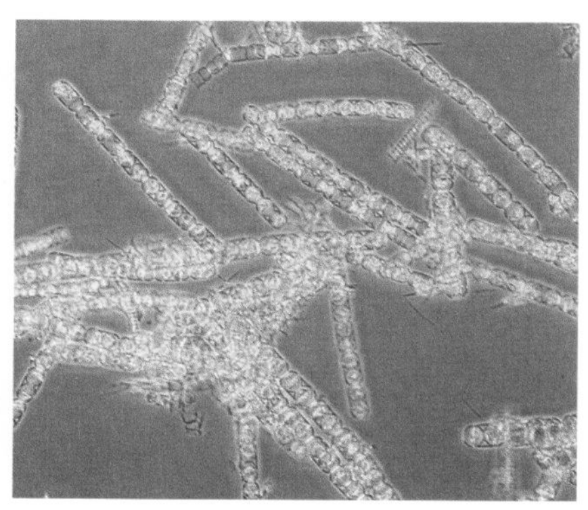

Figure 3.5 *Left*: Colonies of the diatom, *Nitzschia frigida*, often dominate under first year ice, where it may form dense brownish mats (Syvertsen, 1991). *Right*: Colonies of the diatom, *Melosira arctica*, are often found under multi-year ice, where it may form reddish-brown mats of more than 30 cm in thickness (Syvertsen, 1991).

few millimetre up to a centimetre thick, but the most important kind is the sub-ice assemblage, which may reach several decimeters in thickness in the Arctic Ocean, but much less in the Barents Sea. The most important algae under the ice in the Barents Sea are again diatoms, like *Thassiosira spp., Bacterosira spp., Porsira spp.* and *Chaetoceros spp.* Under old annual ice of more than 50 cm thickness the algae communities are dominated by one species, *Nitzschia frigida,* while communities of another diatom, *Melosira arctica,* often dominate under multi-year ice (Fig. 3.5).

Ice fauna

The ice fauna can also be divided into two groups: the sub-ice fauna and the true ice-fauna. The latter again consists of two groups: the *autochthone* (those that live permanently in the ice all their lives) and the *allochthone* (those that are associated with ice only for part of their lives, or for part of the year only). The sub-ice fauna is completely dominated by the hyperid amphipod, *Parathemisto libellula* (Fig. 3.6), which is of great importance as food for polar cod (*Boreogadus saida),* diving seabirds, harp seals and ringed seals, while the autochthone fauna consists of gammarid amphipods. The polar cod (Fig. 3.13) is conspicuous among the allochthone ice-fauna, which also includes numerous species of copepods, like *Calanus glacialis* and amphipods, like *Apherusa glacialis* and *Gammarus wilkitzkii.*

Poikilothermic animals that make their living in ice must be able to avoid freezing and to tolerate dramatic changes in salinity. During spring and summer,

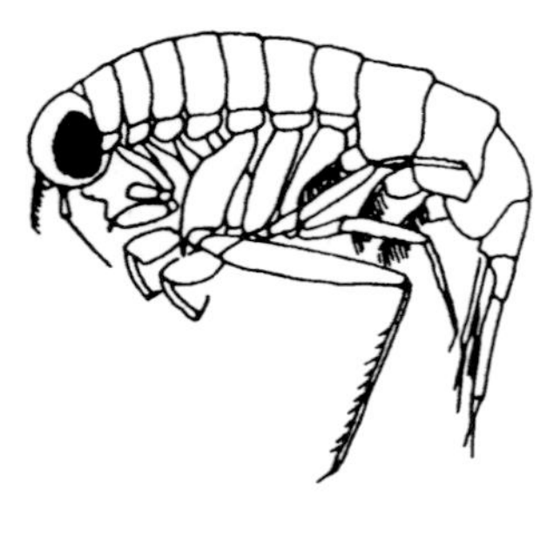

Figure 3.6 *Parathemisto libellula*, a hyperid amphipod, which is found both under the ice and in open water, is important food for many marine mammals and seabirds. *Left*: The stomach of a Brünnich's guillemot stuffed with the amphipod. (Photo: F.Mehlum). *Right*: Drawing of the animal, which may reach a length of 60 mm (usually 10-25 mm), with the large eyes typical of the hyperid amphipods.

pools of meltwater with a salinity as low as 1-2 ‰ are created on top of the ice and this freshwater percolates through the brine channels and hits the autochthone amphipods which are what is called euryhaline, i.e. tolerate great shifts in salinity, down to 3-4 ‰. The amphipod *Gammarus wilkitzkii* acts as an osmo-conformer down to salinities of 25 ‰, but starts to osmo-regulate when the salinity of the water drops below that level. The sub-ice amphipod, *Parathemisto libellula*, on the other hand, only tolerates a change in salinity between 28 and 35 ‰, and has to evacuate when the ice is melting.

Sediment

Sedimented organic material serves as food for benthic organisms. Sedimentation of biogenic material causes, together with grazing by higher organisms, a reduction of the algae population in the *euphotic zone*. It follows, that sedimentation and grazing are competing processes. How much that is consumed by the higher organisms depends on the degree of match, or mismatch, between the populations of phytoplankton and zooplankton in time and space. Sedimentation of biogenic material is most pronounced at the end of an algal bloom. Faeces, particularly from zooplankton, may contribute significantly, and sometimes even dominate in the sediments, since such faecal pills are large and sink fast. When organic matter reaches the bottom it can just sit on the surface and contribute to the sediment deposits, or become bio-deposited. Biodeposition implies that the material is eaten by benthic organisms, like molluscs or tunicates. Larger bottom dwellers may also bury the material in the sediments, but such materials may be resuspended in the water due to currents.

Benthic organisms

In the Svalbard area the red and brown algae contribute each 35-45 % to the algal biodiversity, while the green algae contribute the final 10-30 %. Such macro-algae are found down to 30-40 m whereafter the arctic light no longer can support photosynthesis. It follows, that these algae are concentrated in a narrow belt around islands and along the mainland. It is typical in the Arctic, however, that the macroalgae flora is missing in the tidal zone, since the ice removes all the benthic organisms there.

Approximately 80-90 % of all marine organisms are benthic. Just around Svalbard alone more than 1,000 species have been identified, and if the entire Barents Sea is included the number is likely to become doubled. Sponges (Porifera) and sea anemones (*Cnidaria*) (Fig. 3.7), ragworms (*Polychaeta*), crustaceans (*Crustacea*), molluscs (*Mollusca*), moss animals (*Entoprocta*), echinoderms (*Echi-*

Figure 3.7 Rock bottom location at the west coast of Spitsbergen (Magdalena-fjord), Svalbard (79°N), showing an abundance of filterfeeders, like colonical ascidians (*Synoicum turgens*), solitary ascidians (*Halocynthia pyriformis*), sea anemones (*Urticina crassicornis*), soft corals (*Gersemia rubiformis*) and bryozoans (Photo: B. Gulliksen).

Figure 3.8 The seafloor at 25 m depth in the Rijp-fjord of Nord-Austlandet, Svalbard (80°N) showing typical red crustose corallines and brittle stars (*Ophiura*) (Photo: B. Gulliksen).

nodermata) and sea squirts (*Ascidacea*) (Fig. 3.8) are the most common groups of benthic animals. In addition there are benthic fish. Several benthic species in the Barents Sea have commercial value, in particular, deep water prawn (*Pandalus borealis*), which is also one of the most important food items for the cod (*Gadus morhua)*, scallop (*Chlamys islandica*) and fishes like Greenland halibut (*Reinhardtius hippoglossoides)* and long rough dab (*Hippoglossoides platessoides).*

Demersal and pelagic fish

The fish fauna of the Barents Sea consists of only some 150 species, of which the commercially most important are capelin (*Mallotus villosus*) and cod, and sometimes herring (*Clupea harengus*). Ecologically, capelin and polar cod (*Boreogadus saida*) are most important of the truely arctic species, while cod, haddock (*Melanogrammus aeglefinus*), seithe (*Pollachius virens*), herring, Greenland halibut (*Reinhardtius hippoglossoides*) are the most important of the boreal species.

Capelin

The capelin (*Mallotus villosus*) is a circumpolar salmonid fish (Fig. 3.9) which stay close to the Polar Front during winter, and therefore changes its distribution with the movements of the Polar Front (Fig. 3.10). In northwest, the capelin is located

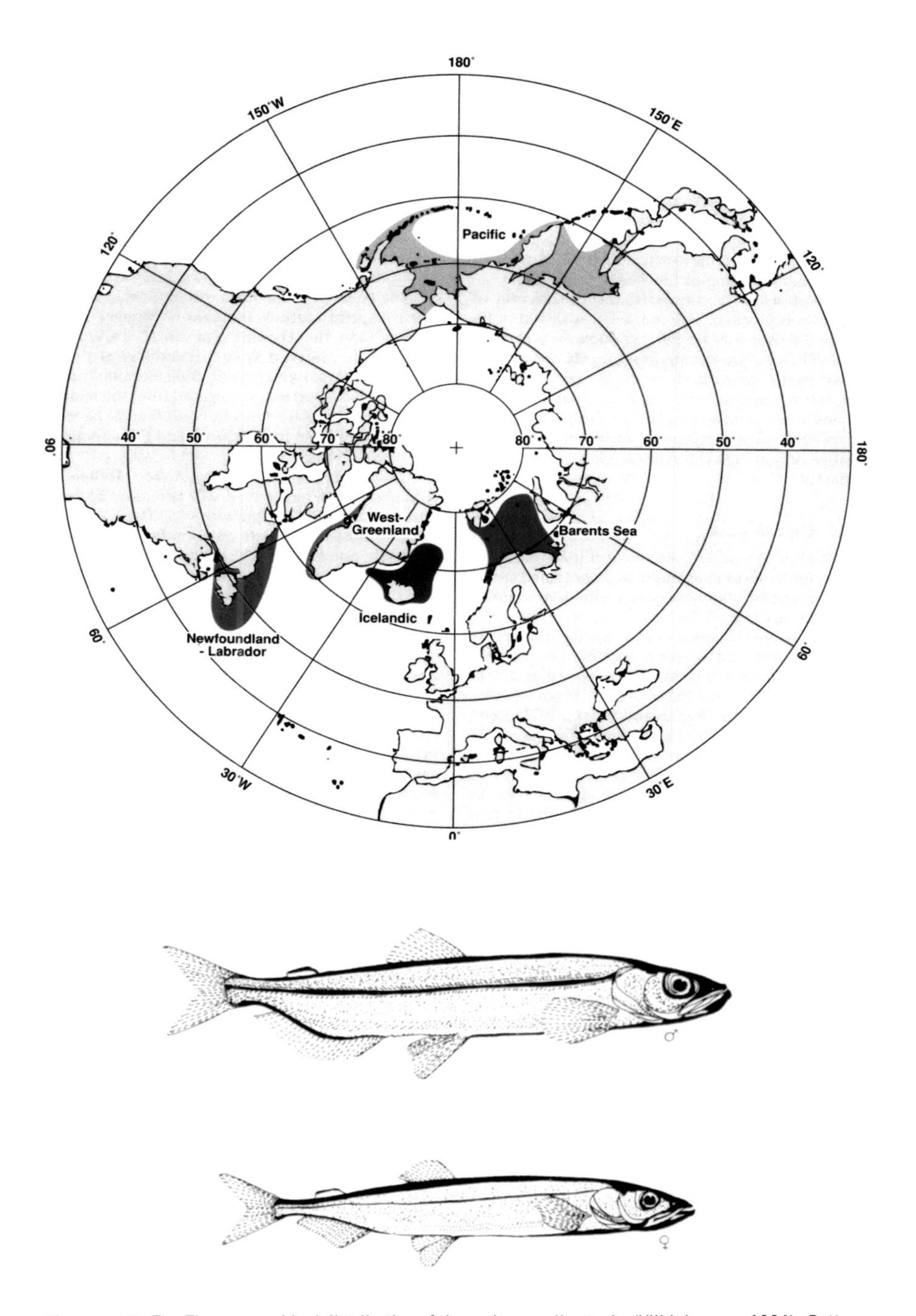

Figure 3.9 *Top*: The geographical distribution of the major capelin stocks (Vilhjalmsson, 1994). *Bottom*: Male and female capelin.

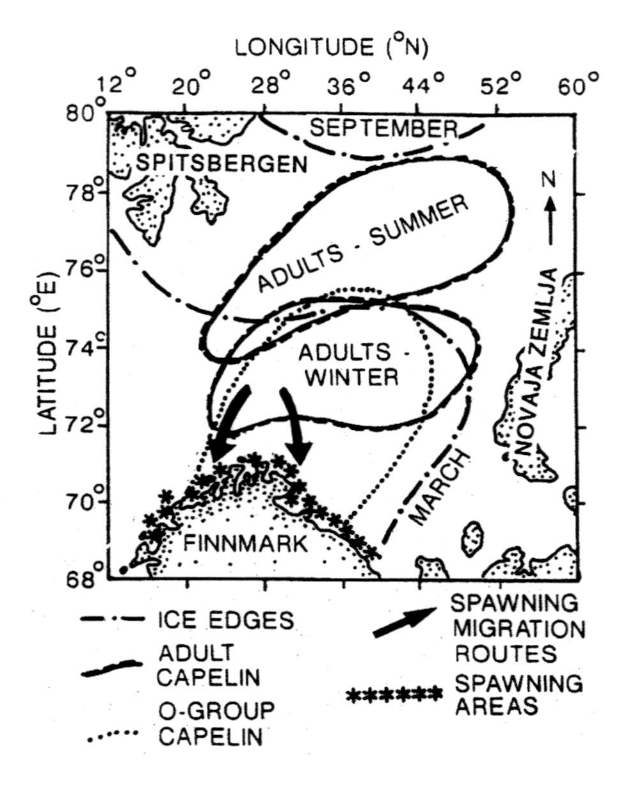

Figure 3.10 The spawning areas and the distribution of adult capelin in summer and winter in the Barents Sea. The position of the ice edge at mean maximum (March) and mean minimum (September) extensions are also indicated (Redrawn from Reed & Balchen, 1982).

along the southern edge of the sea ice. In early January-February the mature fish starts a migration south to the northern coasts of Norway and Russia, where they spawn in March-April. During summer they again migrate north to the feeding grounds in the central and northern part of the Barents Sea. During summer the capelin also migrate vertically during the day, being in the upper part of the water column during the night and at depth during the day (Vilhjalmsson, 1994).

Capelin matures at 3 years of age and usually dies after spawning. Young capelin feed primarily on copepods (*Calanus finmarchius*) and young pelagic sea slugs (*Limacina retroverso*). Adult capelin feed most intensively in July-October, when it primarily eats copepods, amphipods and krill, the latter sometimes contributing 70-100 % of its diet. The Barents Sea population of capelin has undergone dramatic changes from 1973 till present. (Fig. 3.11). The stock collapsed in 1987-89 and began building up again in 1990-92, whereafter it dwindled to minimum size again, but is now in 1998-2001 back on the increase. The reasons for this obviously is a combination of different events. The capelin is a very shortlived species and reproductive success is much dependent on oceanographic conditions such as temperature and salinity, besides predation of larvae by an increasing

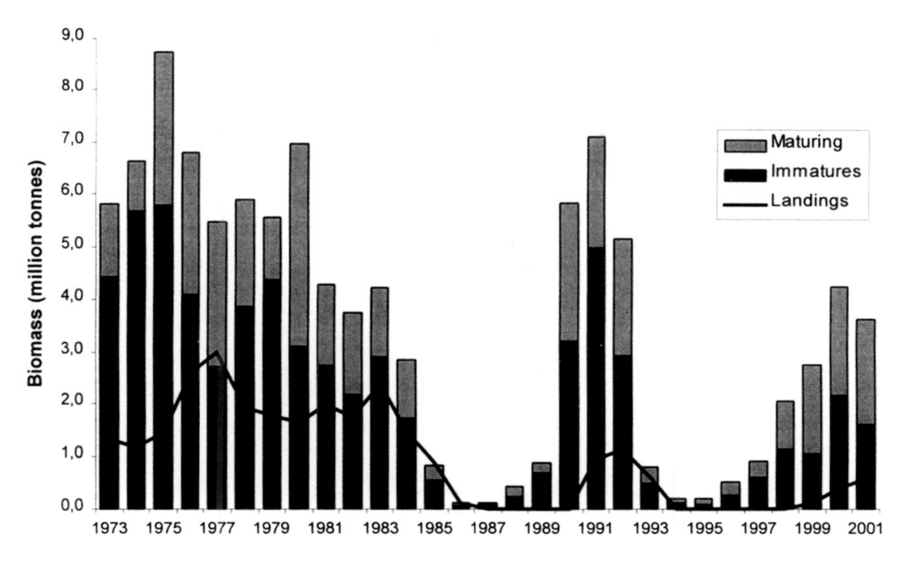

Figure 3.11 Estimated biomass of mature and immature capelin in the Barents Sea (1973-2001) together with the biomass landed during the same period (H. Gjøsæter, HI-Bergen, pers. comm.).

stock of immature herring and cod, as well as increased predation on adult capelin by adult cod, harp seals, and, of course, at times pretty wild commercial fisheries.

Herring

Herring (*Clupea harengus*) is not really an arctic species, but belong to the northern Atlantic and the Norwegian Sea, with separate sub-populations in the White Sea and the Kara Sea of Russia and in the Baltic Sea. The atlantoscandic population spawns along the west coast of Norway and the larvae are taken by the coastal current into the Barents Sea, where it grows till 3-4 years of age. During its sojourn in the Barents Sea the larvae eat primarily copepods and krill, while herring larvae are important food for immature cod. The huge population of atlantoscandic herring collapsed around 1960 and has since been at a very low level (Fig. 4.20).

Polar cod

The polar cod (*Boreogadus saida*) is a circumpolar species (Fig. 3.12) which is common around Svalbard, in the White Sea and the Kara Sea, and this truly Arctic species grows and spawns in waters of around 0 °C (Fig. 3.13). The polar cod is also found over a vast area of the central and eastern Barents Sea during autumn, and spawns west of Svalbard and west of southern Novaja Zemlya from

Figure 3.12 The geographical distribution (blue) of polar cod (*Boreogadus saida)* in arctic waters.

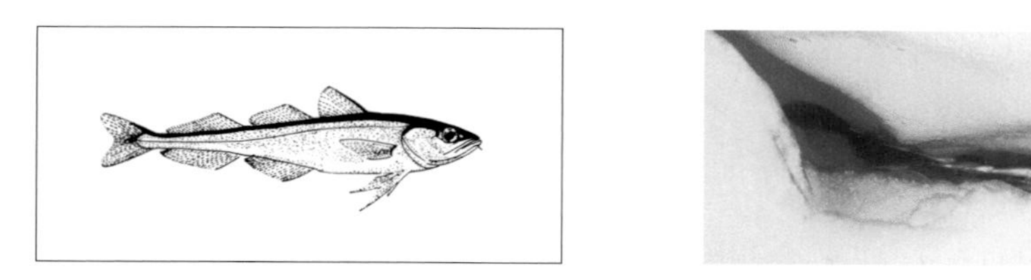

Figure 3.13 The polar cod is a truly arctic fish, which is often found in close contact with ice (Photo: B. Gulliksen).

December till February, usually in ice-covered waters. After spawning in spring it again disperses in the northern part of the Barents Sea. The polar cod larvae feed on copepods and later on hyperoid amphipods, while adults also feed on krill.

The polar cod has a slow growth rate and lives till the age of 5-6 years, at which time they have grown to about 30 cm.

Cod

There are several populations of cod (*Gadus morhua*) in the North Atlantic from Biscaya to the Barents Sea and from Iceland to the east coast of North America. The economically most important population is now the Norwegian-Arctic population in the Barents Sea, where it primarily is distributed in the Southern ice-free part and along the west coast of Svalbard. The (8-10 years old) mature animals start to migrate to the north-west coast of Norway, notably the Lofoten area, where they spawn in February -March. Immature cod migrate instead to the coast of northern Norway in early spring to feed on spawning capelin. The young larvae are taken from the spawning areas by the Coastal Current into the Barents Sea during summer and early autumn. The larvae start feeding on immature copepods and later take a variety of mature crustaceans, but most important for adult cod are capelin, herring and in some years, immature cod. Some cod may live to the age of 40 years.

The cod fisheries along the Norwegian coast have had an enormous economical and cultural importance for coastal Norway from early medieval times till this day, when the stock is relatively low (Fig. 4.20), and salmon farming is most important, at least, as far as the economical side is concerned.

Haddock

Haddock (*Melanogrammus aeglefinus*) is, only second to cod, the most important demersal fish of the Barents Sea. Like cod, it spawns off the coast of Norway in March-June whereafter it goes back to the Barents Sea. Again, like cod, the larva drift with the Coastal Current into the Barents Sea where they grow till mature age. Immature haddock feed on polychaetes, amphipods and fish larvae, later to be supplemented by krill, echinoderms and a variety of fish.

A wide range of excellent underwater photographs of marine organisms that inhabit the sea floor, and the open water as well as those associated with sea ice in the Svalbard area has recently become available (Gulliksen & Svensen, 2004).

The Kara Sea

With its western boundary at Novaya Zemlya and its eastern limit at the western shores of the Taimyr Peninsula and the Severnaya Zemlya Archipelago (Fig. 1.15), the Kara Sea is wide open to the waters of the central part of the Arctic

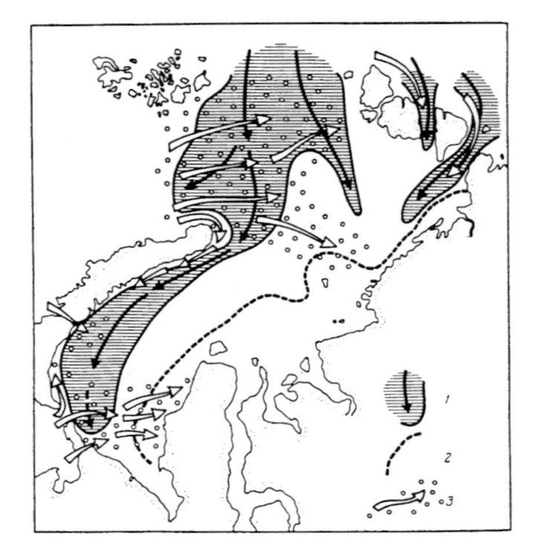

Figure 3.14 Main ways of penetration into the Kara Sea of benthos of different biogeographical nature (according to different workers). 1: Forms of the intermediate warm layer and of the cold deep layers; 2: Northern boundary of forms brought down by the discharge from the mainland; 3: Sublittoral deep-water forms and those of the Barents Sea (marked by circles) (Zenkevitch, 1963).

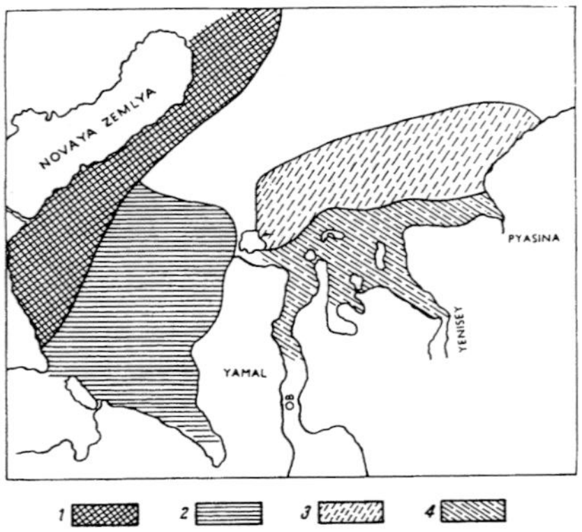

Figure 3.15 Areas of the Kara Sea according to plankton distribution. 1: Area of penetration of atlantic and arctic forms from the north; 2: Area of penetration of Barents Sea forms from the south; 3: Area of predominance of brackish water forms: 4: Area of predominance of fresh-water forms (Zenkevitch, 1963).

basin through the sound between Franz Joseph Land and Severnaya Zemlya (Fig. 3.14). Like other Siberian seas, the Kara Sea loses much of its salinity, especially in its upper layers, from the inflow from large rivers. The abundant brackish areas at river mouths and estuaries also harbour a varied, mostly original fauna, which, in its origin is a high arctic relict brackish water fauna, consisting mainly of fish and crustaceans (Fig. 3.15).

The Kara Sea is exposed, in its western part, to the influence of the warmer and more saline waters of the Barents Sea with its characteristic flora and fauna, while warm and saline atlantic waters, of the intermediate layer of the central part of the arctic basin, carrying its own fauna rich in forms, penetrate from the north through the troughs into the deeper layers of all the four Siberian seas, but mostly into the Kara Sea. The penetration of the boreal and abyssal fauna with the deep cold waters from the north is also characteristic of the Kara Sea.

The shallows off the southern coasts of the Kara Sea, on the other hand, differ greatly both in their conditions and fauna from those of the deeper central part, in that the in-shore areas are well oxygenated, warmer, less saline and populated by a fauna which is rich in variety and at times in numbers. The deeper parts with their lower temperature and higher salinity, has a thick brown mud floor and is populated by a fauna which is poor in diversity and number, with a great preponderance of echinoderms, exceptionally large sizes of invertebrates, very poor in fish, and very low biomass and productivity, as their main characteristics. The biological poverty of the Kara Sea is unquestionably related to its extensive ice-cover and the dilution of the surface layer, which prevents vertical circulation in summer and promotes an increase of ice formation during winter.

The phytobenthos of the Kara Sea is, according to Zenkevitch (1963) represented by only 55 forms, which is less than a third of the species found in the Barents Sea. The highest bottom-fauna diversity is found along the eastern shores of Novaya Zemlya and off the western coast of Yamal, due to the influx of Barents Sea waters, and in the great depth of the northern parts, where the deep water of the North Atlantic and the central parts of the Arctic Basin penetrate into the Kara Sea. As one moves from the northern parts into the south, and from the shores of Novaya Zemlya into the East Siberian Sea one experiences the least summer melt of any of the Russian Arctic seas. Normally, over 50 % of the sea still has at least a partial ice cover at the height of the melt season, spring break-up occurring in June and July while ice begin to form again in September. In the central part of the sea the diversity of the fauna is reduced, while the number of individuals of some species increases markedly. Thus, the dominance of echinoderms on the floor of the Kara Sea is such that this sea may quite rightly be called the "sea of the echinoderms". In fact, in the deep western part of the sea no less than four-fifths of the benthos biomass consists of echinoderms, the rest being mainly bivalves.

The Kara Sea fish fauna includes 53 species, but it is only in the brackish coastal waters off the mainland and along the coast of Novaya Zemlya that there exists any limited local fisheries, based primarily on whitefish (*Coregonus sp.*), frostfish (*Osmeridae*), navaga (*Eleginus navaga*) and polar cod. The exceptional poverty of the fish population of the open parts of the Kara Sea is obvious from the fact that this area is covered with ice for most of the year and plankton production is suspended for 8-9 months.

The Laptev and East Siberian Seas

The Laptev Sea is to the east of the Taymyr Peninsula and Severna Zemlya, extending to the New Siberian Islands, while the East Siberian Sea extends from there to Wrangel Island off the Chukotsk Peninsula (Fig. 1.15). The Laptev and East Siberian Seas have the most severe climate and the lowest salinity of all the seas off the northern coast of Asia. As in the Kara Sea, a deep gully enters the western part of the Sea from the north, and saline and relatively warm water flows into the Laptev Sea through it, but half of the area of the Laptev Sea is less than 50 metre deep, south of 76°N not exceeding 25 metres, and the abundant in-flow of fresh-water from the great Siberian rivers results in extremely low salinities. Thus, from a distance of more than 100 km to the northeast of the Lena estuary the salinity is 5 to 6 ‰ down to a depth of 20 to 25 metres, and in 1893 Nansen of Norway, with the "Fram" recorded a salinity of 14.9 ‰ 500 km from the Lena estuary. The result of this is that the East Siberian Sea experiences the last summer melt of any of the Siberian seas. Normally, over 50 % of the sea still has at least a partial ice cover at the height of the melt season, while freeze-up begins in September and is complete by mid-October. After a long harsh winter, spring break-up occurs in June or July and the surface waters of those parts that are freed from the ice are warmed partly by the river waters, and partly by the sun, to a few degrees above zero. But the pack-ice is not far away even in summer.

Studies of the Laptev Sea flora and fauna are few and far between, but most of the phytoplankton seems to be brackish water forms. The mean biomass of the summer zooplankton of the Laptev and East Siberian Seas may, again according to Zenkevitch (1963), be as low as 72 mg/m^3, and the number of species is as low as 49. So far 405 species of zoobenthos have been found in the Laptev Sea, but the diluted water masses of its southeastern parts only contain 73 species, with high-arctic species being overwhelmingly dominating. A total of altogether 39 species of fish, being predominantly brackish forms, have also been found in the Laptev Sea, while the presumably even lower number in the East Siberian Sea is unknown. Some detailed data, primarily on the land-ocean systems of the Laptev Sea have recently been published (Kassens *et al.*, 1999).

The Chukchi Sea

The Chukchi Sea is located to the east of Wrangel Island and reaches as far as Point Barrow in Alaska, and is connected with the Pacific Ocean by the shallow Bering Strait (Fig. 1.15). The Chukchi Sea is also very shallow, being for the most part less than 50 m deep, its greatest depth being 180 m. A fairly warm, strong current enters the Chukchi Sea through the Bering Strait. In general the movement through the Bering Strait is that of the Pacific Ocean waters into the Chukchi Sea, and only to a very small extent a flow of the Chukchi waters to the south. Still, the conditions in the Chukchi Sea are rather austere. For seven months (November to May) the temperature of even the surface waters remains below

-1.5 °C; in June, September and October it reaches about 0 °C, and only in July and August, does it off the coast, rise appreciably above zero, while the deep layers of water have a temperature of almost 0 °C even in mid-summer. The severe temperature conditions, the shallow depth, the preponderance of hard bottoms, low salinity and ice-cover naturally result in a very low biodiversity. Thus, the fish fauna includes 37 species, of which 75 % are of 5 families. In this context, it is rather amazing to find that some typical atlantic forms, (*Portlandia lenticula* and *Portlandia fraterna*), along with atlantic water, have penetrated as far as the edge of the continental shelf in the northern parts of the Chukchi Sea over the arctic basin, all the way from the Atlantic Ocean.

The Bering Sea

The Bering Sea is framed by the Seward and Chukchi peninsulas on the north, and by the Kamchatka Peninsula and the 1,900 km long Aleutian chain to the south (Fig. 1.15). The narrow, 85 km long passage of the Bering strait connects the Bering Sea in the north to the Chukchi Sea and the Arctic Ocean. The Bering Sea lies in the northern part of the Pacific Ocean. It exchanges water with the Arctic Ocean and with the Pacific Ocean , into the Bering Sea from the Gulf of Alaska through the Aleutian Islands and out of the Bering Sea through Kamchatka Strait (Fig. 3.16). The Bering Sea is distinguished by a very broad continental shelf – approximately half of the geographic area – with an average water depth of only 50-75 m. The remainder is an oceanic basin 3,000 to 4,000 m deep. Continental shelves are in general more productive than most other regions of the ocean, because tidal and wind energy can occasionally interact with sufficient strength to resupply nutrients to the well-illuminated surface layers from reservoirs beneath the thermocline. The wide shelf in the Bering Sea provides an expansive area over which such physical processes can occur.

This shelf edge area, which is often called *the green belt* (Springer *et al.*, 1996), because of its prodigious productivity, apparently extends around the entire perimeter of the Bering Sea continental shelf and across the northern shelf, through Bering Strait, and into the Chukchi Sea (Fig. 3.17). The interaction of strong tidal currents with the abrupt, steep shelf break promotes upwelling at the front, which also helps supply nutrients to the surface layers. As a result, primary production apparently remains elevated throughout summer, long after the termination of the spring bloom.

Among the numerous species of herbivorous zooplancton in the Bering Sea, large calanoid copepods are particularly important links in the food web between

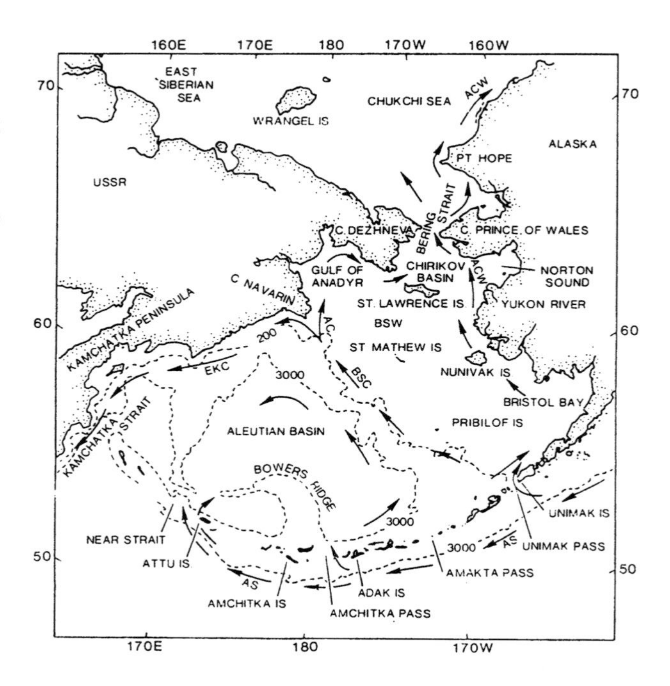

Figure 3.16 Major currents of the Bering and Chukchi seas. ACW is Alaskan Coastal Water, AS is the Alaskan Stream, EKC is the East Kamchatka Current, BSC is the Bering Slope Current, and AC is the Anadyr Current (Shuert & Walsh, 1992).

phytoplankton and higher trophic levels. *Neocalanus cristatus, N. plumchrus, Eucalanus bungii* and *Metrida pacifica* are oceanic species found in deep water across the North Pacific. They are large, efficient grazers of large diatoms typical of the spring bloom flora, and are in turn important prey of most species of planktivorous fishes, birds and mammals. However, they do not occur over the shelf inshore of the middle or coastal areas where they are replaced by *Calanus marshallae,* the only large copepod found higher on the shelf, and by various species of small copepods.

The benthic communities of the Bering Sea are dominated by a large variety of polychaetes, amphipods, gastropods and bivalves. They vary tremendously in space due to food supply, depth, disturbance, sediment type and predation. The infaunal community structure is generally dominated by bivalves and polychaetes, although it becomes dominated by amphipods in the Chirikov basin, north of Saint Lawrence Island. Epifaunal benthos on the eastern Bering Sea shelf is numerically dominated by bi-valves, and to a lesser extent, arthropods and echinoderms; however, nearly 70 % of the total epibenthic biomass is made up by sea stars. Limited studies of fauna on the shelf break and deep basin in the eastern portions of the Bering Sea indicate that the benthic fauna there is dominated by

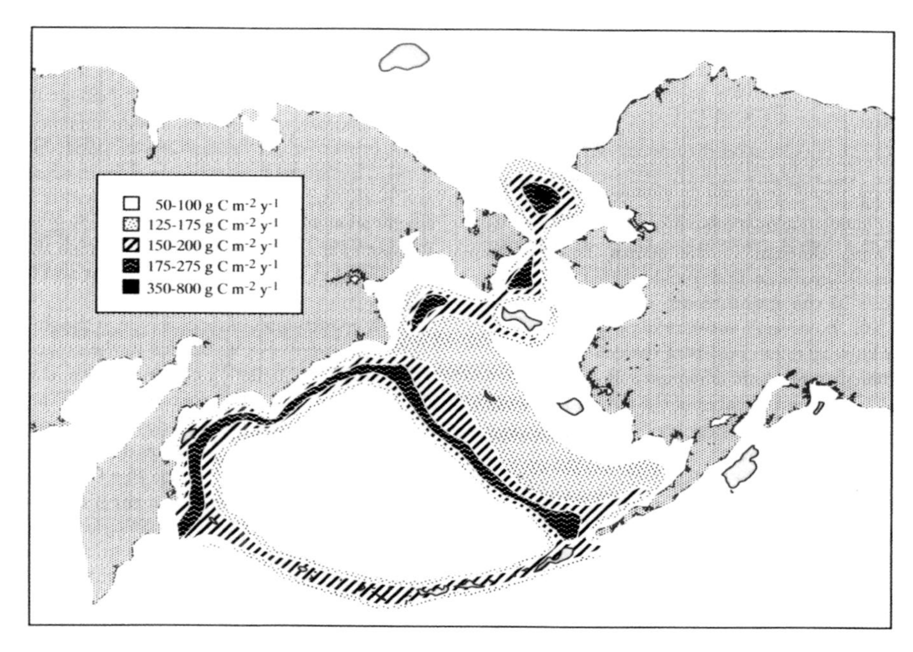

Figure 3.17 Generalized pattern of primary production in the Bering Sea. The Green Belt is the region of high productivity at the shelf edge with branches to the north-west and south-west. Elevated production along the Aleutian Arc completes a circle around the basin (Springer, McRoy & Flint, 1996).

polychaetes of low biomass. Although there are many benthic and pelagic invertebrates in the Bering Sea, only a few are commercially significant, and because of their commercial importance, most of the available data are on squid, crabs, scallops and shrimp.

Several species of squid, of which *Berryteuthis sp.* and *Gonatus sp.* are the most important, inhabit the Bering Sea seasonally and also are widely distributed. Predators of squid are marine mammals and pelagic fishes.

Three species of king crabs are found in the Bering Sea and Aleutian Islands region. The red king crab (*Paralithodes camtschatica*) is the largest, most widespread and most abundant. They are generally found on the continental shelf in the eastern Bering Sea, along the Aleutian Islands and the Gulf of Alaska, generally at depths of 180 m or less. In the western Bering sea, two large populations of red king crab have been identified in the Sea of Okhotsk and Kamchatka shelf area and the second in the northwest Ayano-Shantarskii. The red king crab was transplanted by the Soviets from the Sea of Okhotsk to the Fjord of Murmansk in the Kola peninsula area of the Barents Sea in 1961, and has now established itself and spread westwards into Norwegian waters, where it is now being harvested commercially. The blue king crab (*P. platypus*) has a more limited distribution. Bering Sea populations are found around Saint Lawrence Island, Saint Matthew Island and the Pribilof Islands, while the brown king crab (*Lithodes aequispina*) is the smallest of the three commercial species and inhabits deep water (more than 180 m) along the continental slopes of the North Pacific Ocean and the Bering Sea. In addition to these crabs, tanner crabs (*Chionoecetes spp.*) are found throughout the Bering Sea region. Two species are relatively large and therefore are of commercial importance. The tanner crab (*C. bairdi)* is the larger of the two species and is usually encountered in the same habitat as red king crab. The snow crab (*C. opilio*), which is smaller, but believed to be the most abundant of the tanner crabs, is found in the northern and central Bering Sea on the continental shelf of Russia and Alaska.

Eight pandalid shrimp species representing two genera are taken in commercial and subsistence fisheries off Alaska, but only five are of commercial importance. For many years shrimp fishermen have targeted primarily what is locally called pink shrimp (*Pandalus borealis*), with up to 85 % of the total shrimp catch in some years made up of this single species. One unique aspect of shrimp life history, which is worth mentioning, is individual transformation of sex. Almost all shrimp develop as males, although initial development as females has been reported in a few species. Shrimp reach maturity as adult males two years from hatching. At two and one-half years of age most breed as males, although at three and one-half years a few have already transformed into females. By four and one-half years of age, all shrimp have transformed into sexually mature females. Six months is the average time required for an individual to change sex, but few females live beyond six years.

Five species of pectinid scallops are commonly found in the Bering Sea and Aleutian Islands area. The most common and the primary target of the commercial fishery is the weatherwane scallop (*Patinopecten caurinus*), which are found in the intertidal waters to depths of 300 m on beds of mud, clay and sand.

In the Bering Sea, the partly bottom-water dwelling Alaskan, or walleye, pollock (*Theragra chalcogramma*) is by far the most important finfish. It inhabits continental slope and shelf waters along the northern rim of the North Pacific, extending from southern Oregon into southern Chukchi Sea and south along the Asian coast to the southern Sea of Japan (Fig. 3.18). At present, the walleye pollock stock represents the greatest single-species biomass in the Bering Sea, accounting for approximately 50 % of the total biomass of all groundfish in eastern Bering Sea, Aleutian Islands and Bogoslof area, and they probably account for at least that pro-

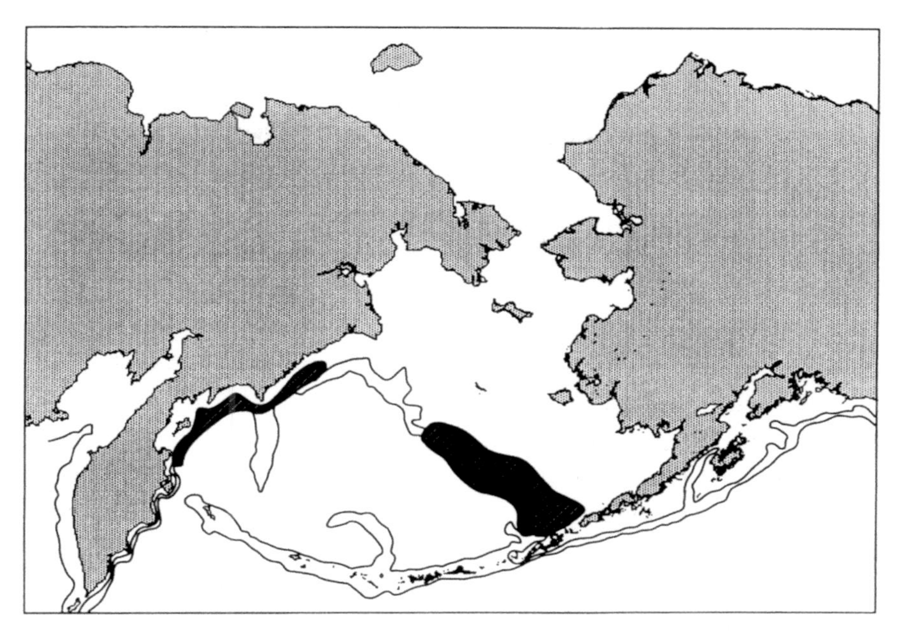

Figure 3.18 Generalized distribution of the main spawning regions of walleye pollock in the Bering Sea (Springer, McRoy & Flint, 1996).

portion of the biomass in the western Bering Sea, as well. Spawning occurs primarily in the first quarter of the year in the Aleutian Basin with major aggregations occurring in the vicinity of Bogoslof Island, and from March through October along the eastern Bering Sea outer shelf, with major spawning concentrations occurring between the Pribilof Islands and Unimak Island. Diets of adult pollock consist mainly of copepods, euphausiids and fish, the majority of which are juvenile pollock. Pollock are a major part of the diets of northern fur seals and other marine mammals, and are also important prey for many seabirds and other fish in the area. The estimated adult biomass of walleye pollock has been fairly stable over many years, amounting to a staggering 10 million metric tons.

Capelin *(Mallotus villosus)* are widely distributed in the Gulf of Alaska, Bering Sea, and Sea of Okhotsk and along the Kamchatka Peninsula, and have been known to be very abundant in the Bering Sea. Spawning occurs in spring in intertidal zones of coarse sand and fine gravel. In the Bering Sea, adult capelin are found near shore only during the months surrounding the spawning run. At other times of the year they are found far off-shore in the vicinity of the Pribilof Islands and the continental shelf break. Their seasonal migrations may be associated with the advancing and retreating pack-ice front in the Bering Sea. Capelin are not harvested commercially for the time being, but are commonly found in the stomachs of seabirds, marine mammals and fishes, and are believed to be important prey when, and where, they are abundant.

All five Pacific salmon species are found in the Bering Sea area. Sockeye salmon (*Oncorhynchus nerka*) is the most common, the Bristol Bay supports the largest sockeye fishery in the world. Chinook (*O. tshawytscha*), chum (*O. keta*), coho (*O. kisutch*) and pink (*O. gorbuscha*) salmon are all found in lesser degrees of abundance. Depending on species, juvenile salmon migrate from their freshwater rearing areas between birth and age two. Once in the marine environment, juvenile fish follow an annual migration pattern that takes them away from the coast and into the Bering Sea and North Pacific Ocean. Once the fish achieve the proper age and size for reproduction, they return to their freshwater streams of origin to spawn, an event that takes place between early summer and early winter. After spawning, all salmon of these five species die.

Predators in the oceans include seals, killer whales, and humans. After spawning, the dying salmon are preyed on by bears, for which they are very important for the pre-hibernation fattening (Fig. 3.19).

In addition to these pelagic fisheries, salmon are caught during the spawning runs by coastal people who often establish villages or temporary fishing camps at

Figure 3.19 Brown bears enjoying a day of salmon fishing in an alaskan river.

the mouths of rivers and along streams to take advantage of this annual event. The importance of salmon to native peoples of the Bering Sea is reflected in their culture, art and song. Sockeye salmon from the Bristol Bay area have always dominated the eastern Bering Sea salmon catch and continue to do so. The Bristol Bay catch for all salmon species in 1993 totalled approximately 42 million fish, of which sockeye salmon comprised 41 million.

The Beaufort Sea and the Canadian Arctic Oceans

Numerous channels connect the Arctic Ocean and the Atlantic Ocean through the Canadian archipelago (Fig 1.15). In the deep channels of the Queen Elizabeth Islands and the western archipelago currents are weak and variable. There is southward flow toward Peary channel and stronger eastward flows on the south side of Barrow Strait. In the broad southern channels of the archipelago and in the passages connecting with the Arctic Ocean, the few available measurements show southward flows, but because of shallow sills only surface waters penetrate from the Arctic Ocean into Baffin Bay.

Sea ice is persistent throughout the year in the Beaufort Sea (Fig. 1.15), except near the mainland coast in summer. Away from the coast, the ice cover is close to 100 % for nine or ten months of the year. In the southeastern sector of the Beaufort Sea there is commonly a well developed north-south lead along the edge of the fast ice off Banks Island. During intervals of strong and persistent easterly winds, a large polynya may also form in this location in May or June (Fig. 3.20). Such polynyas have great importance for both marine mammals and birds (e.g. Stirling, 1977). There are significant contrasts between ice conditions in the northern and western parts of the Arctic archipelago and those elsewhere. For example, Sverdrup Basin is normally completely covered by multi-year ice throughout the year, with ice 2-3 m thick. Usually ice opens and moves in the eastern archipelago only during late summer and early autumn. In contrast, ice breaks and clears early in Lancaster Sound, where the ice is 1-1.5 m thick; and it does the same on the northern side of Viscount Melville Sound. Surface melt begins in the southern channels during April and has by June extended to all of the archipelago. Breakup occurs first in the south (from the Amundsen Gulf to Queen Maud Gulf) and from Baffin Bay westward towards Barrow Strait, with openings appearing on the southern and eastern sides of Melville Island in late July and August, with open water reaching its maximum extent in September.

The biology of the permanently ice-covered waters of the high Arctic Canadian archipelago is poorly known, but probably poor also in biomass and diver-

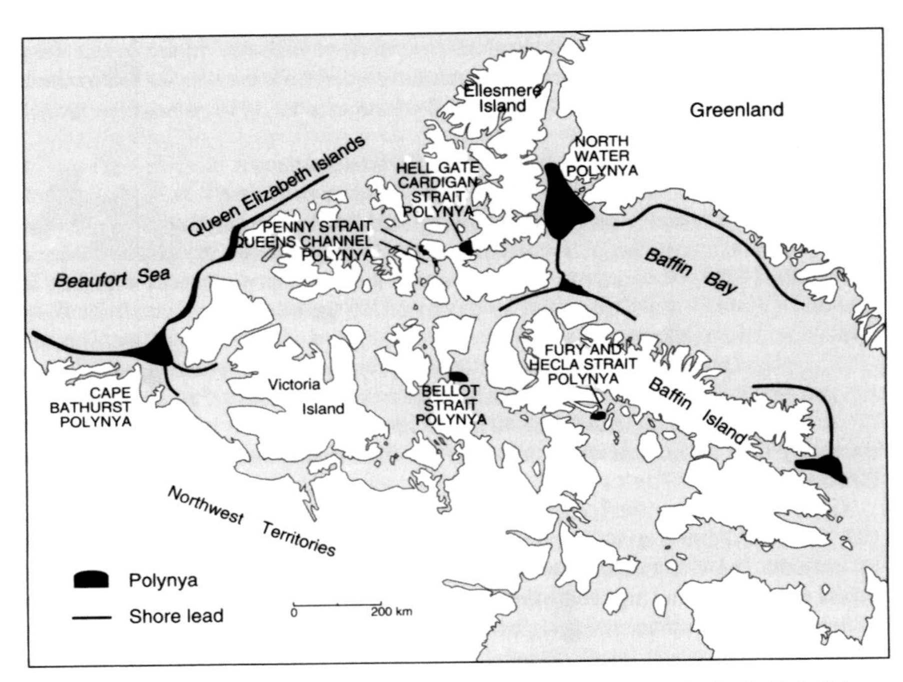

Figure 3.20 Locations of recurrent polynyas and shore leads in the Canadian Arctic (Smith & Rigby, 1981).

sity, much like the East Siberian Ocean. However, Welsh *et al.* (1992) have studied the marine ecosystem of the Lancaster Sound region, which is normally ice-free from the end of July till the end of September, whereafter polynyas and leads sometimes provide areas of open water. They have described an ecosystem of primary producers, of which ice algae and kelp contribute only 10 % and 1 %, respectively. Ice-associated amphipods, like *Apherusa glacialis, Onisimus glacialis* and *Weyprechtia pinguis* are present, but seem to be of minor importance, while *Pseudocalanus acuspes, Calanus hyperboreus, C. glacialis and Metridia longa* seem to account for nearly all the energy flow through the herbivorous copepods. The bivalve *Mya truncata* is important among the benthic organisms, with a high biomass between 10 and 30 metres of depth, but with relatively little growth. Other bivalve clams are *Hiatella arctica*, the large *Serripes groenlandicus* and the small, but numerous *Macoma calcarea.* Sea urchins are other abundant benthic species, and brittle stars are particularly important numerically in deep water. Sea cucumbers, pycnogonids, terrebellid polychaetes and anemones are other noticeable benthos, while the ctenophore *Mertensia ovum* reach high summer biomass in the water column.

Of the vertebrates, the polar cod is the most important among the fishes. This species is generally found everywhere in low concentrations throughout the year, but in summer they occasionally occur in enormous schools, in one case, one such school was estimated to contain 12,000 tons of fish. Such schools most often occur in nearshore waters and are subjected to intense predation by hordes of seabirds, primarily black-legged kittiwakes and northern fulmars, but they are also preyed upon by harp seals, beluga and narwhal.

CHAPTER 4

Marine Mammals

Seals and whales are marine mammals which are very well adapted to their aquatic environment. Whales spend their entire life-cycle in water, while all species of seals haul out on ice or shores to give birth and to moult. Both groups of animals are equipped with a thick subcutaneous layer of fat (blubber) which contributes to the streamlining of the body, and hence, reduces the cost of swimming, and provides insulation against the ice-cold arctic water as well as storage of energy for periods of fasting during breeding and moult.

The arctic seals give birth on ice-floes in early spring, when weather conditions are rather bad. The pup is nursed for a period of a few days to several weeks, depending on species, and then abandoned by the mother who goes off to get bred by the males. Gestation period is about 11 months, but most species have delayed implantation of the blastocyst for about 2 months. Both sexes go to sea and fatten up after breeding for a couple of months and then haul out again to moult for a period of a few weeks, during which period the animals seldom venture into water and consequently live off their body reserves. This results in dramatic seasonal changes in the body composition of most arctic seals (Fig. 4.1).

The three species of truly arctic whales, of course, give birth and nurse their thermally underprivileged calves in ice water from day one. The size of the newborn calf differs wildly, from about 80 kg in beluga and narwhal to over a ton in the bowhead. Newborn seals, like newborn whales receive a very energy rich milk (Table 10.2), but unlike seals, the whales normally nurse their young for more than a year, and the arctic whales consequently only reproduce at intervals of 2-3 years.

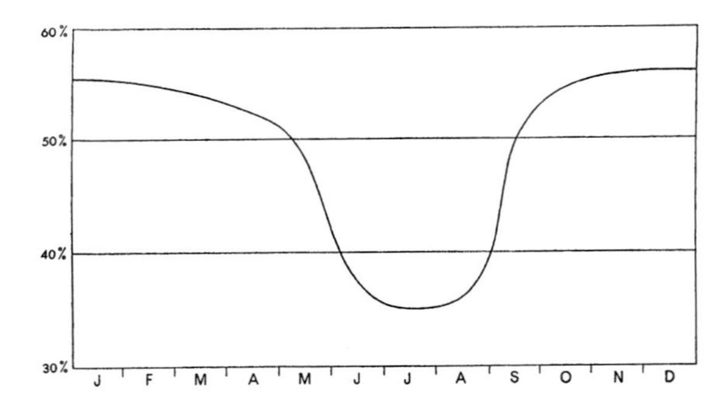

Figure 4.1 Blubber in % of total body mass in 97 adult ringed seals shot at Svalbard throughout the year in 1954/55 (Lønø, 1970).

Most arctic seals give birth to a pup which is equipped with a coat of (lanugo) fur which offers prime insulation against cold in air, but only the hooded seal pup, which sheds its lanugo fur *in utero*, is born with a subcutaneous layer of fat. The other acquire such a layer of blubber from suckling the extremely energy rich milk (Table 10.2). At the time of weaning, when they are abandoned by their mothers, the pups start to moult, and aquire the usual short seal fur in preparation for going to sea. During the moult, which takes in the order of a couple of weeks, the pups live off their fat reserves and are ready to transfer to marine life at the time when food becomes abundant in the arctic seas. The newborn whale (Fig. 10.32) has, of course, no fur at all, but these creatures are born with a subcutaneous layer of fat, which helps to keep them warm in the hostile arctic waters.

There are two suborders of the order of whales (*Cetacea*): the toothed whales *(Odontoceti)* and the baleen, or filterfeeding whales *(Mysticeti)*, while the order of seals (*Pinnipedia*) are divided into three families: true seal *(Phocidae)* which are cigar-shaped, with short fore-flippers, and haul or hump themselves caterpillar-like across the ice, eared seals *(Otariidae)*, with external ears and strong forelimbs and the ability to swing their hind-flippers forward to assist in quadrupidal locomotion, and walrus *(Odobenidae)*.

Arctic seals

Spotted seal (*Phoca largha*)

The spotted seal was previously recognized as a subspecies of the common seal (*Phoca vitulina*), but is now established as a separate species (Fig. 4.2). It is a small seal, the adult male weighing about 100 kg, the female a little less, while the newborn weigh about 10 kg. Unlike the young of the common seal, the young of the spotted seals are born with a whitish lanugo fur, which is shed two to four weeks after birth. The fur of the adult consists of short coarse overhair and shorter finer

Figure 4.2 The geographical distribution of common seal (*Phoca vitulina* ; dark blue) and spotted seal (*Phoca largha* ; light blue) in the Arctic.

under-hairs. The colour is greyish to brown with numerous dark spots, the back generally appearing much darker than the belly.

The spotted seal is littoral in summer when the ice has melted away, but in fall and winter it migrates to the edge of the pack-ice, where it breeds both in the Bering Sea and the Okhotsk Sea. The population in the Bering and the Okhotsk Sea is not well known, but may count in the order of 200-300,000.

The pups feed on small amphipods around the ice-floes. The adults are believed to be rather opportunistic feeders that take crustaceans, cephalopods and a wide variety of fish. These seals are monogamous during breeding, which takes place in the period from February to May, depending on location. After breeding all ages congregate in large moulting patches before the ice melts and the seals move to the coast. While the common seals swim the day they are born and dive at age 2-3 days, the pups of the spotted seal do not swim until after weaning, at the age of 2-6 weeks, at which time they may dive to 80 m, adults diving to 300 m.

The common seal (*Phoca vitulina*) is not considered an arctic species, but nearly at 80°N, at Svalbard there is a small population (approx. 500 individuals) which is about 900 km away from any other population of common seals (Wiig, 1989). These seals live in an area which is ice-free for most of the year and the pups are born without any white lanugo, unlike the spotted seal pups, in the second half of June, which is the same as the breeding time for common seals at the coast of the Norwegian mainland. The distribution and diving behaviour of these animals have been studied (Giertz, Lydersen & Wiig, 2001), but aside from this the biology of this very interesting population of (protected) arctic seals is presently unknown.

In 1988 and again in 2002 a large part of the European harbour seal (*Phoca vitulina vitulina*) population was wiped out due to a very contagious phocine distemper virus (e.g. Heide-Jørgensen *et al.*, 1992), but there are no signs of this disease having spread either to Svalbard or the Bering-Okhotsk Sea population of spotted seals.

Ringed seal (*Phoca hispida*)

The ringed seal is the smallest of all phocid seals and has a circumpolar arctic distribution, with subspecies in the Baltic Sea, and truly freshwater relict subspecies in the Caspian Sea and in the lakes Saimaa, Ladoga and Baikal (Fig. 4.3), where they were stuck when the ice withdrew and the water-level sank after the Pleistocene glaciation. To the peoples along the arctic coasts the ringed seal has always had a great subsistence as well as economic importance. During breeding these animals have a high affinity for stable landfast ice with uneven surfaces where snow accumulates, and in which they construct their breeding lairs (Fig. 4.4). In Svalbard, however, much of the breeding takes place on flat fjord ice where snow accumulation is rarely enough to permit birth lair construction. In these areas pups are often born in the open. In summer all age-classes are found along the edge of the permanent pack-ice and in near-shore ice remnants. For the seals which winter in the Bering Sea this requires a migration of many hundreds of kilometers. The total population of ringed seals is not known and vast arctic areas have never been properly surveyed, but it is still quite certain that ringed seals are the most abundant of the arctic seals, counting several million individuals.

Figure 4.3 The geographical distribution of Ringed seals (*Phoca hispida)* in the Arctic.

The name ringed seal refers to the prominent grey-white rings found on the generally dark grey backs of the adult seals. The belly is usually silver, lacking dark spots. The pups are born in early April with a white woolly coat. At birth the pups weigh only about 4.5 kg and are on average only 65 cm long. Growth continues for 8-10 years with a mean weight of adults of both sexes of about 50 kg. The ringed seals construct birth lairs under the snow (Fig. 4.4), in which pups are born and spend the first 5-7 weeks of life together with the mother. During the nursing period the weight of the pup more than doubles and the lanugo is shed 2-3 weeks after birth. It has recently been shown by Lydersen & Hammill (1993) that, unlike most other arctic seals, the ringed seal pup often ventures into water,

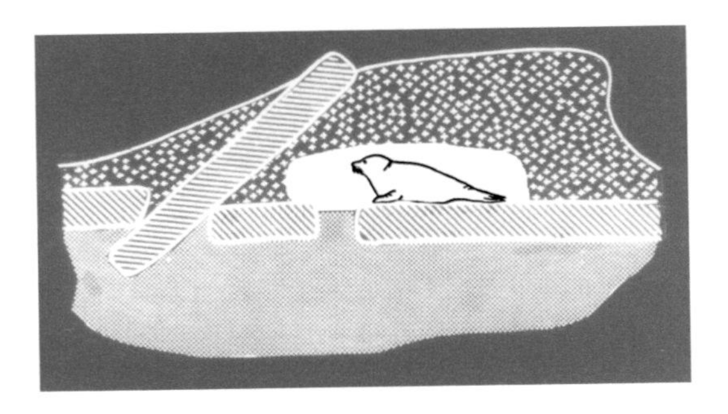

Figure 4.4 The snow lair of the ringed seal, which provides protection both from weather and predators.

where it may dive sometimes down to 90 metres, even during the nursing period. Mating occurs one month after parturition. Ringed seals mature at about 5-7 years of age.

The moult in ringed seals occurs over several months, with a peak in June. During the moult the animals haul out on the ice along cracks and leads and bask in the sun. Average life-span is probably between 15-20 years, but some grow to be as old as 40 years. Predators of the ringed seals include ravens, arctic foxes, wolves, dogs, polar bears and humans.

In late summer, fall, winter and early spring, ringed seals spend most of their time in the water feeding. Foods eaten vary markedly by season and geographical area. Polar cod, pelagic amphipods and other crustaceans make up the bulk of the diet. A seasonal cycle of feeding intensity has been well documented. Feeding is at a minimum during the moult period but does not cease entirely. Ringed seals are able to dive for at least 25 min and reach depths in excess of 200 m (Gjertz, Kovacs, Lydersen & Wiik, 2000).

Ribbon seal (*Phoca fasciata*)

The ribbon seal has long remained an animal about which little is known. This is entirely understandable in view of its distribution in the seasonally ice-covered seas of the North Pacific region and because they seldom occur near shore. The range of the ribbon seal includes the seasonally ice-covered regions of the Okhotsk, Bering and Chukchi Seas, where they are associated with moderately thick smooth clean ice, of the kind present at the inner zone of the ice front, during the late winter, spring and early summer (Fig. 4.5). Their distribution and activities during the late summer and fall are not known. Since they apparently are never seen anywhere in this period it is likely that they become pelagic. It is assumed that the population of ribbon seals is in the order of 200-300,000 individuals.

Births occur over a period of almost 5 weeks from about 3 April to 10 May, the majority in mid-April. At birth the pups weigh about 10 kg and are covered

Figure 4.5 The geographical distribution (blue) of ribbon seal (*Phoca fasciata)* in the Arctic.

by a white lanugo. The nursing period lasts for 3-4 weeks, during which time the pup doubles its weight and the lanugo is shed. Breeding occurs shortly after the pups are weaned. Adults begin their moult during the first half of May. Normal lifespan is probably 20 years.

The food habits of the ribbon seal are little known and the few stomach samples which exist were taken during May and June when very little feeding occurs due to the moult. In this period it appears that pollock (*Theragra chalcogramma)* and cod (*Gadus macrocephalus*) are taken in the Okhotsk Sea, while shrimp (*Pandalus sp.* and *Sclerocrangon sp.*), pollock and arctic cod are taken in the Bering Sea.

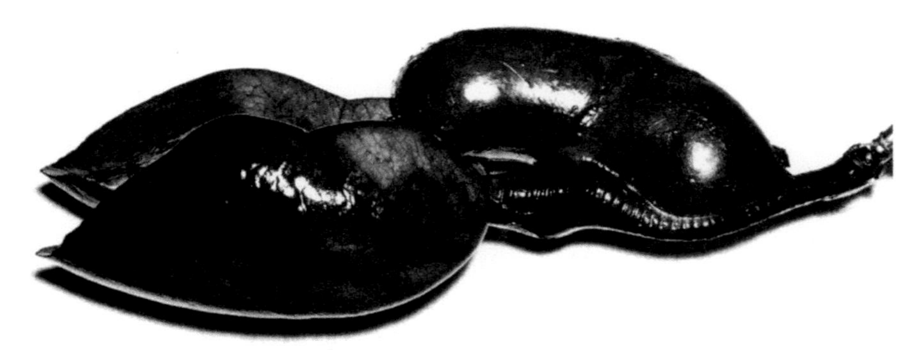

Figure 4.6 Excised and inflated trachea and air sac (right) and lungs (left) of a three-year-old male ribbon seal (Photo: J. Burns).

In ribbon seals a slit-like opening, located approximately three-quarters of the distance from the anterior end of the trachea, is connected to an air sac. This structure occurs in both males and females, although the air sac (Fig. 4.6) is considerably more developed in adult males. The function of this interesting structure is unknown.

These seals have an unusual tolerance of boats and humans and often rest on ice at a considerable distance from water with comparatively long sleep intervals, indicative of lack of predators.

The most striking and obvious characteristic of the ribbon seal is its distinctive coloration and pattern of markings (Fig. 4.7).

Bearded seal (*Erignathus barbatus*)

The bearded seal has got its name because of its long and numerous moustachial vibrissae, which unlike other phocids are usually curled at the tips. The bearded seal is one of the largest of the northern phocids, weighing on average 230 kg, but with a disproportionately small head. These seals have 4 retractable teats (like the walrus), while other northern phocids have two. The colour of the bearded seal is variable, but most adults of both sexes are basically light to dark grey, slightly darker down the middle of the back. Others may be tawny-brown to dark brown. Newborn pups have dark (usually brown), dense lanugo fur (Fig. 4.8), but at the time of weaning the pelage resembles that of older seals.

The bearded seals have a circumpolar range (Fig. 4.9) They avoid regions of continuous, thick shorefast ice (unlike ringed seals) and are not common in regions of unbroken, heavy, drifting ice. These seals utilize instead areas where

Figure 4.7 A ribbon seal, sporting its almost bizarre fur coat (Photo: L.M. Shults).

Figure 4.8 The newborn bearded seal has a dark brown fur, which is rather unusual for a newborn mammal in the high Arctic (Photo : A.S. Blix)

the ice is in constant motion, producing leads, polynya and other openings (e.g. Stirling, 1997).

The pupping period is long, extending from mid-March to the first week of May. Pups are born onto the ice, some times, in my own experience, at ambient temperatures as grisly cold as -40 °C. Nursing period is comparatively short, lasting 12-18 days. During this perid of time the weight of the pup increases from a birthweight of 35 to 85 kg. Bearded seals can swim as soon as they are born, and Lydersen, Hammill & Kovacs (1994) have shown that soon after birth they spend half the day in the water and may dive to 80 m, while actively feeding. Thus, in this respect they are unusual among arctic pinnipeds, but remarkably similar to the ringed seal pups. Both have evolved under heavy predation from polar bears and it is likely that this behaviour is important to avoid predation.

The main breeding period coincides with the end of lactation. The bearded seal may live to the age of 30 years. Moulting peaks in May- June and coincides

Figure 4.9 The geographical distribution (blue) of the bearded seal (*Erignathus barbatus)* in the Arctic.

with the period of maximum haul-out. Major predators on bearded seals are polar bears and man, for which these seals have been important for subsistence for thousands of years. Boats, lines, clothing and other items have been made from their very durable skins. The total population of bearded seal is probably in the order of 1 million. Except for the mother-pup pairs and breeding animals, bearded seals tend to be solitary.

Bearded seals are very vocal and produce a distinctive song. My old friend, John Burns of Alaska, tells that their long musical underwater sounds are well known to Eskimo hunters who in the days of kayak hunting tried to locate them by listening for them. Although parts of the song are audible at close range in air, it is easily heard by placing a paddle in the water and pressing an ear against the butt of the handle. These songs are stereotyped and repetitive and performed when they dive slowly, apparently in a loose spiral, while releasing bubbles. It has been suggested by Rey, Watkins & Burns (1969) that the song is produced by mature males and that it is a proclamation of territory, but it is likely that both sexes take part in it.

The bearded seals are benthic feeders utilizing primarily epibenthos. Feeding depths are usually between 50 and 200 m. The total array of food items consumed by bearded seals is quite large, but relatively few species comprise the bulk of the diet. In the Barents and the Kara Seas primarily crustaceans, like shrimp, and molluscs, like gastropods and bivalves are eaten. Other food include a variety of worms (notably *Priapulus sp.*), cod and other demersal fish. In the Okhotsk Sea the most important food are crabs (mainly *Hyas coarctatus aleuticas*), echiuroids, crangonid shrimps and others. In the Bering-Chukchi Seas the snow crab (*Chionocetes opilio*) is most important, while other significant prey include the crab *Hyas coarctatus*, which together with arctic cod also seems to be the most important food item in the Beaufort Sea.

Hooded seal (*Cystophora cristata*)

The young of the hooded seal shed a creamy white lanugo fur *in utero* and is born with a short pelage which is blue-grey on the back and cream white on the belly (Fig. 4.10), hence this pup is called "blue-back" by the sealers. After the yearling coat is shed at the first annual moult, mottled pelage patterns begin to appear.

Figure 4.10 The newborn hooded seal, named "Blue back" for its steel-blue colour on the back, while the belly is snowy white (Photo: A.S. Blix).

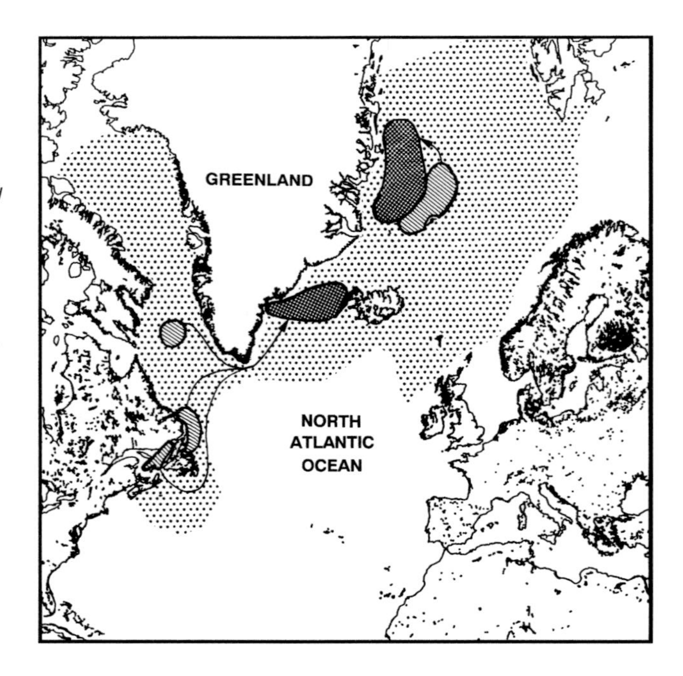

Figure 4.11 The geographical distribution of hooded seals in the North Atlantic (lightly dotted area). Hatched areas: breeding grounds; densely dotted areas: moulting grounds. Arrows indicate migration routes from breeding grounds, to moulting areas (Folkow & Blix, 1995).

These consists of dark brown, brownish- black, or black patches against silvery grey background all over the body, more pronounced on the back and upper sides (Fig. 4.14). Males are much lager than the females at maturity, and weigh from about 200 to 360 kg, while mature females weigh 150-250 kg.

The normal distribution of the hooded seal is limited to the Arctic and sub-Arctic North Atlantic, where the hooded seals prefer rather heavy ice (Fig. 4.11). The population probably consists of three different breeding stocks, with whelping areas in the pack-ice off Jan Mayen, in the Davis Strait and off the Coast of Labrador/Gulf of St. Lawrence. We (Folkow, Mårtensson & Blix, 1996) have recently shown that the animals from the Jan-Mayen stock aggregate in the area north of Jan Mayen for moulting, while it has long been known that the animals from the other stocks migrate to moult in the Denmark Strait. We have also shown by use of satellite technology that the animals of the Jan Mayen stock, which breed in late March, disperse widely, but mainly along the edge of the pack-ice after breeding, and that they concentrate to moult along the ice-edge north of Jan Mayen in June-July, whereafter they again disperse widely in the Norwegian and Greenland Seas (Fig. 4.12). These seals mainly perform meso/bathypelagic dives of 5-25 min duration down to 100-600 m, but particularly some individuals seem to specialize in deep dives down to a staggering 1,000 m with durations up to one hour (Fig. 4.13) (Folkow & Blix, 1999). From our, and other

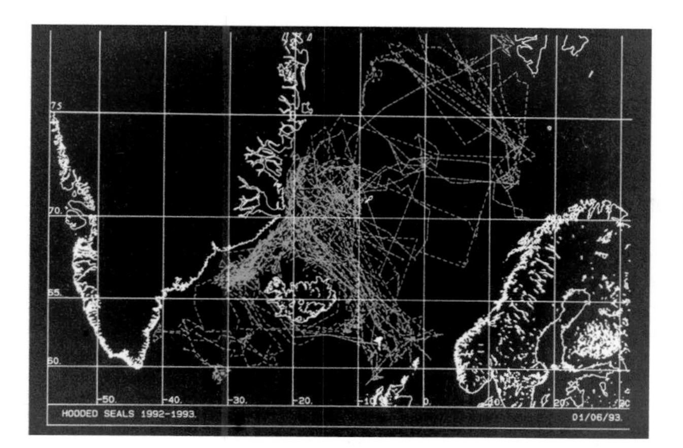

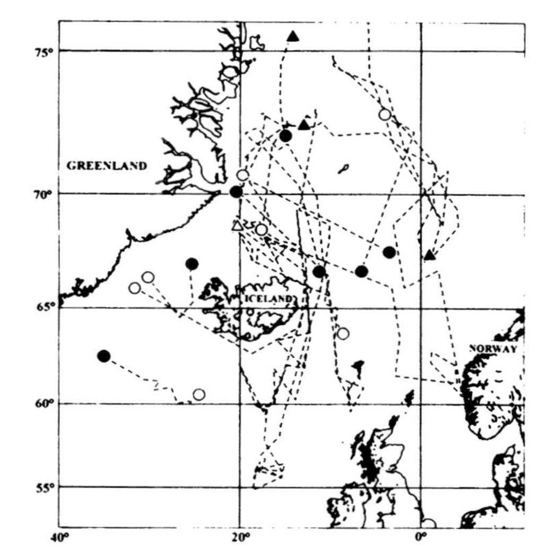

Figure 4.12 *Top*: Distribution and overall movements of 15 hooded seals of the West-ice, or Jan Mayen, stock obtained by satellite telemetry, between moulting in July and breeding in March. Note, however, that these track lines do not correctly reflect the frequency distribution of seals in different areas, since location fixes from hauled out seals in the sea ice were obtained much more frequently than from seals diving at high sea. (Folkow, Mårtensson & Blix, 1996). *Bottom*: Distribution and overall movements of 11 hooded seals of the West-ice, or Jan Mayen, stock, as determined by satellite telemetry, between breeding (April) and moulting (July). Open and filled symbols, tagging location and end (lost) location, respectively (Folkow, Mårtensson & Blix, 1996).

information, it is likely that hooded seals primarily feed on Greenland halibut, redfish, polar cod, blue whiting (*Micromesistius poutassou)* and squid (*Gonatus spp.*).

Hooded seals are considerably less abundant than their close neighbours the harp seals, but the censuses for hooded seals are few and far between and the size of the population is uncertain. It is assumed that the Newfoundland stock numbers some 400,000, the Davis Strait stock in the order of 100,000 and the Jan Mayen stock in the order of 350,000 individuals.

The hooded seals have at times been hunted intensively on the moulting grounds in the Denmark Strait since these were discovered in 1874 by the Norwegian sealer Edvard Johansen of Tromsø, but did not become part of the catch

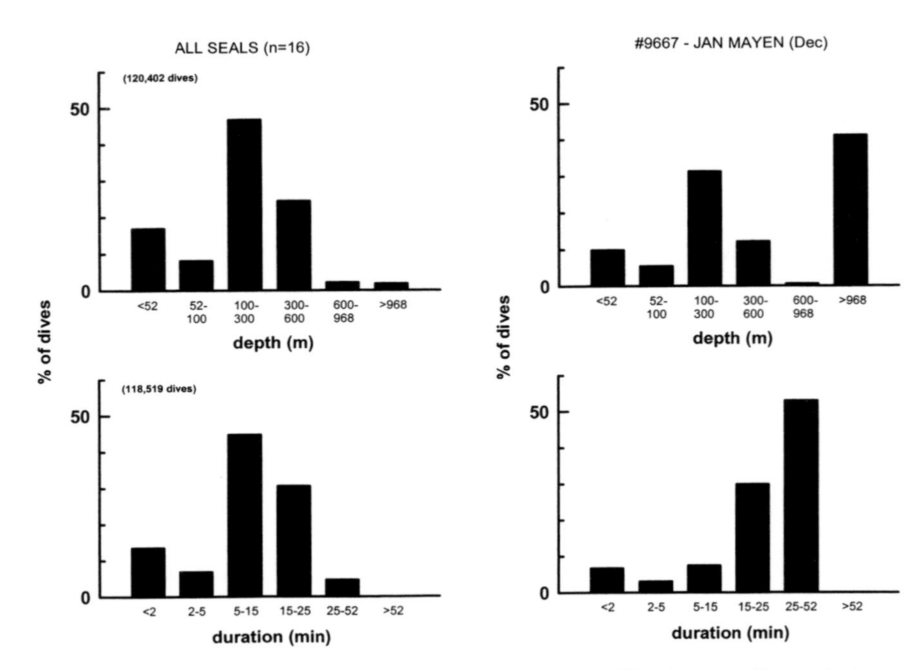

Figure 4.13 Dive depth and duration (*left*) of 16 hooded seals of the West-ice, or Jan Mayen, stock, performing a total of 120,000 dives throughout a year. *Right*: Dive depth and duration of one juvenile male hooded seal, which was diving off Jan Mayen in December, showing a very high frequency of dives down to about 1,000 m with durations up to one hour (Folkow & Blix, 1999).

at Jan Mayen until the 1920-ies. Commercial sealing at Newfoundland became a major late-winter occupation in the late nineteenth century, but the preference of breeding hooded seals for heavy pack-ice made them largely protected on their breeding grounds during the sailing ship area. The harvest has always focused on new-born pups, but the capture of the pup often requires the killing of the aggressive mother, and this has made the proportion of adults in the catches rather high. The hunt in the Jan-Mayen and Denmark Strait areas has been regulated since 1958 and the summer hunt in the latter closed since 1961. The catch at Newfoundland has been regulated since 1961 when a closing date was agreed between Canada and Norway, while an opening date has been enforced since 1968. The Gulf of St. Lawrence was closed to commercial sealing in 1972, and Norwegians were not allowed to hunt in Canadian waters after 1982. Since then, the hunt in Canada has been carried out by local landsmen. Thus, without doubt, man has been the most serious predator of hooded seals (Table 4.1).

The hooded seal derives its name from the inflatable nasal sac (Fig. 4.14), which is developed only in the males. It reaches its fullest size in sexually mature

Table 4.1

Average annual reported catches of hooded seals by Canada and Norway from Newfoundland and the Gulf of St. Lawrence, (CANADA) and by Norwegians and Russians from the Jan Mayen area (NORWAY). The Norwegians were banned from Canadian waters in 1982, whereafter the Canadian hunt was carried out primarily by local landsmen. The great variation in the annual catches, particularly in recent years in Canadian waters, is mostly caused by variations in ice conditions making the animals more or less available to the Canadian landsmen (i.e. people walking out onto the ice and catching seals without the use of ships).

YEAR	CANADA	NORWAY
1946-50	6,062	41,409
1951-55	5,972	54,428
1956-60	6,455	38,228
1961-65	5,043	46,805
1966-70	13,543	31,487
1971-82	2,251	19,863
1983-90	657	3,791
1991-95	1,551	3,479
1996	25,754	811
1997	7,058	2,934
1998	10,148	6,351
1999	201	4,446
2000	14	1,936

bulls which use it for display during courtship fighting. Moreover, such males are also able to blow the soft and well vascularized nasal septum out through one of the nostrils as a red bladder (Fig. 4.14). The extrusion of this bladder appears to be confined to the breeding and mating seasons, when the males are engaged in rather ferocious fights.

Birth takes place at the end of March, usually at the centre of a large heavy ice-floe. The female attends to the pup on the ice and is herself courted by at least one, but usually several, adult males, which often remain in the water. Such groups are often referred to as families although the appropriateness of this term, even in our sexually liberal times, is questionable, since neither of the attending males are father of the pup, and each one appears to attend to several females during the course of the breeding period. At birth the pup weigh 25-30 kg and is equipped with a (rather unusual) subcutaneous layer of blubber. Lactation was shown by Bowen, Oftedahl & Boness (1985) to last only for 2-4 days. During this time the pup can grow at rates exceeding 7 kg/ day increasing to about 45 kg at weaning. This rate of growth is accomplished

Figure 4.14 The hooded seal has its name from the fact that (only) the male has a "hood" on its head, which it can inflate, even under water, in times of excitement (*top*) (Photo: N.R. Lightfoot). In times of extreme excitement it may even blow the nasal septum out through one of the nostrils and produce a spectacular red balloon (*bottom*) (Photo: R.G. Mason).

through the intake of one of the most energy rich milk produced by any mammal (Table 10.2).

Harp seal (*Phagophilus groenlandicus*)

Pagophilus means ice lover from the Greek words "pagos" meaning ice and "philos" meaning loving. There are three major populations of these ice-loving seals, each with its own distinct breeding ground: one in the White Sea, one in the Greenland Sea, north-west of Jan Mayen, and a third utilizing the Gulf of St. Lawrence and the ice off the east coast of Newfoundland (Fig. 4.15). Adult males average 135 kg in weight and females about 120 kg, but there are great seasonal weight changes.

The harp seal pups are born with a soft, curly yellowish coat, stained by amniotic fluids (Fig. 4.16). This turns white during the first three days. When they are

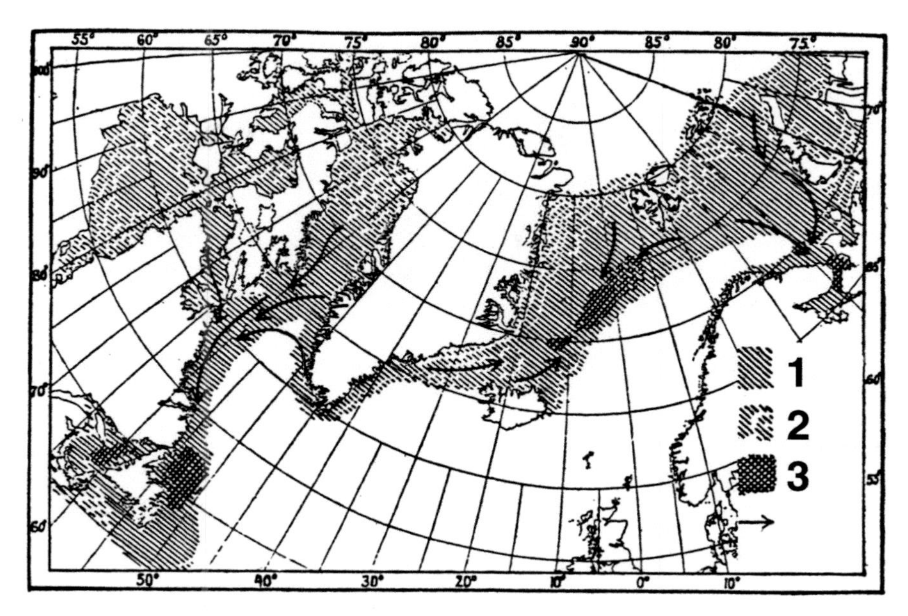

Figure 4.15 The overall distribution and annual migrations of the harp seal, as presented by Fridtjof Nansen (1924) based on his own studies in the field in 1882 and 1888, has remained the state of the art until the introduction of satellite-telemetry (e.g. Figs. 4.20 and 4.21). 1: main distribution; 2: unusual distribution; 3: breeding ground; arrow: migration route to breeding ground.

two to four weeks old this white coat is shed to reveal a short-haired coat of grey, darker dorsally, lighter on the ventral surface, and marked with darker grey and black spots. There is a great variation in the shade of the ground colour and degree of spotting, but all immatures show the same general pattern. For each annual moult the spottings become less apparent and instead a typical "saddle", or harp, starts to develop. Thus, mature animals of both sexes are usually light whitish grey with a horse-shoe-shaped black band running along the flanks and across the back. Its head, to just behind the eyes, is also black (Fig. 4.17).

The newborn harp seals weigh 10-12 kg, but increase their weight rapidly to about 34 kg at weaning. Except for the first few days after birth the pup is often left alone on the ice during the nursing period, which lasts for about 2 weeks, after which the pup is completely abandoned by the mother who goes off to start working on the next pup. During both the nursing period and the following two weeks, during which the pup undergoes its first moult, it reacts to the approach of both man and beast by a typical "freezing"-reaction. This consists of becoming completely motionless, with its head retracted, eyes closed, nostrils closed, respiratory arrest and much reduced heart rate. It is assumed that this "playing dead

Figure 4.16 A newborn harp seal on the ice at an ambient temperature of -18 °C and strong wind, but still doing fine (Photo: A.S. Blix).

Figure 4.17 An adult male harp seal showing the colour pattern which has given this species its name (Photo: A.S. Blix).

reaction" is a defence against polar bear and Arctic foxes, which sometimes may prey heavily on newborn and young pups.

Outside the breeding period, harp seals range from the Kara and Bering Seas through Svalbard, Jan Mayen, Greenland and Baffin Bay to Newfoundland. (Fig. 4.15). The Jan Mayen population, which has been studied intensively by Folkow, Nordøy & Blix (2004), by use of satellite telemetry, stays along the edge of the pack ice after breeding in late March to start to moult in the same general area at the end of May. After moulting, the animals continued to stay in contact with the pack-ice until mid-July, when more than 80 % of our animals migrated into the Barents Sea, where they remained till late in the autumn, when they returned to the Denmark Strait (Fig. 4.18). This summer exodus into the Barents Sea, which holds its own stock of harp seals was at the time completely unknown, and shows the strength of satellite telemetry in this kind of research.

The Newfoundland population breeds in rather dense rookeries in the Gulf of St. Lawrence and along the east coast of Labrador in late February. After breeding the Newfoundland population moves north in early May and reach south-west Greenland in mid-June. During summer this population is distributed from Thule and the Canadian archepelago in the north to the northern Lab-

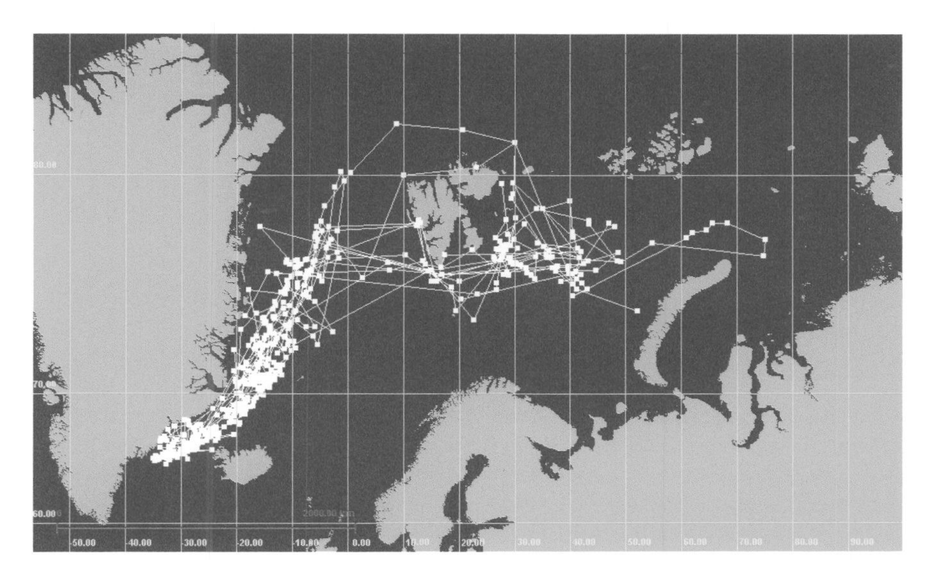

Figure 4.18 Distribution and overall movements of 10 adult harp seals of the West-ice, or Jan Mayen, stock, as determined by satellite telemetry, from moulting in late May till contact was lost in the spring of next year. Note that all but one seal left the Greenland Sea in mid-July to stay in the Barents Sea until October-November, when they all moved back (Folkow, Nordøy & Blix, 2004).

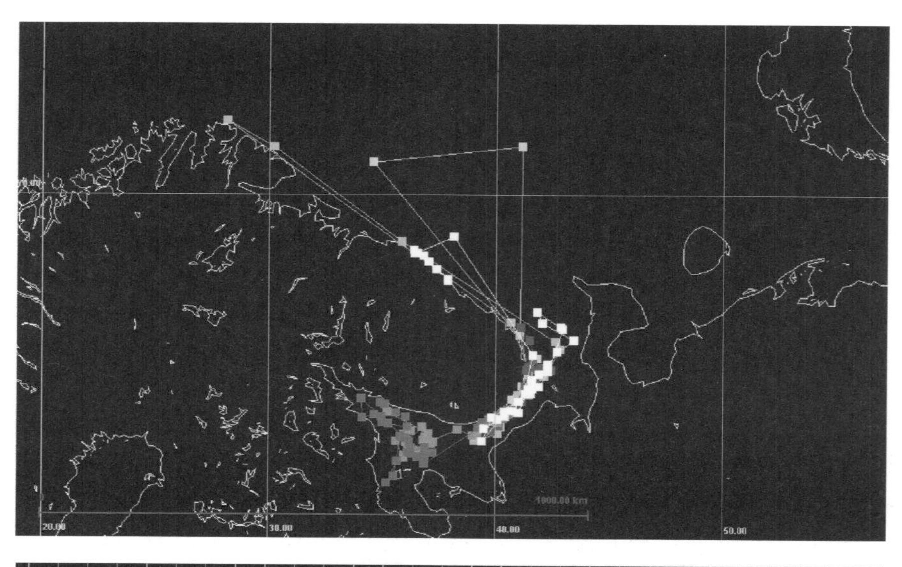

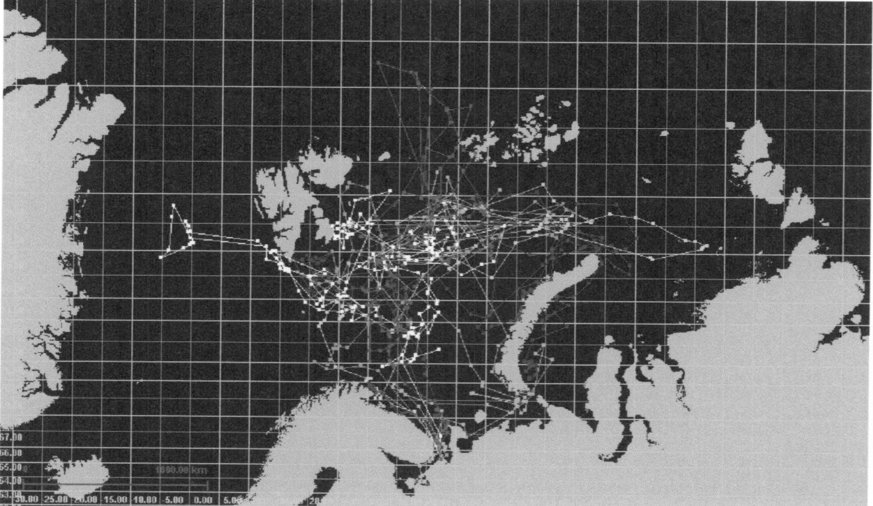

Figure 4.19 Distribution and overall movements of the White Sea/Barents Sea stock of harp seals, as determined by use of satellite telemetry. *Top*: 8 adult females from breeding (at the mouth of the White Sea) in late February until moulting in April/May the same year. *Bottom*: 10 adults of both sexes from the completion of moulting off the mouth of the White Sea in early May until contact was lost at the time of breeding the next year (Nordøy, Blix, Potelov & Folkow, unpublished).

rador in the south. The movement south begins in late September and by early November large numbers have passed northern Labrador to reach the Strait of Belle Isle in late December.

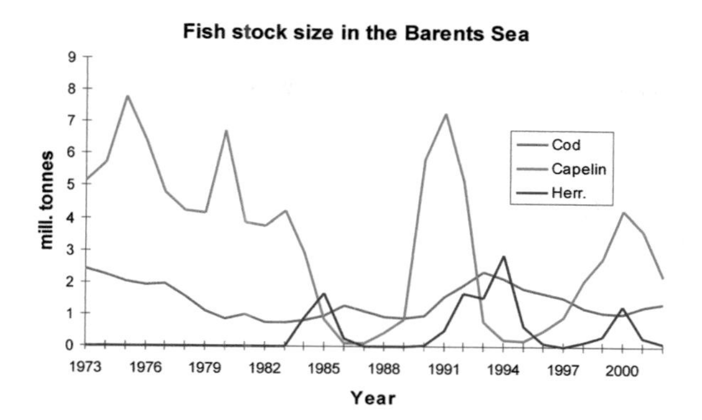

Figure 4.20 Stock size estimates of the commercially most important fish in the Barents Sea: cod, capelin and herring: 1973-2002 (H.Gjøseter, HI-Bergen, Pers.comm.)

The Barents Sea population which breeds in the White Sea of Russia has also been studied in detail by us, again by use of satellite technology. This group of animals leave the White Sea after breeding in early March and move along the coast of the Kola Peninsula up to and beyond the Norwegian border to fatten up until they return to the mouth of the White Sea off Cape Kanin to moult in late April. After moulting in early May the animals are pelagic and fan out in north-west and northerly direction towards the retreating sea-ice, until they start moving back towards the breeding grounds in November, mostly along the west-coast of Novaja Zemlja (Fig. 4.19). During the entire annual cycle it appears that hardly any of the animals go west of the shelf break into the deeper waters of the Atlantic. Instead, they seem to exploit the relatively shallow waters of the Barents Sea, where they during the period May-August mostly perform shallow dives to less than 50 m, while in the period between September and October they mostly dive to depths of 100-200 m. This, as well as other information, indicates that the harp seals feed primarily on the rich sources of capelin, herring and cod in the area, in addition to crustaceans, notably *Parathemisto libellula* along the edge of ice.

The huge Barents Sea capelin stock collapsed in the years between 1983 and 1985, probably due to a combination of overfishing and a series of natural causes (Fig. 4.20). Following this collapse, which coincided with a period when the population of immature herring and cod already were at a low, the coasts of northern Norway were invaded by hordes of harp seals, to the extent that compensation was provided for local fishermen for almost 60,000 harp seals caught in gill-nets in 1987, while the number was reduced to 20,000 in 1988 (Fig. 4.21). After this time the situation has normalized, with, at least, a temporary return of the capelin and increased stocks of herring and cod, but surveys seem to indicate that a severe reduction in harp seal recruitment may have occurred after 1985.

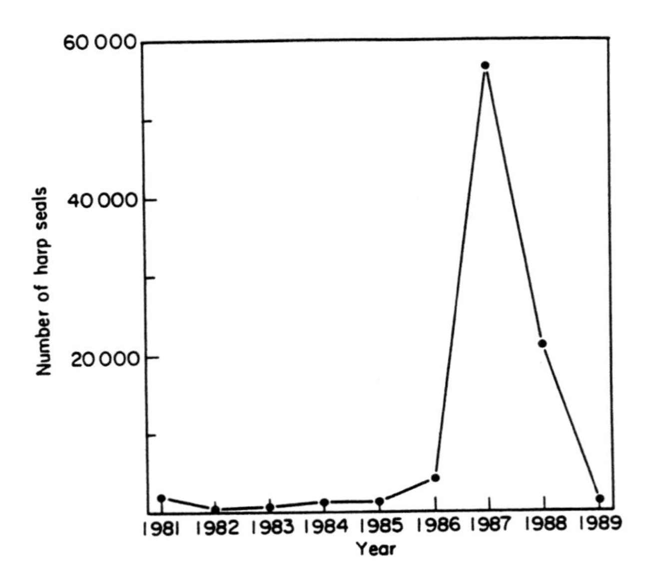

Figure 4.21 Number of seals taken as by-catch in Norwegian gillnet fisheries and reported for financial compensation throughout the 1980s (Haug *et al.*, 1991).

The present (2001) population estimates are about 2 million in the Barents Sea/White Sea, about 360,000 in the Jan Mayen area and no less than about 5 million in the Newfoundland area. All these populations have been exploited commercially for more than 150 years. The catch statistics for the Canadian, the Jan Mayen and the Barents Sea stocks after the Second World War are shown as examples of this bonanza of the past (Tables 4.2 & 4.3).

Now, lets assume (what we pretty well know) that an average harp seal needs about 3 kg of fish (depending a bit on what kind of fish it eats) a day. Let's further assume that a harp seal feeds (not counting breeding and moulting) for 300 days a year. That implies that a harp seal needs 900 kg of fish material each year. Now, let's assume (what we again pretty well know) that about 2/3 of its food requirements is actually met by consumption of fish. If so, each harp seal takes away about 600 kg of fish, which for the Barents sea population of some 2 million seals amounts to a staggering 1.2 million tons of fish! It goes without saying, therefore, that an uptake of this magnitude (and forget for the moment some 100,000 minke whales that feed primarily on fish in the same waters), will greatly effect the energy flow in the ecosystem, and ultimately destroy the rich fisheries in the Barents Sea, unless properly managed.

Walrus (*Odobenus rosmarus*)

Walruses are almost circumpolar in their distribution (Fig. 4.22). Those inhabiting the North Atlantic region are regarded as the subspecies *O. rosmarus rosmarus*,

Table 4.2

Average annual reported Canadian and Norwegian harp seal catches off Newfoundland and in the Gulf of St. Lawrence, Canada, 1946-2000. The Norwegians were banned from the hunt in 1982, whereafter the catches were made by Canadian landsmen. (1+: animals aged one year and older).

YEAR	PUPS	1+	TOTAL
1946-50	152,981	64,995	217,976
1951-60	218,489	97,870	316,359
1961-70	223,218	59,073	282,291
1971-82	137,083	28,544	165,627
1983-90	37,909	12,180	50,089
1991-95	32,395	22,685	55,081
1996	184,856	58,050	242,906
1997	220,476	43,734	264,210
1998			282,624
1999			244,603
2000	85,485	6,583	92,068

The great variation in the annual catches, particularly in recent years, is mostly caused by variation in ice conditions, making the animals more or less available for the landsmen.

Table 4.3

Average annual reported Norwegian and Russian harp seal catches in the WEST ICE (Jan Mayen area) and EAST ICE (White Sea area), 1946-2000. (1+: animals aged one year and older).

	WEST ICE			EAST ICE		
YEAR	Pups	1+	Total	Pups	1+	Total
1946-50	26,606	9,466	36,070	-	-	170,373
1951-60	25,250	8,266	33,506	-	-	127,022
1961-70	17,524	3,365	20,889	38,268	18,889	57,157
1971-80	11,543	1,744	13,287	35,594	6,529	42,123
1981-90	5,095	3,394	8,489	48,530	18,860	67,390
1991-95	281	6,968	7,249	29,951	9,526	39,477
1996-00	3,251	1,473	4,724	30,638	3,646	34,284

while those of the North Pacific region are bigger and distinguished as *O.r. divergens*, the two subspecies differing principally in general body size, but also in the size of the tusks, which are the trade mark of the walrus. The average weight of Pacific walrus, being the only population for which a proper sample is available, is 1,200 kg for males and 800 kg for females, the newborn calf weighing about 50 kg (Fay, 1982). The exposed part of the tusks of the pacific walrus may reach

Figure 4.22 The geographical distribution (blue) of walrus (*Odobenus rosmarus)* in the Arctic.

a length of 70 cm in males and 60 cm in females, the tusks of the atlantic walruses usually being much shorter. Unlike the true seals, the walrus is able to rotate the hind-flippers forward for support of the body in quadrupedal locomotion. The walruses are covered with short, coarse cinnamon-brown hair, usually no more than 1 cm in length. The hair is most dense on the youngest animals, while old males may appear to be naked.

Aside from their generally bizarre appearance, the walrus is also special in having an extreme elasticity of the pharynx, the lateral walls of which are expandable as a pair of large pouches. These pouches are situated posterior to the glottis and involve a small portion of the anterior end of the oesophagus. Pouches appear to

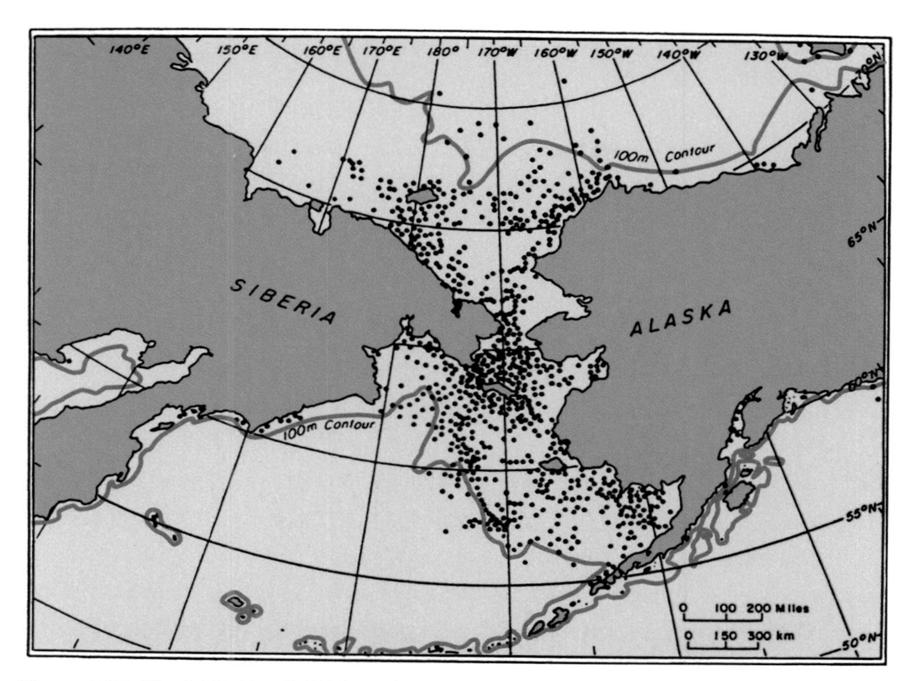

Figure 4.23 The distribution of sightings of walruses (all months 1930-79) in relation to the position of the 100-m isobath in the Bering-Chukchi region. Each dot represents one sighting, regardless of the number of animals (Fay, 1982).

be present in all adult males, but only in a few females, young animals not having them at all. The size of the pouches, which are rather thinwalled when inflated, differ substantially in size, but may hold as much as 50 liter each in adult males. It was supposed by Fay (1982) that the pouches are used for buoyancy, especially when the animals sleep in the water.

The calves undergo a prenatal moult, about two to three months before birth, shedding a fine, white lanugo while *in utero*. At birth the calves appear ashen grey to grey-brown overall, becoming tawny-brown with distinctively black flippers within one to two weeks.

Walruses inhabit the moving pack ice over the shallow waters of the continental shelf, where they feed on benthic invertebrates at depths down to 100 m (Fig. 4.23). A study in Svalbard has shown that their deepest dive there was 67 m, with a mean depth of only 22m, while their longest dive lasted 24 min, with a mean duration of only 6 min (Gjertz *et al.*, 2001). Most of the populations appear to be migratory, moving southward with the advancing ice in autumn and northward as the ice recedes in spring. Thus, the Pacific walrus inhabits the pack-ice of the north-central and south-eastern Bering Sea in winter (December-March), migrate northward through Bering Strait in spring (April-June), spend the summer along the ice edge in the Chukchi Sea (July-September), and return to the Bering Sea again in autumn, some individuals travelling more than 3,000 km per year. Some walruses stay in the Bering Sea throughout the summer, however, utilizing small islands in the Bristol Bay and the Gulf of Anadyr as haul-out grounds. The distribution of the population of walruses at Svalbard has recently been studied by Wiig, Gjertz & Griffiths (1996) by use of satellite transmitters, and they found them to be rather stationary with movements limited to the area between Svalbard and Frans Josef Land.

The world population of walrus is not well known, but probably in the order of 250,000 animals. Of these at least 80 % are in the Bering-Chukchi Sea region. The Laptev Sea is believed to hold 4,000-5,000 animals (1971) and the eastern Canadian Arctic and West Greenland (1966) about 25,000. All segments of the world population, except that in the Laptev Sea, were severely depleted by commercial exploitation in the eighteenth and nineteenth centuries. There is no commercial harvest from any population at present, but limited use for subsistence by

Figure 4.24 Walruses are extremely gregarious, hauling out in vast aggregations, normally on ice, but when not available, on the shore (Photo: I. Gjertz).

local Eskimos. Even so, only the Pacific population appears to have been restored to its former, pre-exploitation numbers.

The walrus (in the Bering Sea) mates in February and gives birth over a year later in May. The single calf is suckled for about two years. After weaning the food comprises mainly of benthic organisms, especially bivalve molluscs, but a wide variety of other invertebrates are also taken, and, occasionally, fishes and marine mammals are also eaten. From most of the molluscs, only the muscular foot or siphon is eaten; the remainder is discarded. Walruses to some extent compete with bearded seals for food.

The walrus has a penis with an exceptionally large *os penis*, or baculum, which measures up to 62 cm and may weigh as much as 1,000 grams, being the largest both actually and relatively of any living mammal. The reproductive organ of the females is also rather exceptional, in that also the clitoris of this species is equipped with a small mass of bone which is the homologue of the baculum.

Walruses are among the most gregarious of mammals, tending to travel almost always in small groups and to haul out in herds of up to several thousand individuals. On land or ice, in any season they tend to lie in close physical contact with their neighbours (Fig. 4.24).

Arctic whales

Beluga (*Delphinapterus leucas*)

Beluga, which is a corruption of the Russian name belukha, or white whale, are toothed whales, with a conspicuous white colouration and without any dorsal fin. Instead, belugas have a rather tough ridge on the mid-line of their back which may be used to break thin sea ice to allow the animals to breathe. Belugas reach a length of about 5.5 m and a body weight of a little less than 2 tons. There are clear indications, however, that average body weight is different in different populations. Belugas are equipped with an unusually thick (2-7 cm) layer of blubber, which provides both insulation and energy in times of need. They are social animals and appear sometimes in flocks of as many as 1,000 animals (Fig. 4.25), but more often than not females with calves and adult males form distinct groups.

The distribution of the beluga is circumpolar ranging into the subarctic (Fig. 4.26). There is a near continuous distribution of belugas across the Russian Arctic, limited in the Atlantic to the north coast of Norway and in the Pacific to the Okhotsk Sea. Belugas are also present along the east and west coasts of Greenland and in North America, extending from Alaska across the western Canadian Arctic to a large population in Hudson Bay, along the islands of the eastern Canadian

Figure 4.25 An aerial view of Canada's Cunningham River estuary, filled to capacity with beluga whales (Photo: Gunter Ziesler (Bruce Coleman)).

Arctic and into the Gulf of St. Lawrence. The north limit appears to be north Greenland, Ellesmere Island and Svalbard. Movements of the herds are seasonal (e.g. Smith & Martin, 1994). In winter most belugas inhabit ice-permeated off-shore waters. In spring, as leads form in the pack-ice and ice retreats from the coast, most belugas migrate to nearshore areas. During the open water months, nearly all major concentrations occur in shallow bays or in estuaries of large rivers north of 40°N. The whales then return to the wintering areas before, or with the advancing ice. It is assumed that the animals wintering in Cook Inlet, Bering Sea, Okhotsk Sea, Barents Sea, West Greenland, St. Lawrence Estuary, West Hudson Bay and Hudson Strait represent different discrete populations.

Belugas are first able to reproduce at an age of about 4-7 years in females and 6-9 years in males. There is no evidence of delayed implantation. Gestation is estimated to last 14-15 months. Breeding takes place at sea in April -May. Calves are born from April to July, depending on location, at intervals of about three

Figure 4.26 The geographical distribution (blue) of the beluga (*Delphinapterus leucas)* in the Arctic.

years. At birth the calf is dark slate grey, and the animals only become white upon sexual maturation. At birth the calves have a length of about 1.5 m and weigh about 79 kg (Fig.10.32). The presence of many belugas in estuaries, where they have been harassed by man for centuries, but where the water is shallow and comparatively warm, has been interpreted as behaviour aimed to conserve energy and to minimize the thermal costs for newborn calves. The diet of the calves appears to be milk during the first year, supplemented with easily captured prey (molluscs, annelids and crustaceans), during the second year.

Belugas are probably unique among cetaceans in that they undergo an intense period of moult each summer. In so doing they lose their old outer skin (epidermis) layer, which has by this time become yellow, and regain their pure whiteness. The whales hasten the moult by rubbing themselves on gravel and sand, usually in shallow fresh water in river mouths and estuaries.

Polar bears, walrus and killer whales are natural enemies of belugas. In fact, belugas are somewhat prone to becoming entrapped in shallow water at low tides as well as in small openings in the ice, and such circumstances make them easy prey for human hunters as well as polar bears.

Belugas are of the most vocal of odontocetes and have an excellent echo-location system. The animals project broad-band pulses with peak frequencies in a narrow beam forward from the forehead and listen for the echo returns. These whales are even able to produce a lot of noise above water. Passing air through the "lips" of their blowhole, they can blow whistle and scream over large distances.

Belugas are known to feed intensively during April through September. (Tom) Smith of Canada and (Tony) Martin of England have studied the summer distribution and feeding behaviour of belugas of both eastern and western Canada by means of satellite telemetry for many years (e.g. Martin, Smith & Cox, 1988). They have shown that these whales, at least in summer, swim straight down to the seabed at depths of 250-500 meters, remain at the bottom for 5-10 min and return directly to the surface. This particular behaviour may indicate that the whales are feeding on seabed invertebrates of some kind, but even so, fish are likely to be the most important prey throughout most of their range. In fact, more than 100 kinds of organisms have been identified in the summer diet of belugas. These include arctic cod, herring, smelt, char, sculpin, suckers and eelpout, as well as benthic invertebrates and squids. The winter diet of belugas is not known.

The world population of this species is poorly known, but probably counts in the order of 100,000-200,000 animals.

Belugas are utilized by the Eskimos throughout their range and are in some areas heavily hunted both for their meat and their skin, which is used both for food and for ropes and lines.

Narwhal (*Monodon monoceros*)

This species is unique in that the male bears a long forward pointing tapering tusk. Like its close relative the beluga, the narwhal shows marked colour changes with age. Newborn narwhals are uniformly grey, while young suckling animals, 1-2 years old, are uniformly dark grey. With increasing age, white patches appear around the genital area and gradually extend to the rest of the ventral surface and then up the flanks to the back. The narwhal, like the beluga, has no dorsal fin.

Figure 4.27 The most common geographical distribution (dark blue) of the narwhal (*Monodon monoceros)*, and areas where the species is sporadically observed (light blue) in the Arctic.

At birth, usually in August, the calf weighs about 80 kg, while adult females may weigh about 1,000 kg and males as much as 1,600 kg, and reach a length in excess of 4 m. Mating takes place mainly in April, with a gestation period of about 14 months.

The general distribution of the narwhal is shown in Fig. 4.27. In the eastern Canadian Arctic, narwhals occur regularly in Jones Sound, Lancaster Sound, Baffin Bay, Davis Strait, northern Hudson Bay and Foxe Basin. Narwhals also live in the waters of western Greenland from Disco Island to as far north as Smith Sound and Kane Basin, and in the seas of east Greenland. In the Eurasian sector of the Arctic, they are rare, but sometimes seen in the Barents, Kara and Laptev

Seas, while they are seldom, if at all, seen in the western Canadian Arctic, Alaskan or Siberian waters.

Narwhals display a pronounced seasonal migratory cycle. For example, after spending the winter in Davis Strait, they move northward through the pack-ice, congregating in May and June at the edges of the fast ice in the fjords of northern Baffin Island and northwestern Greenland. They remain there in August and September. Prior to the formation of new ice in October, narwhals leave the fjords and migrate southwards. They spend the winter in the pack-ice of Davis Strait, west and southwest of Disko Island.

The world population of narwhals is not known, but is likely to be between 50,000 and 100,000 animals. Narwhals are hunted by the Eskimos and about a thousand animals are probably taken annually. Besides providing meat the tusks are important trade items to the Inuit.

The dentition of the narwhal has received considerable attention, and the presence of the remarkable tusk that may exceed a length of 3 m (Fig. 4.28) undoubtedly has contributed to much public and scientific interest in the species. In fact, the narwhal is most likely responsible for the myth of the unicorn. Male narwhals occasionally develop two tusks, and sometimes even a rare female narwhal may have a tusk. Both tusks of bi-dental narwhals are invariably spiralled in the same direction (sinistrally), contrary to the general rules of bilateral symmetry in the animal Kingdom. Numerous and often fanciful suggestions for the use of the tusk of narwhal have been put forward. The most fanciful aside, it is likely that the tusks both are used to uncover and root out benthic food organisms, since their ends invariably are worn smooth, and it is equally likely that since narwhals are polygamous animals the tusks are important for sexual display between dominant males. This is further supported by numerous observations of males actively fencing with their tusks (Silverman & Dunbar, 1980).

During summer the narwhals in the northwestern Greenland area make daily dives to depths of more than 500 m and frequently dive to 1,000 m, but most of the time is spent in the water column at depths of between 8 and 52 m, and more than half of the dives last less than 5 min, and only a few more than 20 min (Dietz & Heide-Jørgensen, 1995; Heide-Jørgensen & Dietz, 1995). Similar behaviour was also observed in Canadian waters, where water depth excluded dives of more than 300 m. (Martin, Kingsley & Ramsay, 1994).

The most important prey eaten by narwhals during summer seems to be polar cod (*Boreogadus saida*), which sometimes occur in enormous schools in the high Arctic, and Greenland halibut (*Reinhardtius hippoglossoides*), but arctic cod (*Arctogadus glacialis*), redfish (*Sebastes marinus*), squid (*Gonatus fabricii*) and shrimps

Figure 4.28 A Greenland Eskimo holding on to the head of a narwhal with two tusks. Only the males develop tusks, and normally, only the left one is developed, but some 1-3 % of the males have two, in which case, both spiral counter-clockwise; a rare example of bi-lateral asymmetry.

(*Pasiphaea tarda* and *Hymenodora glacialis*) are also eaten (Finley & Gibb, 1982). The shrimp (*P. tarda*) is bathypelagic, occurring at depths of 800-1,200 m. The polar cod is usually pelagic, living close to the undersurface of the ice, while the Greenland halibut is demersal (200-600 m). The squid *(G. fabricii*) is eurybathic, occurring from near surface to deeper than 4,000 m. These observations suggest that narwhals feed over a great range of depths.

Narwhals are usually found in groups of three or four, occasionally up to ten. The animals are usually segregated sexually during summer into groups of immature males, groups of mature females and calves, and groups of large adult males.

Narwhals, like belugas, are very vocal animals and their underwater sounds can often be heard above the surface. The sounds are mostly narrow-band, regularly spaced pulses, ranging from 1.5 to, at least, 24 kHz. It is likely that this sound is used both for orientation and communication.

Bowhead whale (*Balaena mysticetus*)

Bowhead whales are baleen whales, which are restricted in their range to frequently ice-bound seas, which make them unique among mysticetes (Fig. 4.29). Bowheads are also large arctic whales, which may weigh as much as 100 tons. They have proportionately much larger heads than other baleen whales, being two-fifths of body length in the adult, more than one-third of the entire bulk of

Figure 4.29 The geographical distribution of the bowhead whale (*Balaena mysticethus)* in the Arctic in blue, with post WW-II observations of animals in the Atlantic, Barents, Kara and Laptev Seas indicated with blue dots.

the animal (Fig. 4.30). Within this enormous head there is a brain which in a fully grown animal, of say 60 tons, weighs about 3 kg. The bowhead has no dorsal fin. Its colour is basically black or brown, with limited, well defined areas of white or light grey. Much of their chin and lower jaw can be white, and some bowheads have a light grey to whitish band around the tail stock up to the origin of the flukes. The young ones tend to be lighter in colour than adults.

The baleen of the bowhead are considerably longer than those of other species (Fig. 4.32); the longest plate measured in Alaska in recent time is 3.13 m, but plates as long as 5.18 m were reported during the early commercial fishery. The

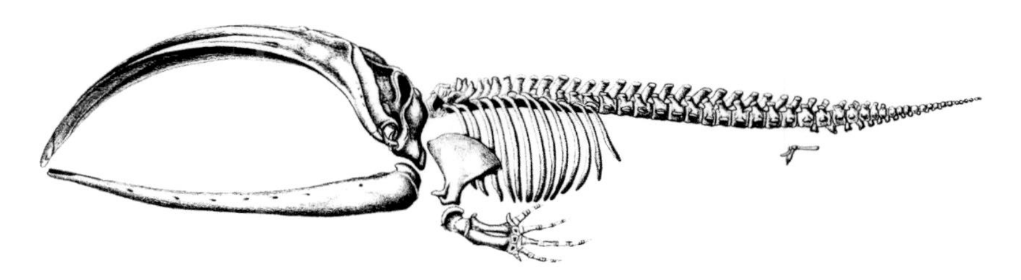

Figure 4.30 Lateral view of skull and skeleton of a bowhead whale, showing the huge head in relation to the rest of the body (Drawn by B.S. Irvine after Eschricht & Reinhardt, 1866).

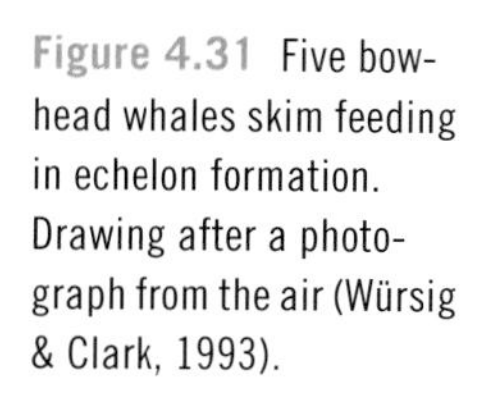

Figure 4.31 Five bowhead whales skim feeding in echelon formation. Drawing after a photograph from the air (Würsig & Clark, 1993).

baleen is generally bluish black, and the number of plates on each side ranges from 237-346.

The bowheads were heavily hunted all over their range, starting in the North Atlantic in the early 1600s until terminated due to depletion of the stocks in the late 19th century, while the North Pacific stock was only discovered in 1848, but

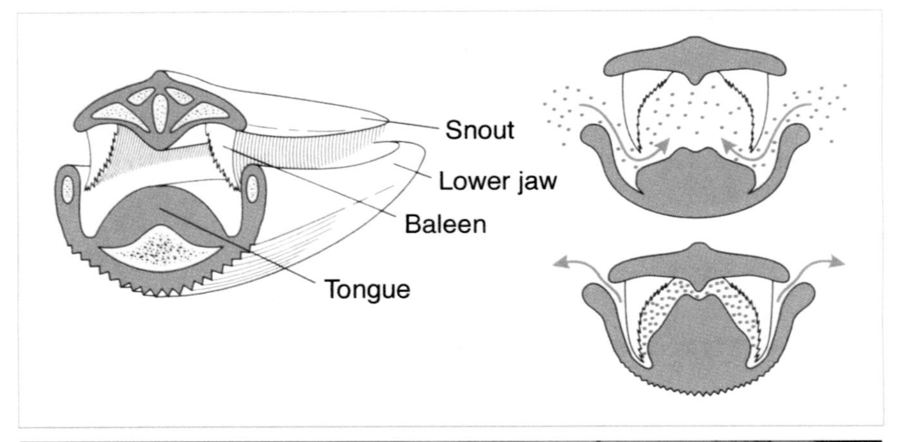

Figure 4.32 *Top*: Cross section through the mouth-part of a baleen whale showing the rows of baleen plates hanging down from the "ceiling" of the mouth cavity. The whales feed by filling their mouth with plankton-containing water, whereafter they close their mouth and press the water out through the row of baleen plates whereby the plankton get stuck on the hairy inner surface of the baleen. *Bottom*: Head of a bowhead whale (on its back), showing an impressive row of black baleen plates (Photo : unknown).

already at the turn of the century also this stock was completely depleted – all due to the demands from the fashion industry for narrow waists – and hence the need (and great prices) for "whalebone" corset stays. In all honesty, however, what started the whaling industry in the first place was the need for and high yield of oil, which might amount to 275 barrels (32,000 liters) in one animal.

Bowheads produce a V-shaped blow which is characteristic of the species. Four different stocks of bowhead are recognized: the Svalbard stock, which is still at rock bottom; the Davis Strait and Hudson Bay stock which is also at a low level, counting in the order of a few hundred animals; the Bering Sea stock, which is in pretty good shape, counting in the order of 10,000 individuals; and the Okhotsk Sea stock with a couple of hundred individuals. In spite of all this the bowheads are still hunted by aboriginal whalers in Alaska in what appears to be one of the most controversial and serious whale conservation problems in the world to day. This hunt is carried out by use of traditional methods with high (25 %) "struck and lost" rates, and is at best maintained for cultural identity reasons. But, even so, I have had the rare opportunity to serve as crew member on board the "umiak" – skinboat of the famous captain John Apangalook of Gambell, Saint Lawrence Island, as my debut to a 3-year sojourn in Alaska in 1976, and I can tell you: The surreal experience of chasing 50 ton whales from skin boats under sail, armed with only hand-thrown lance-harpoons, is an event of a life-time, even for the most experienced Arctic traveller!

Anyway, the Bering Sea stock which is the best studied stock today overwinters in polynyas and along the edges of the pack-ice in the western and central Bering Sea. During spring they migrate north and are distributed in a corridor which extends from the wintering areas through the Bering Strait to the feeding grounds in the eastern Beaufort Sea from April through June, and remain in the Canadian Beaufort Sea and Amundsen Gulf until late August or early September. From September through November the animals are distributed from eastern Beaufort Sea to the northern Bering Sea (Fig. 4.29).

The winter distribution of the bowheads in the Sea of Okhotsk is not known, but they are most likely wintering near the ice-edge off the Kuril Islands and western Kamchatka. In the spring they migrate to Tugurskiy Bay until the ice disappears in late July or early August when they move on to Udskaya Bay, where they seem to remain till late October.

The Davis Strait stock is wintering along the ice-edge from south-western Greenland to Labrador and move north into northern Hudson Bay and along the west coast of Greenland in the spring, and reach the Canadian high-Arctic archipelago in August-September, only to return to the wintering grounds in October-November.

The relict, but hopefully viable Svalbard stock was previously distributed from Iceland along the coast of northeast Greenland via Svalbard and Frans Josef Land to the Kara Sea and counted whales in the ten-thousands, but as a result of ruthless over-harvesting, it now probably count in the tens. The migration pattern of this stock is not known, but whales were present in historic times off Jan Mayen as early as the end of March, and large numbers used to be present along the west and north-west coasts of Svalbard at the end of May, but whaling was also very active off the east coast of Greenland in June-August.

Bowheads spend an estimated 5 (spring migration) to 16 % (autumn migration) of the time at surface, while individual dives seldom last for more than 2 min, while harpooned whales have been reported by Scoresby (1820) to dive for 56 min. Among the more conspicuous activities of bowheads are breaching. Occasionally, migrating animals perform long series of breaches: one whale is reported to have breached 57 times during a 96 min period; another 39 times in an unbroken series.

Bowheads often feed at the surface by so called "skimming" in groups of 2 to 10 individuals, frequently swimming in echelon formation (Fig. 4.31). However, most of the time the bowheads feed in the water-column. They sometimes also forage at or near the bottom in shallow areas. The most important food seems to be copepods (*Calanus sp.*), euphausiids (*Thysanoessa sp.*), mysids (*Mysis sp.*) and several species of amphipods, which they filter out of the water by use of their huge baleen plates (Fig. 4.32).

The principal mating period is unknown, but is believed to occur in the period from March-September. Most calvings seem to take place in April-May, but calvings also takes place at both sides of this period. The length of gestation is uncertain, but probably 12-14 months. Calving interval is supposed to be 3-4 years. Lactation period is not known but probably 5-6 months. The mean length of the newborn calf is 4.5 m, which makes it likely that it weighs more than a ton!

Bowheads, like humpbacks (*Megaptera novaeangliae*), are very vocal animals that produce a great variety of sounds. In general, the majority of bowhead vocalizations are low (<400Hz), frequency modulated calls, but the repertoire includes a rich assortment of amplitude-modulated and pulsed calls with energy up to at least 5 kHz. Bowheads also sing! An average song lasts about one minute, but song bouts can last anywhere from a minute to many hours. The song seems to be different every year. It is likely that the song may serve a reproductive function – either to attract females or to dominate rival males. Source levels for bowhead calls in the spring have been estimated as high as 180-189 dB re 1 μ Pa at 1 m. It is hypothesized that the various calls serve several purposes. First, calling at regular intervals helps maintain the cohesion of the migrating herd. Second, by adopting signatures, members of the group are able to keep track of specific individuals in the herd. Third, the stereotypic calls may allow bowheads to monitor changes in the ice conditions. It has also been suggested that bowheads actively use surface echo of low-frequency calls to navigate through ice (Würsig & Clark, 1993).

Stellers sea cow (*Hydrodamalis gigas*)

The sea cow was discovered in 1741 when a ship under the command of the Russian captain Vitus Bering became wrecked on an island which was later named Bering Island in the Komandorsky group of islands in the North Pacific Ocean. Almost all our information on this rather peculiar animal is contained in the notes of the naturalist Georg W. Steller, who was a member of Bering's shipwrecked crew, or from studies of bones, a complete skeleton of which is on display in the museum of natural history in St. Petersburg in Russia.

The sea cow which upon its discovery occupied cold sub-arctic waters probably reached a length of about 8 m and adult animals may have attained a weight of 4 tons, or more. It lacked teeth and finger bones and its very heavy skeleton consequently looked rather bizarre (Fig. 4.33). The animals were abundant in the shallow water off the island, but it has been estimated from the harvest records from the area that only about 2,000 animals existed at the time of its discovery. The sea cow apparently lacked the ability to dive and they showed no fear as they floated at the surface feeding on the large brown algae called kelp. In this context it is worthwhile to ponder why no vertebrate animal, other than the sea cow, utilizes this rather abundant food resource. Anyway, according to Steller, grazing occupied much of the sea cow's time in summer, and the animals then became rather fat. In winter, however, little food was available and the animals were starving with their ribs showing clearly through their very thick skin (e.g. Reynolds & Odell, 1991).

Now, why bother about an extinct animal in this book? I think that the saga of the Steller sea cow has something to tell us all. Due to the fact that the meat of the sea cows was very tasty, said to resemble beef, and the fearless animals being very easy to catch, it was taken, not only by fur hunters on the island, but also by a great many ships on route from Russia to Alaska that stopped to lay in stores. The harvest was extremely wasteful and only 27 years after its discovery this most intriguing of species was irreversibly gone due to greed and ignorance by man. It

Figure 4.33 Russian hunters kill a Steller's sea cow for food, while Arctic foxes wait for scraps. Due to wholesale slaughter, the species became extinct 27 years after its discovery (Painting by Alfred J. Milotte).

is likely, however, that slow breeding, small population size and limited range contributed to its demise. The very restricted range of this large and very distinct animal is intriguing in itself. Why should such a large animal have adapted to utilize the abundant kelp resources, pretty much without competition, be restricted to the Bering and Copper Island (Os Mednyy) at the end of the Aleutian chain in the North Pacific? It is tempting to think that what happened on Bering and Copper Island only was the last act of an ongoing process of removing this species from the surface of the earth. Sad as this is, on the brighter side, we have to appreciate that the Steller sea cow is the only marine mammal which has been exterminated in historic times, and hopefully that is the way it will be.

CHAPTER 5

Land Invertebrates

Damp polar desert soils and vegetation mats contain a small but active micro-fauna of consumers that feed on bacteria, algae, fungal hyphae and plant debris. Soils of Devon Island, for example, yield according to Ryan (1977), both protozoa, rotifers, tardigrades, turbellarians, nematodes, enchytraeid worms, copepods, ostracods, cladocerans, mites, spiders and insects. The class Myriopoda, which is of importance in forest ecosystems, is almost absent in the tundra, and is represented by only a few *Chitopoda* species. There are also only a few species of molluscs in the tundra fauna, and they are all slugs *(Limacoidea)*, and the elsewhere important earthworms *(Lumbricidae)* are represented in the Eurasian tundra by no more than 5 species, while they are absent in the American tundra. In wetter areas platyhelminths, collembolae and larvae of chironomids and other flying insects become dominant.

The evaluation of species richness of arctic insect fauna is problematic. Danks (1981) listed 1,650 species. According to his estimate, 50 % of the entire arctic insect fauna consists of *Diptera*. Springtails (*Collembola*) are another important part of the tundra fauna (Fig. 5.1). Among invertebrates they show perhaps the greatest potential for adaptation to arctic conditions. About 184 species of springtails are known from the northern tundra zone alone. The entire arctic fauna of this group may include 4-500 species, or 7-8 % of the world fauna of about 6,000 species.

All insects are most active in summer, and although seldom apparent on cold sunless days, those of the high tundra must be able to hunt, feed, evade

Figure 5.1 The most common springtail (*Hypogastrura tullbergi; Collembola)* in Svalbard (Photo: H.P. Leinaas).

predators and reproduce at temperatures which would immobilize species from lower latitudes.

Only bumble-bees in flight are know to maintain higher than ambient body temperatures, from heat generated by the flight muscles that is kept from escaping by the furry surface of thorax and abdomen. Basking is highly characteristic of Arctic insects. Many diptera bask in sunlight inside flowers, warming themselves and at the same time facilitating pollination. Flying insects keep close to the ground and out of the wind, haunting sheltered valleys and seeking direct sunlight as much as possible.

Life cycles are adapted to low temperatures. Where similar temperate species take a year or less to complete their life cycles, polar species may take several years (Fig. 5.3). Individuals feed and grow slowly, spend longer at each growth stage and ultimately produce fewer eggs (Bale *et al.*, 1997; Strathdee & Bale, 1998).

Insects are, of course, the most numerous of arctic animals, and both migratory and resident arctic animals, as well as man, are often plagued by hordes of biting and stinging insects during summer. Still, very few insects are restricted to the Arctic. Most of the available information on arctic insects has been obtained in the North American Arctic, but the great majority of the high Arctic insects are supposed to have a circumpolar distribution. In this chapter I will therefore primarily limit myself to species which occur in the high Arctic, or are endemic to the arctic region. In so doing, I will have to lean heavily on Kevan & Danks

(1986), who have studied this topic much more than I, who, in fact, has limited my contact with arctic insects to a huge number of mosquito bites during my more than 30 years of arctic residence.

The mayflies *(Ephemeroptera)*, dragonflies *(Odonata)*, and stone flies *(Plecoptera)* have no known representatives in the high Arctic. The true bugs *(Hemiptera)* occur in low densities, the aphids *(Aphididae)* being represented with three species which feed on plant saps in the high Arctic. The leafhoppers *(Cicadellidae)*, seed bugs *(Lygaeidae)*, leaf bugs *(Mirdae)* and psyllids *(Psyllidae)* are small plant-sucking insects of which a few inhabit the Arctic. Predatory bugs are only represented with a single species, *Anthcoris melanocerus*, a minute pirate bug. Aquatic bugs are also few, with two species of water boatman *(Corixidae)* and a few species of shore bugs *(Saldidae)*.

Water striders *(Gerridae)*, damsel bugs *(Nabidae)*, assasin bugs *(Reduviidae)*, ambush bugs *(Phymatidae)*, stink bugs *(Pentatomidae)*, and many others which occur abundantly in temperate zones are entirely absent from the Arctic. Even thrips *(Thysanoptera)* are poorly represented.

Beetles *(Coleoptera)* form the largest order of insects in the world, but most families are entirely missing from the high Arctic. Of the ground beetles *(Carabidae)* only one species, *Amara alpina*, is found in the high Arctic, while the rove beetles *(Staphylinidae)* are represented by two, as are the predatory diving beetles *(Dystiscidae)*.

The true flies *(Diptera)*, on the other hand, are well represented in the Arctic, and some 150 species have been described. Of these the crane flies (*Tipulidae*) are large and often numerous and are important food for birds. The mosquitoes *(Culicidae)* are represented by few species, like *Aedes impinger* and *A. nigripes*, but anybody who has travelled in the Arctic during summer will know that they sometimes encroach the tundra in uncountable numbers. By far the most abundant and diverse of arctic insects, however, are the non-biting midges. In some places they may contribute up to half the insect diversity, and some species of chironomids may emerge in such vast numbers that they may cloud one's vision. The hoover flies *(Syrphidae)* are conspicuous flower visitors, of which 5 species penetrate into the high Arctic. Among other flies which extend into the high Arctic are few species of leaf mining flies (*Agromyzidae*), dung flies *(Scathophagidae)*, blow flies *(Calliphoridae)*, which are conspicuous from their size, and although most species feed on decaying matter, some, like *Protophormia terraenovae*, like living meat and may cause subcutaneous lesions in caribou, while bot and warble flies (*Oestridae*) are notorious for their appetite on reindeer and caribou. Females of *Cephenemyia trompe* lay their first-instar larvae in the nostrils of caribou. These

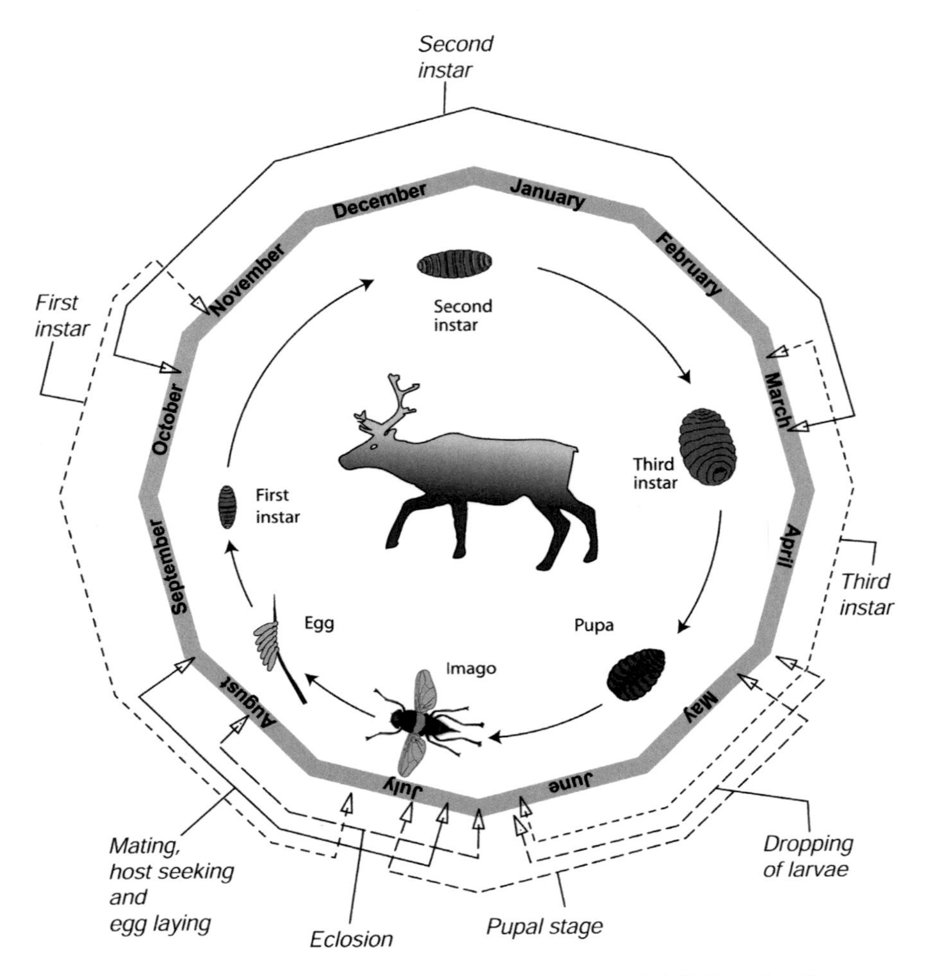

Figure 5.2 The life cycle of the warble-fly (*Hypoderma tarandi*). The adult fly (imago) produces eggs which are attached to the hairs in the reindeer pelt in August. The eggs hatch in September/October and 3 stages (instars) of lava develop under the skin of the reindeer until May, when the third stage larvae emerge and drop to the ground, where they go through metamorphosis into a pupal stage. Metamorphosis of the pupa into the adult fly (eclosion) finally occurs in July (A.C. Nilssen, Tromsø, unpublished).

larvae develop and overwinter in the nasal cavities, and drop to the ground the next year. The warbles *Hypoderma tarandi* start life as eggs attached to caribou hair (Fig. 5.2). After hatching the larvae penetrate the host's skin and migrate subcutaneously to near the mid-line of the back. There the warbles develop, making an opening through the skin by which to breathe. Equally parasitic are the

Figure 5.3 The woolly bear caterpillar of the moth (*Gynaephora groenlandica*). This species is known for its long life cycle, which can vary from 7 to 14 years (Anon., 2001).

bristle flies (*Tachinidae*), but these flies, of which four penetrate into the high Arctic, use other insects as hosts.

Butterflies and moths (*Lepidoptera*) are represented in the Arctic with several species, of which most are small and inconspicuous, and they tend to stay close to the ground and to seek sheltered creek courses. The caterpillars, which probably eat a diversity of vegetation, are particularly interesting, in that they live in the larval stage for prolonged periods. It has been estimated that the "wolly bear" caterpillars (Fig. 5.3) of *Gynaephora* spp. of the *Lymantriidae* in the high Arctic may live for 14 years or more before pupating and emerging as adults in their last season.

The *Hymenoptera* (sawflies, parasitic wasps, ants, wasps and bees) are also well represented in the Arctic. The sawflies and parasitic wasps are represented with several species, but little is known about their biology. Among the stinging *Hymenoptera (Aculeata)*, the ants *(Formicidae)* are notably absent in the Arctic, and only two species of social wasps *(Vespidae)* have been recorded just north of the treeline. The bees *(Apoidae)* are represented primarily with bumblebees, of which three species of *Bombus* reach into the high Arctic. Of these, *Bombus polaris* and *B. hyperboreus* are the most common ones.

In addition to the above mentioned insects an abundance of ectoparasitic lice *(Phthiraptera)* and fleas *(Siphonaptera)* occur in the Arctic in association with their hosts.

Insects obviously play a part in the Arctic ecosystem by cycling of nutrients and energy, but invertebrate herbivores probably consume less than 2.5 % of primary production, and insectivores probably consume a comparable fraction of insect productivity. Thus, insects in the Arctic appear to metabolise relatively little organic matter directly, yet their influence in stimulating decomposition and in pollination may be highly important to the ecosystem. In any case, the populations of insects are difficult to estimate because habitats are heterogeneous and may change from year to year. In Svalbard, for example, it was found by Leinaas and associates that dry lichen tundra had springtail (*Collembola*) populations of 38,000/ m^2, whereas in wet sites 243,000/ m^2 were found. For references, see: Hertzberg, Leinaas & Ims (1994). Thus, the role of insects in the arctic ecosystem is presently poorly understood.

CHAPTER 6

Fresh Water Communities

The high arctic desert has, as we know by now, very little water available during summers, and none in winter. However, snow melts and rain gathers in streams and even small rivers, that flow for a few days or weeks in summer, but are stilled by freezing for the rest of the year. Ponds and lakes form in poorly-drained rolling country, accumulating debris and nutrients, freezing partially or completely in winter but providing a limited habitat for plants and animals in summer.

Still water/lakes and ponds

Freshly fallen snow surfaces are virtually sterile, though bacteria are present in detectable quantities. In late spring, snow surfaces within a few 100 m of the coast often show a flush of colour, usually pink, green and brownish yellow caused by patches of uni-cellular or colonial algae. Snow algae are widespread in polar regions, but require persistent banks of melting snow to form spectacularly coloured patches. They are rarely seen for example on the colder coasts, interior ice-caps, or on iceshelves close to the sea, where ablation exceeds snowfall in summer. Wind-blown dust and rock fragments encourage local melting. The resulting shallow pits and pools, often extending over many hectares, accumulate salts and nutrients from year to year. Many are rich enough to support a distinctive cryoconite flora of algae and cynano-bacteria, which survive winters encapsulated in the ice and flourish briefly during the summer thaw.

Over much of the Arctic surface drainage is poor; permafrost seals the subsoil, and snow-melt provides a flush of water in late spring. In the high north there is rarely enough snow for melt- water to accumulate, and ponds are rare. When winter snow is plentiful, however, ponds and shallow lakes occupy large areas; over much of the tundra they form networks of static waterways that make walking difficult after the thaw. On ice-scoured uplands and coastal flats there may, in fact, be more ponds and lakes than dry land.

Arctic ponds and lakes freeze to depths of 1 to 2.5 m from September on, and accumulate a layer of compacted snow on the top of the ice during winter. Char Lake (74°, 42'N) on Cornwallis Island with an area of 52.6 ha, is, for example, typically icefree from early August to mid-September. Biological activity starts long before the ice has melted. Aquatic algae and bacteria begin photosynthesis as early as February when snow still covers the ice and less than 1 % of available light penetrates. Often the algae accumulate in a thin layer on the underside of the ice sheet, dispersing when the ice melts. Thus, phytoplankton is already plentiful under the ice by mid-May and reaching peak values in late May and June. Algae that are capable of photosynthesis at very low levels of light intensity proliferate early in the season to be replaced by a succession of other species as light intensity increases, but primary productivity is generally low. Lack of nutrients, rather than low temperature or incident radiation, is the single most important factor restricting primary production. Thus, production is often highest in early summer when nutrient levels are maximal, though the water is still close to the freezing point. The fauna of these lakes is usually restricted to rotifers, tardigrades and other small detritus-feeding invertebrates, a few species of copepods and other crustaceans. The planktonic copepods, *Limnocalanus macrurus*, produce eggs from September to November; the nauplii, which hatch after about one month, mature in the following June. Mycids (*Mycids relicta*), by contrast, have a two-year cycle; females require the same amount of energy for breeding as individuals of the same species in warmer climates, but take a year longer to accumulate it (for refs.: Stonehouse, 1989). Chironomids, including several species which bite man and other warm-blooded animals, are plentiful in the lakes, forming the main food of the only truly arctic freshwater fish, the Arctic char.

Populations of arctic char *(Salvelinus alpinus)* that are stationary in the lakes and rivers throughout their lives, as well as populations that are anadromous and only spend their first few years in fresh water, are found throughout the arctic region, sometimes even in the same lake (Fig. 6.1). After some 4-6 years of fresh-water residence the anadromous populations begin the process of transition

Figure 6.1 The geographical distribution (green) of the arctic char (*Salvelinus alpinus*).

to the marine environment, and eventually spend each summer feeding in the sea and each winter in fresh-water.

Arctic char are long-lived and may reach the age of 30 years. They reach sexual maturity as late as age 10, and in northern areas, where productivity is low, they may reproduce only every second year, or even less frequently. Spawning takes place in fresh-water in the late autumn when, particularly the male, attains spectacular colours, with a brightly red belly and a metallic blue back. The anadromous char may reach a weight of 10 kg, while the much less spectacular stationary forms seldom grow to more than 1 kg (Fig. 6.2). The arctic char has always been very important as food for arctic man of every denomination.

Figure 6.2 The arctic char may develop in a variety of ways: It is assumed that the most aggressive juveniles grow rapidly to become large and spectacular during breeding, while the less successful grow slowly to become small and inconspicuous throughout life. If their lake of origin has river connections to the ocean the large ones will become anadromous, while the social losers will spend their entire life in the lake. If the lake of origin is landlocked, however, the aggressive ones become cannibalistic and may again grow fast, while the less successful ones are eaten, or grow slowly and stay small. The two specimens pictured here are a large 50 cm/1,500 g cannibalistic male, and a small inferior 10 cm/10 g female, both caught in a landlocked lake in Bear Island in the North Atlantic, both being mature and ready to reproduce at the time of capture (Photo: G.N. Christiansen).

Running water

Few rivers start in polar regions, though some very large ones, as we know, flow across Arctic and subarctic lands to discharge into the arctic oceans (Fig. 1.17). But, more characteristic of the region are small rivers and streams fed by springs and surface runoff, and seasonal short- lived torrents that flow from glaciers. In these up to half the annual output may occur in the three or four weeks of springtime thaw. They carry heavy loads of sediment, abandoning it along their course and in extensive deltas; the finer deposits are often colonized by vegetation, providing some of the richest and most productive communities in the tundra mosaics. More characteristic still are the thousands of thinly melt-water streams that trickle from ice-sheets and snow-banks for a few days or weeks each summer, often following the same channels year after year, and sometimes building up mats of vegetation along their course.

CHAPTER 7

Amphibia and Reptiles

Amphibia and viviparous reptiles reach the northern edge of the forest but rarely cross the tree line, and they are therefore not considered in this text.

CHAPTER 8

Terrestrial Birds and Mammals

The old Lord and Master of arctic biology, Lawrence Irving of Alaska, used to say that an arctic animal is an animal which is commonly found north of the 10° July – isotherm in January. The terrestrial mammals which qualify according to this definition are listed in Table 8.1, and the few resident land birds in Table 8.2. In addition to these truly arctic animals, there are a great many migratory birds and a few migratory mammals that visit the Arctic during a few hectic summer months which will be dealt with later. In this chapter we shall familiarize ourselves with most of the resident animals before we look into how they have adapted to their austere arctic environment in Chapter 10.

Resident birds

Because birds are visible and audible in their activity by day, a great deal is known about their distribution and activities as they move and change in their annual programs. What we see and hear of birds seems understandable to us in terms that represent significant individual and social operations. The unconcealed and often noisy behaviour of birds, much like our own, serves to characterize their populations better than we find possible for the usually secretive, obscure, and frequently nocturnal mammals, which are so strongly guided by the sense of smell, that finds so little use by birds and men (Irving, 1972).

Myriads of birds reside in the Arctic during the short hectic summer, but only four are found in the high-arctic in January.

Table 8.1
Arctic land mammals

	Weight (kg)	Nutrition
Man	60.0	C
Lepus (Arctic hare)	2.0	H
Dicrostonyx (Lemming)	0.1	H
Canis lupus (Wolf)	50.0	C
Alopex (Arctic fox)	4.0	C
Mustela (Weasel)	0.1	C
Ursus maritimus (Polar bear)	500	C
Rangifer (Caribou)	100.0	H
Ovibos (Muskox)	250.0	H

C = carnivorous ; H = herbivorous

Table 8.2
Birds found in January in the high-Arctic

Species	Weight (kg)	Occurence on Land (L) or Sea (S) ice	Latitude of southward nesting limit
Lagopus mutus (Rock ptarmigan)	0.5	L	45°
Nyctea scandiaca (Snowy owl)	1.5	L S	60°
Corvus corax (Raven)	1.5	L S	0°
Acanthis hornemanni hornemanni (Arctic redpoll)	0.015	L	70°

These are: Snowy owl (*Nyctea scandiaca*), Raven (*Corvus corax)*, Rock ptarmigan (*Lagopus mutus*), and Arctic redpoll (*Carduelis hornemanni hornemanni*), with a weight range from 1,500 grams to only 15 grams (Table 8.2). If the birds which winter on the tundra are included, however, the list becomes much longer.

White feathers characterizes some of these birds, like the snowy owl, and some, like ptarmigan, adopt white feathers in winter. But whiteness is by no means a requirement for arctic life. Thus, the truly resident arctic birds come in all shapes and colours from the ivory white snowy owl to the ebony black raven.

Snowy owl (*Nyctea scandiaca*)

The snowy owl is a large bird, weighing from 1,500 to 2,500 grams, with conspicuous bright yellow eyes in an almost completely white plumage in adult males (Fig. 8.1), while adult females have brownish-grey horizontal stripes on white,

Figure 8.1 Incubating female snowy owl, with the all-white male in attendance (Photo: O. Gilg & B. Sabard).

with a completely white throat, face and legs. The strong curved beak is almost concealed by the long feathers of the facial disc and the feet are completely hidden in feathers with only their strong claws being visible.

The snowy owl has a circumpolar distribution in the high Arctic (Fig. 8.2), but like the ermine, it is missing from the south-eastern and the entire western coast of Greenland where there is no lemming. The snowy owl is almost completely dependent on lemmings for its survival and reproduction during summer, while arctic hare and rock ptarmigan are eaten, in addition, during winter. This implies that the snowy owl may be abundant during a lemming high and almost disappear due to emigration and reduced reproductive activity when the lemmings are low.

The snowy owl starts breeding in early May. It usually lays 3-6 snowy white and almost completely spherical eggs, and when the lemmings are at a high sometimes 9 eggs, in a nest in which it invests very little effort on the open rolling tun-

Figure 8.2 The geographical distribution (green) of the snowy owl (*Nyctea scandiaca)* in the Arctic.

dra in mid-May to early June. The eggs hatch after 35 days and the newly hatched chicks are coated in white fluffy down. They are fed by both parents and are ready to fly in 35 days, within which time they have moulted into a plumage with brownish horizontal stripes on a white background.

The snowy owls are mostly residents in their area year round, unless a collapse of the lemming population causes mass movements, but the birds in the most northern Greenland move to a more southern position along both the east and west coast in August and return to the high north in April. Sometimes the birds which live close to the coast also venture onto sea ice in winter, as far as we know in search for sea-birds. I have, in fact, myself a treasured memory of meeting a large male snowy owl standing at attention on an ice-floe in the pack-ice of the northern Bering Sea in May, when I was hunting bowhead whale and walrus with the Inupiat Eskimos from St. Lawrence Island in Alaska.

The energetics of the snowy owl has been studied by Gessaman (1972). He estimated that an adult owl needs 4 lemmings a day in October, increasing to 7 during the coldest December. Factors in the comparatively low requirement are the rather low basal metabolic rate and the sedentary habits of this bird. A good guess made by Irving (1972), is that free living owls generally stand quietly and, if undisturbed, are on their wings only some 5 % of the time during spring and summer with an abundant supply of lemmings. It is likely, however, that this rather relaxed lifestyle must change appreciably in the cold winters when lemmings are concealed by the snow.

Arctic redpoll (*Carduelis hornemanni*)

The arctic redpoll is a small passerine finch of 11-18 grams (Fig. 8.3) with one sub-species (*C.h. exilipes*) in Alaska and along the northern part of mainland Canada and another (*C.h. hornemanni*) in Greenland and Ellesmere and Devon Islands of the Canadian North West Territories. The exact geographical distribution of this species is not easy to determine, since it interbreeds with its close relative, the common redpoll (*Carduelis flammea*) at the southern edges of its distribution, but its core area of operations is indicated in Fig. 8.4. The arctic redpoll is much lighter in colour than the common redpoll and has a shorter and stronger beak. The bird is a permanent resident of the high Arctic, but undertakes annual migrations on a local scale in many regions. In Greenland the arctic redpoll is rare along the coast and seems to be closely associated with the highland of the inner parts all the way north to Thule and Peary Land.

Figure 8.3 The seed-eating arctic redpoll, with a weight of only 15 grams, is the smallest of the arctic winter residents (Photo: F. Falkenberg).

Figure 8.4 The geographical distribution (green) of the arctic redpoll (*Carduelis hornemanni).*

The nest of the arctic redpoll is built of straw in the low willow brush usually at altitudes above 200 m, and the up to 7 eggs are laid at the beginning of June and hatch after only 11 days of incubation. The chicks which are attended to by both parents are ready to fly after another 11 days.

In Greenland it is assumed that the arctic redpoll feed primarily on seeds of grasses and sedges, which during winter are taken from wilted plants that stick out through the snow. One can probably safely assume that that is tough work for a 15 gram bird in the darkness of the high-arctic winter, but unfortunately, precious little has been done to understand how this hardy and courageous bird has adapted to its environment. In fact, it appears that the chosen few who have

Figure 8.5 The geographical distribution (green) of the raven *(Corvus corax)*.

ever been close to this bird have spent their time on endless discussions over the systematic differences between birds in different regions.

Raven (*Corvus corax*)

The raven is a large (1,000 to 1,500 grams) ebony black passerine bird with a long and strong beak. This bird is distributed almost all over the northern hemisphere, with the exceptions of northern Ellesmere Island and northern Greenland, Svalbard, Novaya Zemlya and Taimyr Peninsula (Fig. 8.5).

The raven is territorial during the breeding season, but roams widely for the rest of the year. They arrive at the nesting site in rocky lowland terrain, often in the vicinity of large sea-bird colonies, already in February, when they build a rather elaborate nest of dry branches, twigs and sticks lined with moss, straw and hairs, on ledges in steep mountain walls. Their courtship-display has the form of a veritable air-show, reminiscent of the by-gone Spitfire days. The 3-6 eggs are subsequently laid, usually in early April at a time when the ambient temperature at the most northern locations every so often drops to -30 °C, or even lower. The eggs are incubated by the female for 21 days. The chicks are hatched naked and are fed by both parents for 5-6 weeks where after they fly from the nest, but are still fed by the parents for a long time during summer, and leave on a southern migration in August-September.

The ravens will eat almost anything, but feed primarily on eggs and chicks of seabirds, lemmings, arctic hare, reindeer calves and carrion from other animals. During winter they are gregarious, and flocks of 10-20 birds are often found close to human settlements, where they scavenge on all sorts of remains. The ravens also frequent the beach to a large extent to feed on marine organisms in the littoral zone, and even venture far out into the pack-ice to feed on carrion from polar bear and arctic fox.

Rock ptarmigan (*Lagopus mutus*)

In spite of the prominence of the rock ptarmigan in the arctic fauna, surprisingly little research has been done on this species, but Salomonsen (1950) of Denmark has given detailed descriptions of their natural history based on a life-long association with them in Greenland. The rock ptarmigan is a gallinaceous bird with a typical hen-like appearance (Fig. 8.6). It is normally non-migratory throughout its range (Fig. 8.7), but leaves the most northern parts of Greenland during the dark period between November and early February. In Greenland several sub-species have been recognized, e.g. *L. mutus reinhardti*, which is greyish in the south, *L. m. saturatus*, which is more brownish in the north, and the large *L. m. captus* in the far north, which is yellowish brown in the autumn.

The Svalbard variety (*L. m. hyperboreus*), on the other hand, is stuck on its high-Arctic archipelago and has to endure the full brunt of the long winter night. To do so it has developed some very distinct adaptations which involve both behavioural changes and deposition of huge stores of fat, which are under sophisticated control. These phenomena have been studied in detail over a long period in our laboratory at Tromsø, and will be dealt with in chapter 10 of this book.

The adult birds sport three annual plumages. In the winter both sexes are pure white with black tail-feathers, the cock having in addition a broad black stripe run-

Figure 8.6 A Svalbard rock ptarmigan roosting in the snow (Photo: H. Parker).

ning from the beak through the eyes. In summer the cock has brownish-black feathers with coarse, irregular vermiculations and spots of yellowish-brown colour on the flanks and back, while the breast and the abdomen are usually white. The hen, on the other hand, has a summer plumage of broad and regular yellowish transverse bars on a brownish-black background. In the autumn, both sexes have a more delicate colour pattern, being closely and irregularly barred and dotted with blackish-brown on a brownish, yellowish or greyish background, which varies considerably with geographical location. At all times the white wings are very conspicuous during flight. It was first shown by the Norwegian adventurer-zoologist Per Høst already in 1942 that these changes of plumage are regulated by photoperiod, and not by ambient temperature, as held at the time (Høst, 1942).

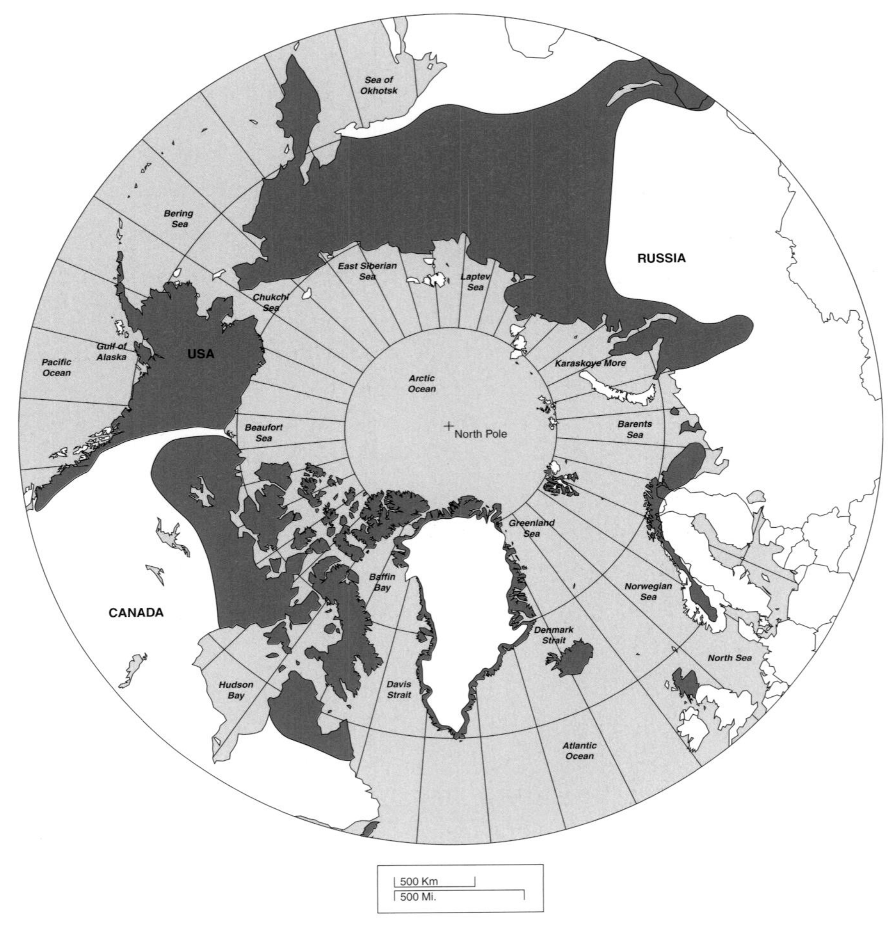

Figure 8.7 The geographical distribution (green) of rock ptarmigan (*Lagopus mutus)* in the Arctic.

We (Mortensen & Blix, 1986) have shown that the insulative value of the plumage is significantly better in winter compared with summer in the Norwegian rock ptarmigan, and that only the snowy owl has better insulation in winter than the Svalbard rock ptarmigan (*L.m. hyperboreus*), where subcutaneous deposits of fat seems to contribute to the insulation. Thus, the Svalbard rock ptarmigan has a lower critical temperature of about -5 °C in winter, but, due to its excellent insulation, it only has to double its metabolic heat production to maintain body temperature at -45 °C, which is the lowest ambient temperature this species ever experiences.

The rock ptarmigan prefers heath-covered ground interspersed with stony patches in the form of large rocks or boulders, which provide shelter from wind and drifting snow and afford suitable places for look-outs, but it also occurs on rolling low-land tundra, and in the blockfields on the mountain sides. It is greatly affected by the weather, especially by snow-fall, and therefore undertakes extensive vertical and horizontal movements depending on availability and accessibility of its food. It feeds in winter not only on the snow-free patches, or on bushes and herbs protruding above the snow, but is also able to dig in the snow in order to uncover its feed. Moreover, it often spends the winter nights in the shelter of boulders or in holes it digs in the snowdrifts (Fig. 8.6) to keep warm at temperatures which often fall below -40 °C. This behaviour may also increase its chances of not belonging to the large group of fellow birds which fall victims to predation by Arctic foxes, and snowy owls during the night.

The courtship-display starts sometimes as early as the beginning of April at lower latitudes and a bit later in the high north. At this time the gregarious habits of the winter are abandoned, and the cocks choose their territory. Simultaneously the supraorbital combs develop in the cocks. The cocks fight each other vigorously, especially in years with a population high. In such years the territories are small, only with a distance of about 3-400 m between two neighbouring cocks. The nest is a mere hollow in the ground, sparsely lined with dead leaves and straw, placed on dry ground, usually on sloping hillsides, sheltered by a rock or hidden in the vegetation. Egglaying (5-12, usually 9-11 eggs) takes place in the first half of June. The incubation lasts 21 days. Only the hen broods. The cock guards her during the greater part of the incubation period, keeping a look-out from the top of a rock or boulder at some distance from the nest.

At Tromsø, we (Gabrielsen, Blix & Ursin, 1985) have shown that when incubating willow ptarmigan hens are approached by animals, or man, they display a "playing dead", or freezing reaction, which implies that they become completely immobile with both their respiration rate and heart rate much reduced. (Fig. 8.8). This obviously is a predator defence reaction which is likely to increase survival of the eggs, not only in willow ptarmigan where it has been documented, but also in the rock ptarmigan where the freezing reaction is regularly seen, during the incubation period.

The very precocious chicks hatch in the first part of July, but do not reach adult size until August, or even later. During the last part of incubation and in the first days after hatching, the hen will display injury-feigning when an enemy approaches, crouching, the plumage puffed up and the wings thrashing. The very precocious chicks run away and hide in the vegetation, lying dead still, pressed against the ground with necks stretched out. After the age of 10 days, however,

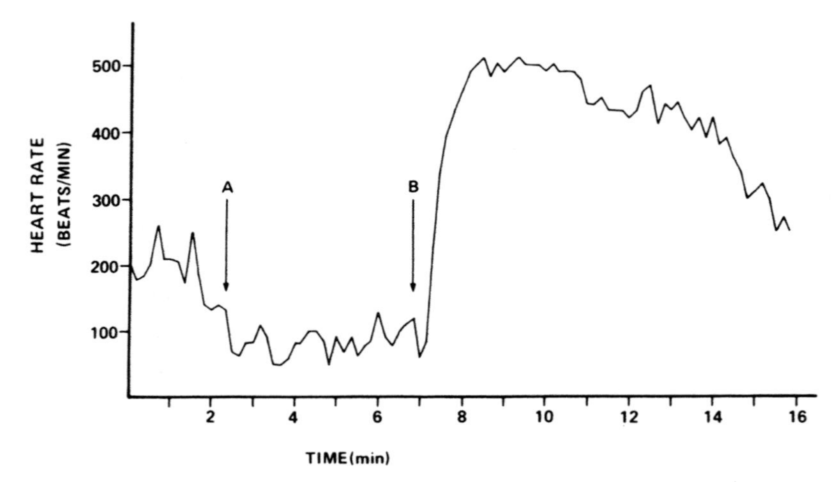

Figure 8.8 The heart rate response in an incubating willow ptarmigan *(Lagopus lagopus)* when approached by a person. A: Hen can see the approaching person. B: The same person turns and leaves the area (Gabrielsen, Blix & Ursin, 1995).

the chicks are capable of flying. When the young are grown up, in the latter half of August, the covey often moves higher up the slopes of mountains, and flocks have been recorded as high up as 1,100 m in northern Greenland. This upward movement is no doubt due to the fact that the food-plants of the ptarmigan appear later at high altitude and that they consequently will be more nutritious than those lower down at this time.

There is a pronounced seasonal variation in the composition of the food of the rock ptarmigan, both in Greenland, Iceland and Svalbard, where its diet has been studied in detail (Unander, Mortensen & Elvebakk, 1985). In summer it consists chiefly of *Polygonum viviparum*, supplemented with a few insects, in the autumn to a great extent of the bulbils of this plant, while in winter it eats mainly *Salix arctica* and some *Dryas octopetala* and *Saxifraga oppositifolia*. Mortensen *et al.* (1983), moreover have shown that, even at Svalbard where complete darkness prevails for more than two months in winter, the rock ptarmigan is able to find high quality food throughout the year, and that inadequate supply of nutrients is more likely to result from restricted availability than poor quality of the diet. It is in this context interesting to note that the most high-arctic birds in Greenland sometimes seem to eat *Cassiope tetragona*, which is hardly eaten by any other mammal or bird due to its high content of resins. The reasons for this unconventional choice of diet is unknown. In Svalbard the chicks of the rock ptarmigan grow very rapidly from about 15 to 400 grams in 40 days. This rapid growth is based on a diet of almost 100 % bulbils of *Polygonum viviparum*, which obviously is an excellent protein substitute for the insects in the diet of the willow ptarmigan (*Lagopus lagopus*) chicks.

The number of rock ptarmigans is subject to large fluctuations. At least in Greenland and Iceland peak years appear to occur with reasonable regularity at intervals of about ten years. The reasons for these fluctuations are not fully understood, but it has been suggested by Gardarsson (unpuplished) of Iceland that high predation in combination with overgrazing of major food plants may trigger the population declines which results from increased winter losses, especially of juveniles. A similar hypothesis has been put forward by Andreev (1988) for the regulation of the willow ptarmigan of eastern Siberia which undergo 10-year population cycles, apparently due to overgrazing of arctic willow.

Resident mammals

Muskox (*Ovibos moschatus*)

To the untrained eye a resting muskox looks amazingly like a rock – both summer and winter, but when it stands up in all its splendour with its almost ancle deep black guard-hairs, much like a solemn funeral horse of past days, you gaze in awe on an animal more pre-historic and obviously arctic looking than any other creature.

The muskox is a ruminant belonging to the order of even-toed (*Artiodactyla*) ungulates, of the family *Bovidae.* Fully grown a bull may reach a weight of 350 kg, while the average for bulls is 275 kg and for females 190 kg. The body of the muskoxen without the fur has a close resemblance with that of a goat, but all live muskoxen are blessed with a fur-coat of considerable dimensions. Very long and dark, almost black guard hairs are hanging down the flanks of the animal covering the inner soft and light brown (Qiviut) wool of outstanding insulative value. On the back there are no guard hairs and hence, a "saddle" of yellowish wool is apparent. Both sexes have horns of a very characteristic shape with a massive base, separated only by a narrow cleft in the middle, and curving, first down and then forward, on both sides of the head. The horns are used as weapons against predators and in fighting between males. The horns are visible in the calves already in the first fall, and the trained eye can tell the sexes apart from the shape of the horns from the age of 2-3 years. The hooves on the front legs are much larger than those of the hind-legs and assist the animal when it is cratering for food during winter.

The muskox is supposed to have originated on the tundra of north-central Asia about one million years ago, and reached the American continent over the

Bering Land Bridge some 90,000 years ago. Today, natural populations occur only in northern Canada and north- and north-east Greenland (Fig. 8.9), where some 50,000 and 20,000 are found, respectively. In 1930 muskoxen from Greenland were successfully transplanted to Nunivak Island off the west coast of Alaska, and from there they have been reintroduced into their former range along the arctic coast of Alaska. Muskoxen from both Greenland, Canada and Alaska have also been introduced into West Greenland, Norway, Wrangel Island and Taimyr Peninsula in Russia and the Ungava region of northern Quebec in Canada, and some of the Norwegian animals have moved out and established themselves in Sweden, while attempts to introduce them in Svalbard have been unsuccessful.

Figure 8.9 The geographical distribution (green) of the muskox (*Ovibos moschatus*). Introduced populations are indicated by green dots.

In addition to these wild populations there are research stations with captive muskoxen at Fairbanks, Alaska, Saskatoon, Canada and Tromsø, Norway.

The calves of the muskox are born from April till mid-June in the high Arctic, where ambient temperatures may fall to -40 °C at the time of calving, but the calves are very well protected against the grisly environment, and soon seem to be enjoying life under the midnight sun. And, of course, if you are born wet, on snow at -40 °C and have managed that, the rest of life is likely to appear rather simple.

After calving the adults shed their under-wool, while the guard-hairs remain, giving the animals a very shaggy appearance. After moulting, the muskoxen often engage in different sorts of play, an activity which is rather uncommon among bovids, particularly among the adults. Such play may include running, chasing, and jumping and whirling in ponds, on land and on snow.

In the high-Arctic the rutting season extends from mid-July to early October, with breeding taking place in August-September. During the rutting season the males indulge in aggressive behaviour. This includes deep rumbling roaring, charging and gland-rubbing, which involves rubbing the preorbital gland, which is located just in front of and below the eye, against the forelegs. Gland-rubbing is probably the most distinctive of all muskox behaviour patterns. The exact significance of the behaviour, aside from the release of some secretion, is not clear. Other activities include: pawing and pit-digging and head-swinging, all to indicate that you better make yourself scarce, unless you are a female muskox of some special attraction. If that does not happen the whole show culminates in both bulls galloping forward and clash head-on with a bang which "can be heard a mile". In fact it has been calculated that clashing speeds reach at least fifty kilometres an hour, the horn bosses receiving the full impact of the clash. Why so few animals die in such collisions, which would certainly cause great damage to modern cars, is probably explained by the bunker construction which envelopes their rather diminutive brain (Fig. 8.10).

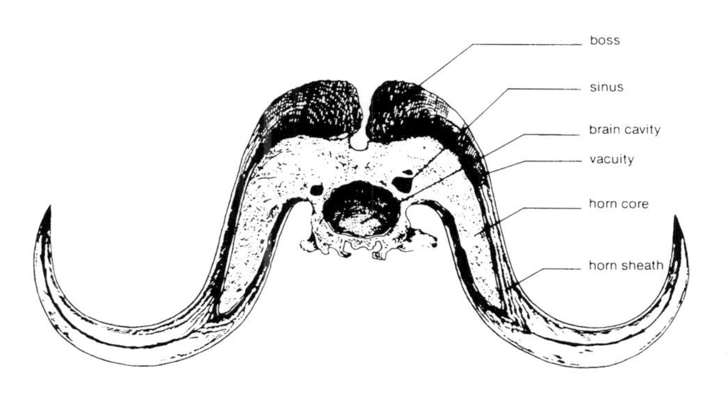

Figure 8.10 A cross section through the skull of a male muskox, which is constructed to withstand the tremendous impact of the head-on collisions with other bulls during the rutting season. The cavity left for the brain is less impressive (Gray, 1987.)

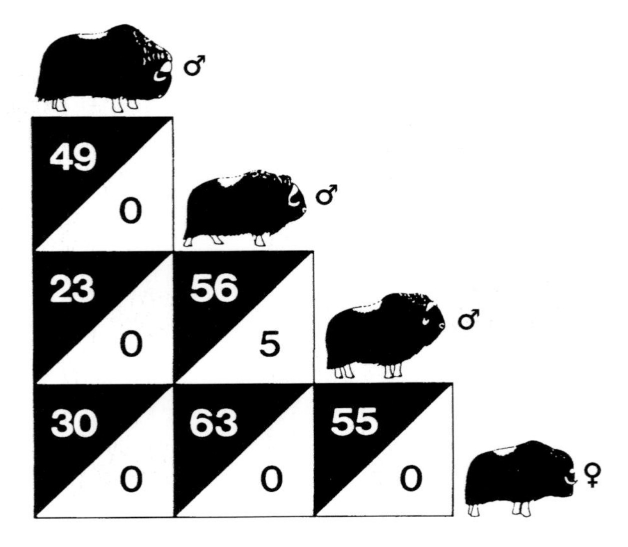

Figure 8.11 The dominance hierarchy of muskox based on the outcome of 281 interactions. The black part of each square indicates the number of times a particular muskox won conflicts with the muskox to the right of the square. The white part of the square shows the number of wins achieved by the muskox illustrated to the right of it in conflicts with the muskox seen above it (Gray, 1987).

After the rutting season the muskox herds settle down for the vigil of the darkness of the high-Arctic winter. Late winter is the time when old animals die, usually from starvation. The muskox herds are organized socially based on a dominance system, or a "pecking order", where adult males dominate adult females, while individuals of each sex dominate others of the same sex (Grey, 1987). This kind of interaction is particularly apparent in displacements from feeding craters in the snow during winter (Fig. 8.11).

In winter it appears that *Ledum decumbens, Empetrum nigrum, Vaccinum uliginosum* and *Betula glandulosa* are important food items for muskox in the Canadian high Arctic, while in summer, willow (*Salix spp.*), grasses like *Puccinellia angustata, Festuca brachyphylla, Alopecurus alpinus* and *Poa spp.*, sedges (*Carex spp.*) and *Oxyria digyna* and *Polygonum viviparum* among many others are important. *Cassiope tetragona* is not eaten in spite of its abundance.

When muskoxen are threatened by wolf or man, their only real enemies, they first react with snorting, then they often run away toward the dominant bull and to high ground where they stop and form a tight group. Usually the dominant bull turns first and stops, the others wheeling around behind him. The protection of the rear of each animal seems to be the basis of the defence formation. In such a formation a great amount of pushing usually ensures close contact and calves would be in a difficult spot in the centre of such a shifting mass. They usually press tightly to the mothers flank and do not huddle in the centre as is often stated in the popular literature. From this tight mass of flesh, individual cows or bulls

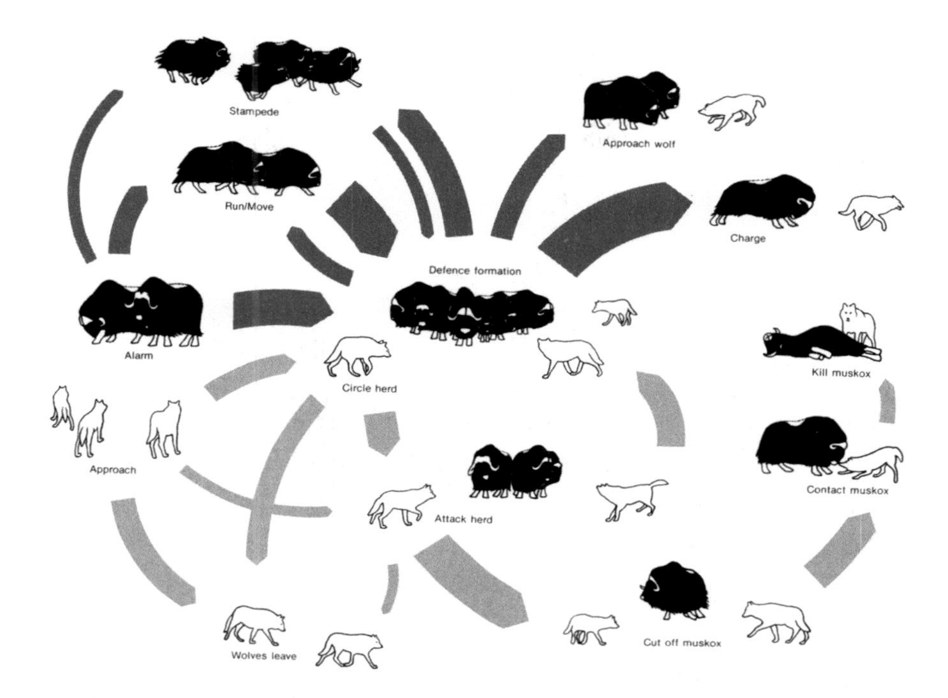

Figure 8.12 Wolf attack and muskox defence. The diagram (starting at left) shows the more common sequences of events in 21 observed encounters. Order of events is shown in the lower half for wolves (paler arrows) and the upper half for muskox (darker arrows). The wider the arrow, the more often the sequence occurred (Gray, 1987).

will charge a wolf that approaches too closely, then return to the herd. Encounters between wolves and muskoxen can last from two minutes to several hours. Wolves are often successful in killing, even strong and healthy muskoxen, particularly after having succeeded in isolating single individuals from the herd (Fig. 8.12) (Grey, 1987).

It has been assumed that solitary bulls were outcasts; old animals that were no longer sexually active, and chased from the herd by younger, more virile males. In concurrence with this we see that bulls in our herd of semi-domesticated research animals at Tromsø invariably are beaten up by younger bulls and become solitary and unconcerned with breeding some time before the age of 10 years. However, recent research indicates that solitary bulls later may join the herd and even become dominant bulls, so obviously this is not always the case.

Reindeer (*Rangifer tarandus*)

The reindeer, like muskoxen, are ruminants belonging to the order of even-toed (*Artiodactyla*) ungulates, but unlike muskoxen, the reindeer are members of the *Cervidae* family, which among other things implies that they have antlers instead of horns. All reindeer belong to the same species, but reindeer have a circumpolar distribution in the north and several sub-species are recognized in different regions (Fig. 8.13), and in spite of belonging to the same species, the North American reindeer are called caribou, while the Eurasian ones are properly called reindeer!

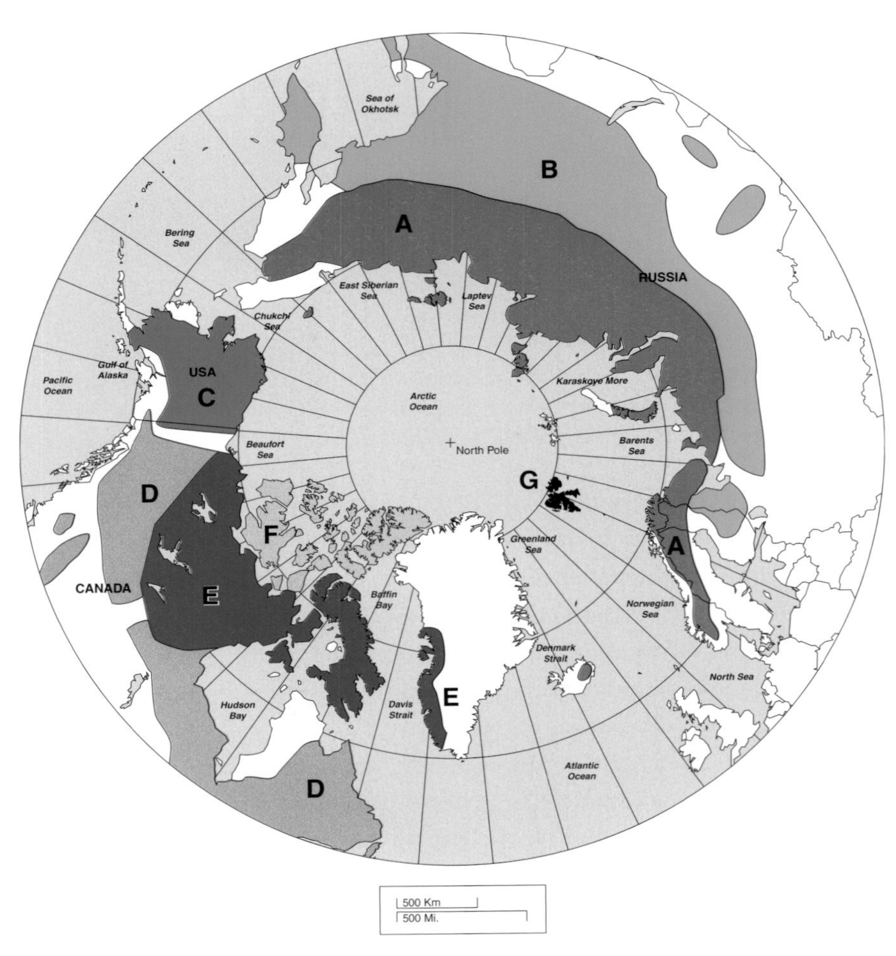

Figure 8.13 The geographical distribution of different subspecies of reindeer (*Rangifer tarandus*).*A*: Eurasian tundra reindeer (*R. t. tarandus*); *B*: Eurasian forest reindeer (*R. t. fennicus*); C: Alaskan caribou (*R. t. granti*); *D*: woodland caribou (*R. t. caribou*); E: barren-ground caribou (*R. t. groenlandicus*); F: Peary caribou (*R. t. pearyi*); G: Svalbard reindeer (*R. t. platyrhynchus*).

The reindeer are found from the very high barren-ground Arctic (Svalbard and Peary reindeer), on the arctic tundra (Eurasian tundra reindeer, Alaskan caribou and barren-ground caribou) and in the boreal taiga (eurasian forest reindeer and north american woodland caribou). The first six of these races are believed to have survived the last Pleistocene glaciation in the Beringia refugium and from there colonized Eurasia and North America, while the North American woodland caribou most likely lived south of the great glaciers (Knut Røed, Norway; *personal communication*).

Reindeer are average sized animals, but the external features of the different races differ in accordance with their habitat : The Svalbard and Peary reindeer are small and clumsy-looking, with short legs and snout, while the woodland caribou are large, longlegged and elegant (Fig. 8.14). The weight of the animals differ from an average of 55 kg in a male Svalbard reindeer to an average of 200 kg in a male woodland caribou, the females weighing some 30 % less.

Reindeer are equipped with a heavy winter pelage of long hollow hairs (Fig. 8.16) which offers prime insulation. This pelage, which may differ considerably in colour, is shed in spring and early summer, when a new one is grown, with the hairs lengthening as the season progresses. The reindeer are also blessed by a number of additional mechanisms for thermal protection which will be dealt with in chapter 10 of this book.

Reindeer are the only members of the cervid family where both sexes carry antlers. Unlike horns, which grow from the base, just like grass, and keep growing throughout life, antlers grow from the tip, like branches on a tree, and they are shed and renewed every year. Antler growth in Norwegian reindeer starts usually in April and is completed in August-September. The antlers start growing from two bone pedicels on the skull of the fore-head and develop in a characteristic branched fashion. During the growth period the antlers are covered by fur, or "velvet", and are well perfused by blood to support the rapid growth. It is not until the antlers are fully developed that they are calcinated and harden into bone. When this has happened the blood supply to the antlers is stopped and the "velvet" dries up and is subsequently brushed off against brushes or other vegetation. The dry antlers are shed at the base, where the growth started, the whole cycle of events being controlled by hormones, notably testosterone in males and oestradiol in females. The bulls use the antlers as weapons during the rut, when the fights between rivalizing bulls sometimes attain herculean proportions. Thus, the bulls shed their antlers after the rut, in November-December, while barren females keep theirs till March-April and pregnant females till the time of calving. It is assumed that the antlers signal status within the herd and that the antlers are

Figure 8.14 All reindeer belong to the same species, but they come in very different shapes. *Top*: the slender and elegant (female) Eurasian forest reindeer. (Photo: M. Nieminen). *Bottom*: the short-legged and bulky Svalbard reindeer (Photo: T.H. Aa. Utsi).

Figure 8.15 It is not only man that is harassed by insects during the arctic summer (Photo: unknown).

important instruments for the females in their competitions for food craters in the snow, not only between themselves, but also with respect to the males, which unlike other cervids are not separated from the females during winter.

Reindeer are, with one or two exceptions, migratory animals. Thus, barren-ground caribou both in Alaska and northern Canada embark on a northern migration between late April and early June, the locus for the movement primarily being a suitable calving area. But as the saying goes: there are no free lunches, and although grub may be readily available the animals are often harassed by insects (Fig. 8.15). After calving, in late July or early August the animals start to move back toward the winter range area. These summer migrations span distances between 150 and 1,000 km, making them the longest migrations known in any terrestrial mammal. In Scandinavia, where most of the reindeer are semi-domesticated and managed by the Sami people, the animals venture on annual migrations from the inland lichen-ranges where they spend the winter to the grass-clad islands off the coast in the spring, and back to the inland winter ranges in the autumn. Many of the traditional migration routes both in Scandinavia and Canada involve crossing of rivers, sounds and fjords, but reindeer are, probably in part because of the airfilled hairs in their pelts, excellent swimmers which willingly negotiate large distances of water (Fig. 8.16).

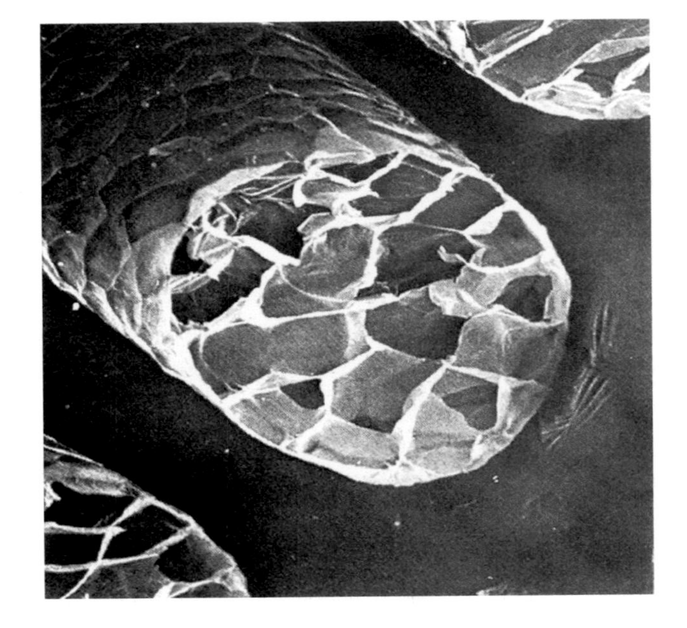

Figure 8.16 Reindeer are good swimmers. *Top*: A herd crossing a fiord in Finnmark, Norway, following the leader, which is towed by a boat, during the annual migrations to the summer pastures at the coast. (Photo: P. Høst). *Bottom*: The hairs of reindeer are air-filled and provide good floatation during swimming, as shown in this scanning electron micrograph of a cut reindeer hair (Timisjärvi, Nieminen & Sippola, 1984).

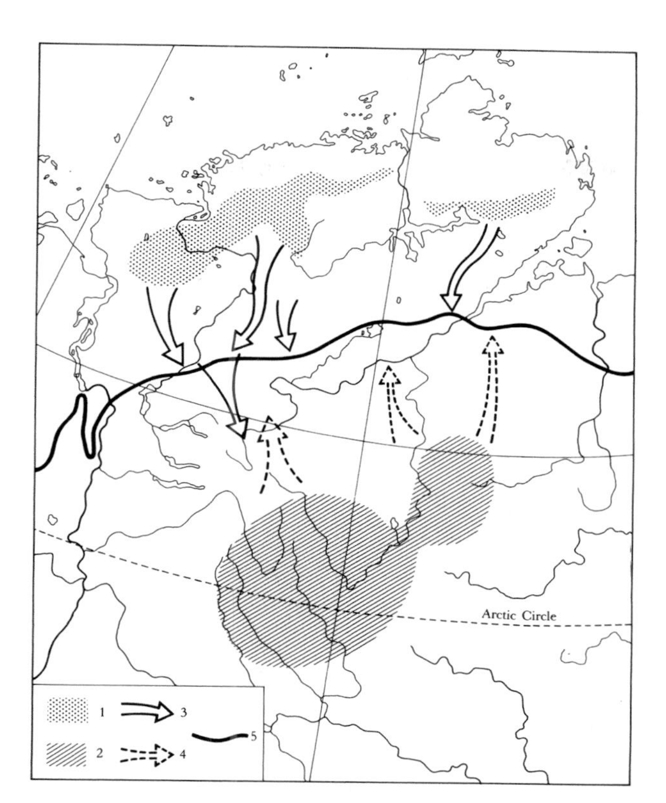

Figure 8.17 The annual life cycle of the Taimyr Penninsula wild reindeer herd. 1. Calving grounds and concentration during summer. 2. Basic wintering area. 3. Autumn migration route southwards. 4. Departure from wintering grounds. 5. Limit of the tundra (Chernov, 1985).

Unlike Scandinavia, where the winter pastures are the limiting factor for the reindeer populations, the supply of lichens in Siberia is often almost unlimited, while vegetation rich in protein is in short supply. The 500,000 reindeer in the Taymyr area, being the largest Russian population of wild reindeer, consequently embark on very long annual migrations almost to the coasts of the Arctic Ocean (Fig. 8.17).

The Svalbard reindeer is different from the other sub-species of reindeer in that it does not migrate. In fact, this animal is rather solitary and only occasionally occur in small loose groups of two or three and usually spend the entire year within an area of a few square kilometres.

Now, why is it that most reindeer migrate over large distances, while the Svalbard reindeer do not? We will examine two possible explanations: nutrition and predation. In an animal with a circumpolar distribution extending over 35 degrees of latitude it is of course hopeless to go into any detailed analysis of the diet of the many different populations of animals. Even so, it is probably common knowledge that reindeer depend to a great extent on lichen ranges during winter and migrate north to thrive on grasses and sedges during summer. This is indeed the case for the Eurasian tundra reindeer, the Alaskan caribou and the barren ground caribou of Canada, except we have already heard that lichen is not a limiting factor during winter in most places in northern Russia. It is not true, however, that reindeer feed exclusively on lichens during winter. Lichen is very rich in carbohydrates and hence, energy, but it is almost devoid of protein, and it is therefore impossible to survive on a pure lichen diet for extended periods. The continued presence of reindeer at South Georgia in the Southern Ocean, brought there in 1911 and 1925 by Norwegian whalers, is further proof of reindeer being independent of lichen, since the animals there feed almost exclusively on evergreen tussock grass (*Paridiochola flabellata*) during winter, while their diet is supplemented by graminoids, forbs and woody plants in summer (Fig. 8.18). Moreover, in the barren-ground caribou in Greenland and Peary caribou and Svalbard reindeer, lichens are not important food items during winter. In Svalbard it is even completely missing. Here instead grasses and sedges, a few woody plants and mosses are important in winter. It has been shown that the digestibility of mosses

Figure 8.18 Reindeer in Grytviken, South Georgia. Reindeer do not depend on lichen during winter. The herd in Grytviken, which was introduced by Norwegian whalers in 1911-1912 and 1925 has thrived, without any access to lichen, mainly on tussock grass (*Paridiochola flabellata*) (Photo: S.D. Mathiesen).

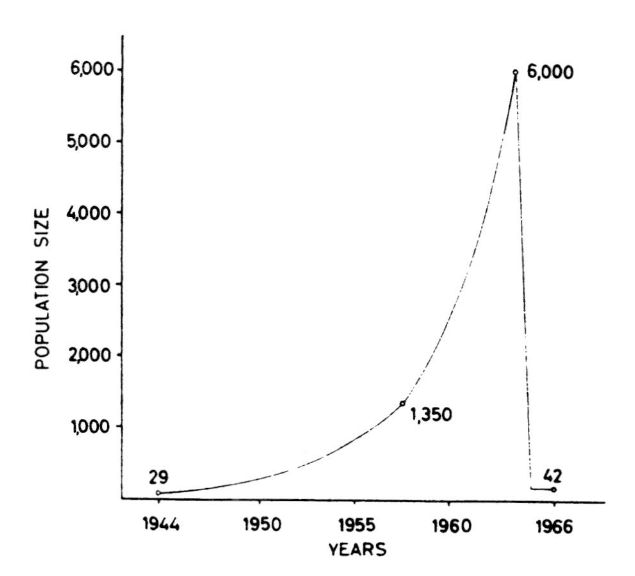

Figure 8.19 Assumed population growth of the St. Matthew Island (Alaska) reindeer herd. Actual counts are indicated on the population curve (Klein, 1968).

is rather low, but this rather unusual food item is readily available and may be very important when little else is to be found. In summer, these animals feed primarily on grasses and sedges. For the woodland caribou, on the other hand, arboreal lichens, supplemented by woody plants are by far the most important food items in winter, while a great variety of food, like mushrooms, lichens, woody plants, grasses, sedges and forbs, are eaten during summer. So it is quite likely that dietary needs may be one cause for the migratory behaviour in some populations of reindeer, but certainly not in others, like for example the Svalbard reindeer which, we have already heard, show no migratory behaviour whatsoever.

Two cases of reindeer population dynamics, where food is likely to be a key factor are worth noticing. One is the celebrated case of the rise and fall of the St. Matthew's Island population in Alaska, as observed and described by Klein (1968), and the other is the case of the population of Svalbard reindeer in Adventdalen, Svalbard, which is being studied by Tyler. On St. Matthew's Island 29 reindeer were introduced in 1944, when the island offered lush vegetation of both lichen and vascular plants. Under these conditions the population increased very rapidly and reached 1,350 animals in 1957 and 6,000 in 1963. Subsequently, the population crashed and in 1966 there were only 42 reindeer left on the island, which by that time had become severely overgrazed (Fig. 8.19). The drama and simplicity of this story has brought it into many ecological textbooks over the years and to many naive readers it has become *the* picture of reindeer population dynamics. This is only too bad, because one has to keep in mind that

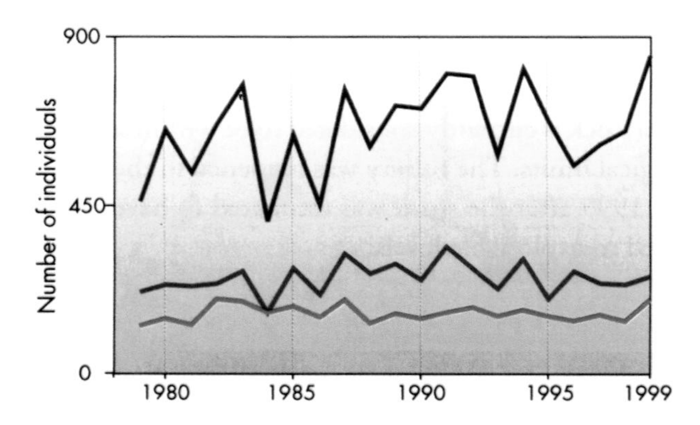

Figure 8.20 Population dynamics of Svalbard reindeer in Adventdalen, West-Spitsbergen, 1979-1998. *Top*: total count in summer (after calving). *Middle* and *Bottom*: females and males, aged two years and older, respectively (Tyler, unpublished).

this is a very special case in that the original reindeer were introduced into a lush virgin range, and that the population due to some ill-conceived ethical barriers were never allowed to recover due to the fact that it was decided to remove the remaining bulls after the crash. This was very unfortunate since the further development of the population would probably, not to say most likely, have progressed different. In this context the population in Svalbard is interesting. Here the local population within a valley, Adventdalen, maintains itself at a fairly stable level, the number of animals varying between 300 and 800. Starvation, caused by overicing of the range in spring, is almost the only cause of death, and only calves and old animals die. The cause of death by starvation in old animals seems to be tooth-wear. When the population, after a series of "good" years with high calf production, which may vary between 0 and 80 %, and low mortality reaches about 800, some 250 animals will migrate and establish themselves in other populations on the island (Fig. 8.20).

Several species of predators are at times preying to some extent on reindeer, but wolves are by far the most important. Wolves are found throughout the range of barren-ground reindeer and caribou, except in Svalbard, but their densities are variable down to very low in Siberia and almost zero in northern Scandinavia, where they have been exterminated due to the local emphasis on reindeer herding. Tundra wolves are wide-ranging and nomadic in habit, and may invade the forest in winter for hundreds of miles in pursuit of caribou herds. Thus, the movements of wolves from summer ranges on the tundra to winter ranges far within the forests, between Great Slave Lake and Lake Athabasca in Canada are almost as well recognized by people who know that country as are the movements of the caribou. However, during the denning season, and during summer and autumn, the wolves are territorial and stationary, and, as pointed out by Skogland of Norway and others, it is quite possible that avoidance of wolf predation during the calving season has also contributed to the migratory behaviour of reindeer and caribou. In this context it is again interesting to look at the non-migratory Svalbard reindeer and note that this animal knows, in fact, no predators.

Arctic fox (*Alopex lagopus*)

The arctic fox is a small canine carnivore, weighing 3-4 kg, and sporting a shining snowy white fur (Fig. 8.21), which many women have fallen for over many years, in winter, while in summer it is brown on the back and yellow on the belly. There is also a relatively rare blue variety of arctic fox which is uniformly blue-grey all year round. Arctic foxes are distributed circumpolarly in the tundra zone (Fig. 8.22) and are opportunistic and generalistic predators as well as scavengers, but microtine rodents are their major food source throughout most of their range. Arctic foxes are monogamous and territorial. Their home ranges vary between 10 and 20 km^2. They are sexually mature at an age of 9-10 months and give birth to pups in underground dens in May-June after a gestation period of 50-52 days.

The size of arctic fox populations in Canada, northeast Greenland, Siberia and Scandinavia fluctuates with the cycles in numbers of microtine rodents. The high fecundity of these foxes, with a mean litter size of 10, makes them capable of making a rapid numerical response when the abundance of microtine rodents

Figure 8.21 The arctic fox in winter costume *(left)* and during summer *(right)* (Photo: A.S. Blix).

Figure 8.22 The geographical distribution (green) of arctic fox (*Alopex lagopus*).

increases. In Iceland and Svalbard, where small microtine rodents are absent, there are no apparent short-term fluctuations in the number of arctic foxes, but the litter sizes are smaller than in areas with rodents.

The diet of arctic foxes varies regionally and seasonally, but may consist of neonate seals, hares, lemmings, seabirds, water fowl, passerines, waders, insects and marine invertebrates. Carrion of whales, seals and reindeer are also important food in many areas. It is also commonly believed that seal carcasses left by polar bears are important to the survival of arctic foxes in winter, and mass migrations onto sea ice in autumn have been reported from North America, but the significance of this food source has not been quantified. The feeding behaviour of the arctic foxes in

Svalbard, where there are no rodents of any kind is particularly interesting. These foxes have been studied in some detail by Prestrud and associates. They found that the seasonal variations in food availability are considerable. In spring and summer, great numbers of sea birds, water fowl and waders occur on the west and north coasts, and the abundance of reindeer carcasses increases abruptly in February-March when 90 % of the annual reindeer mortality occurs. In this period, ringed seals give birth to their pups, and carcasses from bear kills can be found on the fiord ice. From October to February, however, only reindeer and ptarmigan (*Lagopus mutus hyperboreus*) remain. In addition, 24-hour darkness, low temperature, strong winds and drifting snow often make life miserable, one would think, for a lonely hunting fox. Thus, as is to be expected, juvenile overwinter mortality is commonly between 50 and 80 %. The fox may improve its chances of survival by a combination of reduced energy expenditure and storage of energy, either by fat deposition and/or caching (e.g. Prestrud & Nilssen, 1992).

There are numerous observations of arctic foxes caching food (e.g. Sklepkovych & Montevecchi, 1996) and several caches containing more than 100 birds have been found. Prestrud calculated that a cache of 10 little auks (*Alle alle*) and 4 Brunnich's guillemots (*Uria lomvia*) represent an energy storage equivalent to the mean fat deposits of a fox. Still, all foxes in Svalbard deposited fat in autumn, and the mean fat deposits contained enough energy to sustain an average fox at its basal metabolic level for 25-30 days, but the deposits were not depleted before late March. This supports our idea of fat in arctic mammals and birds being crisis rations, the size of which being under photo-periodic control. This concept is dealt with in chapter 10 of this book. Thus, large food items such as carcasses of reindeer and seals and food caches are the major food sources, while the fat deposits add little to the *daily* energy budget for arctic foxes in Svalbard. It has been reported, however, that the subcutaneous fat deposits of arctic foxes contain more unsaturated fatty acids than the abdominal ones. Since unsaturated fatty acids have melting points which are lower than those of saturated fatty acids, it is possible that this indicates that the subcutaneous deposits also contribute to the insulation of the animal.

Arctic foxes in Svalbard have high growth rates in July-September of their first year. More than 90 % of mature size in linear dimensions is reached by October and the full adult size in December. Arctic foxes, moreover, appear to allocate energy, first to reach a certain size, then to development of muscle mass, and finally to deposition of fat. This is exactly the opposite of what we find in the Svalbard reindeer, which strive to deposit fat from the moment of birth and straight through the first summer and autumn, and appear the next

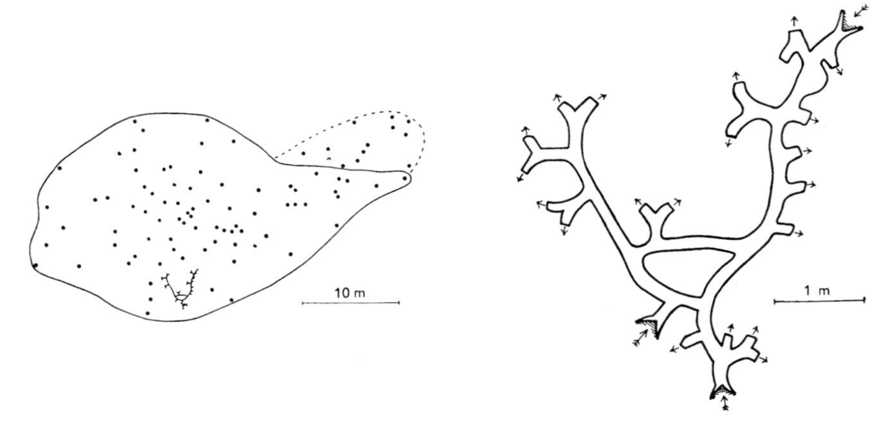

Figure 8.23 Map of an arctic fox den in alpine Norway (Hardangervidda). *Left*: Each dot represents one opening. A small area in the lower part was excavated and the tunnel system accurately determined, as shown in higher magnification at the right. The three inward – pointing arrows in the *right* drawing shows entrances, while the outward-pointing arrows indicate continuation of the tunnel system. During the excavation the foxes enlarged the den with the area inside the dotted line at the left with an extra 13 openings (Drawing by Per Høst).

spring with pretty much the same frame as when they were born (Reimers & Ringberg, 1983).

The standard arctic fox den is a complex structure which may extend over 50-200 m^2 covered with lush vegetation and excavated in sand, gravel or soil with 20-50 entrances (Fig. 8.23), usually in riverbanks, moraines, eskers or hummocks. In Svalbard a typical den is located under boulders, and it is supposed that such dens give better protection against pup predation, which may be substantial in other places.

In carnivore species experiencing cyclic variations in prey abundance, such as coyote, red fox and lynx, large variations in pregnancy rate of yearlings contribute significantly to the annual variations in the overall pregnancy rate. In Canada variable pup survival has been reported to be a major factor in determining the fluctuations in arctic fox numbers, and it is assumed that siblicide and abandonment of litters in years with bad nutrition are the main causes of pup mortality, but there are also data to suggest that variation in intra-uterine litter size is responsible for the variation in reproduction among years.

Rabies is endemic in the Arctic and arctic fox is the main vector of the virus. The Arctic fox virus is a distinct form of the virus, and it is probably not as virulent to humans as some of the other strains. The prevalence of rabies is also low in arctic fox outside epizootics, when the prevalence in trapped foxes often is as high as 70-80 %. The arctic foxes are able to move over enormous distances, even over pack-ice and to undertake mass migrations. Thus, one tagged fox moved 130 km in 6 months in Svalbard and another fox more than 2,300 km in North America. Such long distance migrations will, of course, contribute to the spread of the virus.

Lemmings

Lemmings have a well established place in Nordic history and folklore, and ever since the 14th century books have appeared with stories of lemmings raining down from the sky in Norwegian Lappland, and "everybody know" that myriads of lemming sometimes march into the sea and off cliffs to kill themselves and that they can become so angry that they explode – to the same end. These are all

Figure 8.24 A grumpy male collared lemming from Alaska (Photo: G. H. Jarrell).

myths, but it is true that there sometimes are mass movements of lemmings and also that there sometimes are mass occurrences, as first pointed out by the Swedish Archbishop and naturalist Olaus Magnus in 1555. It is also worth noticing that the Norwegian priest and naturalist Erik Pontoppidan, already in 1751, noted that the magnitude of these mass occurrences vary greatly, and that the real great ones only occur locally. This is indeed consistent with the present view on the lemming cycle.

Lemmings are short-legged, short tailed, short-eared northern mice of several different species, which together have a circumpolar distribution. Within the arctic region, we find, according to the most modern classification, the Norwegian lemming (*Lemmus lemmus*) in northern Scandinavia and north-west Russia, the Siberian lemming (*Lemmus sibiricus*) all along northern Siberia, except Chukotka, and the brown lemming (*Lemmus trimucronatus*) in Chukotka, Alaska and northern Canada, except the Queen Elisabeth Islands, and Collared lemmings in Alaska, northern Canada, including the Queen Elisabeth Islands and northern and north-eastern Greenland (*Dicrostonyx groenlandicus*) and the entire northern Siberia (*Dicrostonyx torquatus*) (Fig. 8.25). Several of these species may be split in local races, and several other species of lemmings extend into the southern arctic region, without being considered true arctic species.

The Norwegian lemming has a beautiful colouration with black head and shoulder and the rest a yellowish brown, and the collared lemmings are unique in that they turn whitish grey in winter (Fig. 8.24), while the rest of the species are boring brownish year around.

Lemmings are characterized by, and known for, their pronounced density cycles combined with shifts in habitat use both at a local and a regional scale. In fact, most lemming populations studied have been found to have cyclic population dynamics. In particular, the Norwegian lemming seems to have population cycles that are erratic with respect to amplitude, but rather regular with respect to periodicity. We also know that the periodicity is variable in time and location (Fig. 8.26). The cycle is not symmetrical; rather, there tends to be a rapid explosion after two or more low years. This rapid increase is then succeeded by a sudden crash, followed by a number of low density years. But, there may be spatial variation in both amplitude and intervals and some populations seem to alternate between cyclic and non-cyclic periods. Reproductive rate is lowest during the peak phase of the cycle, when the animals tend to be more aggressive than those from the low and early increase phases. Scientists have tried to explain the reasons for the lemming population cycles for almost a hundred years. Of the many suggestions are the number and intensity of sunspots, adverse weather, social stress

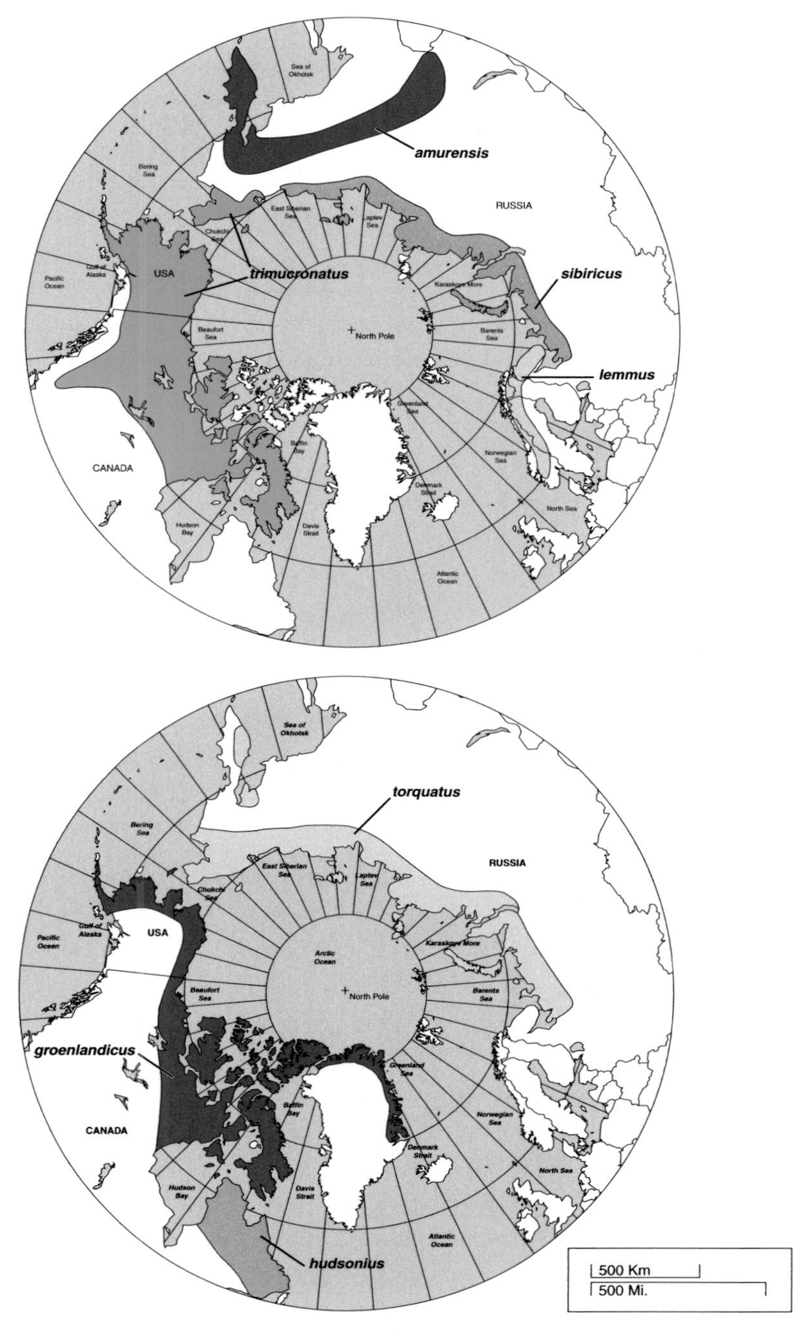

Figure 8.25 *Top*: The geographical distribution of *Lemmus. Bottom*: The geographical distribution of *Dicrostonyx* (Jarrell & Fredga, 1993).

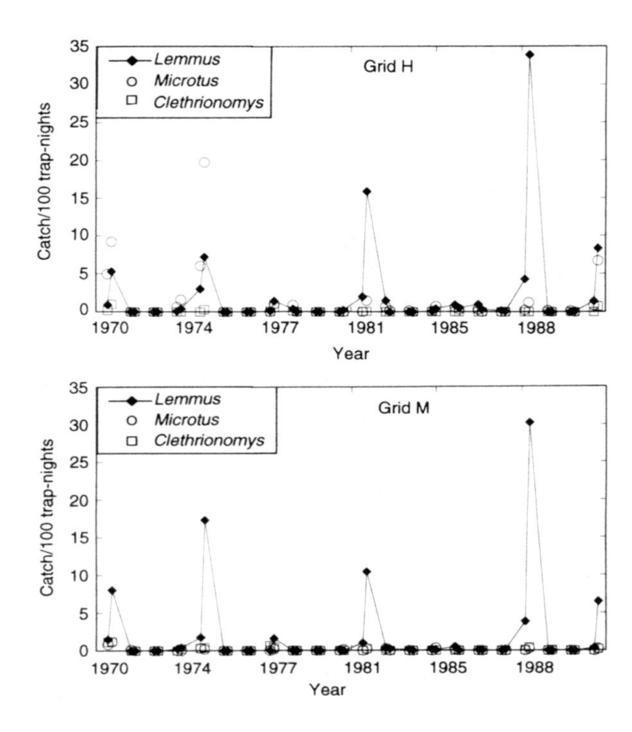

Figure 8.26 Population fluctuations of Norwegian lemming (*Lemmus*), *Microtus sp.*, and *Clethrinonomys sp.* in two grids (H: high productivity and, M: medium productivity) at Hardangervidda in southern Norway, 1970-1991. The data indicate that while *Microtus sp.* and *Clethrinonomys sp.* did not fluctuate much, the lemming population cycled with 7-year intervals (Framstad, Stenseth & Østbye, 1993).

during high densities, genotypic differences with respect to spacing behaviour and reproduction, quantity and quality of available food, production of chemical defence in heavily grazed plants, dispersal, parasites and diseases, habitat quality and heterogeneity, and finally predation. The possibility of predator build-up being responsible for the crash, and consequently the cyclicity of lemming population was first advanced in a crude and simple form by Hagen of Norway almost 50 years ago. It is still likely that resident specialized predators, such as ermine, may generate cycles, if they are sufficiently efficient in finding prey even at low prey densities, and it is quite obvious that they may cause high mortality in the crash phase. Some predators, like snowy owls, are highly mobile and may roam over extensive areas in search of peak microtine populations. It has been suggested that such nomadic predators synchronize microtine fluctuations over large areas. It is finally worth noticing that the microtine cycle disappeared during a large scale predator extermination programme which took place in Norway during 1905-1920.

There is much anecdotal information suggesting that lemmings, and in particular Norwegian lemming, perform long distance movements. It has now been documented that this is indeed the case for Norwegian lemmings living in mountainous Fennoscandia during peak years, while none of the other lemmings do. It is moreover documented that Norwegian lemming in northern Finland exhibits both seasonal migrations and pronounced multiannual mass movements, whereas other lemmings only exhibit relatively short-distance movements during the snow melt. As to why lemmings perform mass movements we can still only speculate, and in any case it seems to be an enigma, at least from an evolutionary point of view, that lemmings have evolved a behaviour that apparently drives large numbers of animals into habitats in which they do not survive, nor ever will return from. However, as pointed of by Stenseth & Ims (1993), there may be a hidden snag: due to the very high densities of lemming during peak years, even a small and seemingly insignificant fraction of the population may in total numbers seem enormous. Be that as it may, and as Irving (1972) of Alaska once said: "We are not inclined to believe that the ways of nature are capricious and suspect that failure to agree upon an explanation of cyclic fluctuations is likely the result of lack of information and understanding". I say, that it is also more than likely that these questions will employ many scientists for many years to come.

The single most spectacular biological accomplishment of lemmings is their ability to breed even in winter under snow, and winter breeding is always associated with cyclic increases and a lack of winter breeding with the decline phase.

Lemmings have a high specific metabolic rate due to their small body size. The metabolic stress on lemmings is further enhanced by exposure to a cold climate. The particularly low conductance of their thick fur and their spherical body shape and small extremities are assumed to be traits evolved to reduce metabolic rate under the harsh conditions typical of the environments in which lemmings live. In this context it is intriguing that some lemming species, such as the Norwegian lemming, which has a high metabolic rate is able to reproduce and survive on very low quality diets consisting sometimes predominantly of mosses, with reported digestibilities of only 13-23 %. Moreover, at least brown lemmings have been shown to have the ability to store huge amounts of fat (40 % of body mass) in captivity, while they hardly ever store more than 5 % of body mass in the wild, and it is likely that the lack of significant fat deposits is related to the poor diet and higher activity levels in the wild (Batzli & Esseks, 1992).

Lemmus primarily eats monocot shoots of grasses, like *Dechampsia, Nardus, Festuca, Poa*, and sedges, like *Carex* and *Eriophorum*, and mosses, like *Dicranum, Pleurozonium, Polytrichum*, but not *Sphagnum* spp., and occur usually in lower, wetter areas, while *Dicrostonyx* eats different grasses on Wrangel Island in Russia, but primarily it eats shrubs, particularly *Salix* and forbs, like *Dryas* and *Saxifraga*, and occur usually in higher, drier areas.

Arctic wolf (*Canis lupus arctos*)

Man has always been under the spell of wolves, and ever since the publication of "The jungle book" by Kipling in 1894, which was subsequently used by Lord Baden-Powell as the basis for the wolf cub division of his boy scout movement, and the probably equally fictitious "Never cry wolf", which was published by Mowat in 1963, man has probably felt closer to the wolves than to any other wild animal.

Arctic wolf is a sub-species or geographic race of the grey wolf (*Canis lupus*), which is almost circumpolar in its distribution (Fig. 8.27), and is basically white year around. The weight of an average adult male is about 50 kg, an average adult female 45 kg.

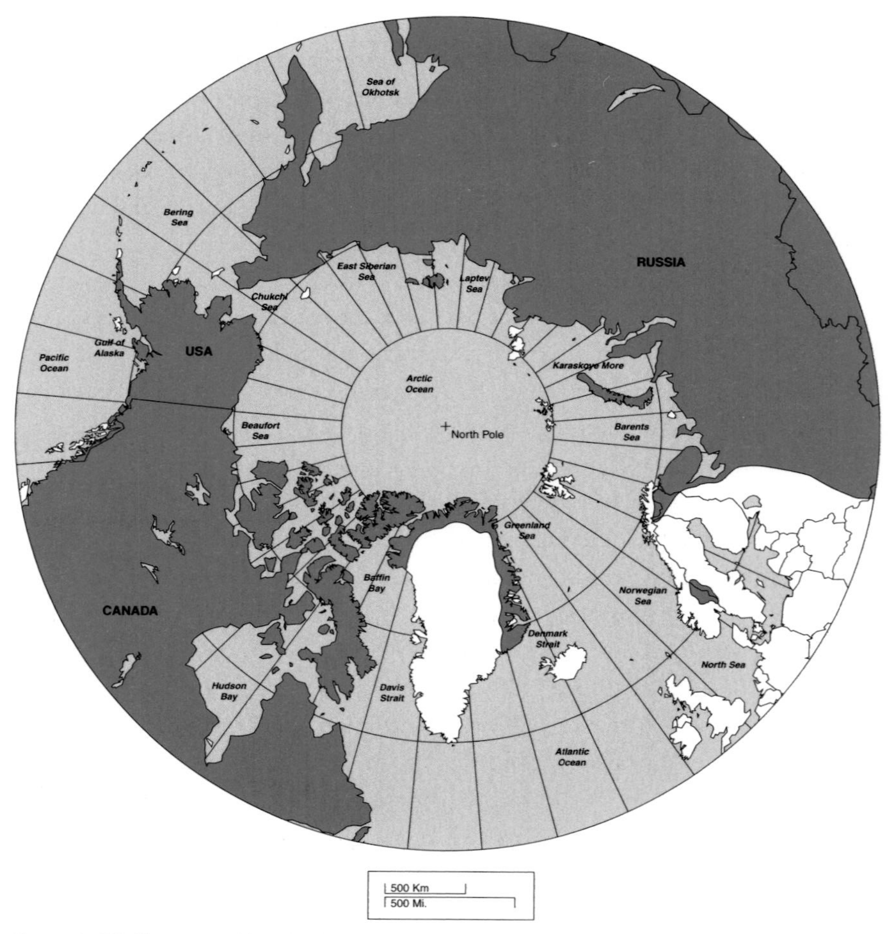

Figure 8.27 The geographical distribution (green) of wolf *(Canis lupus)*.

Wolves are nomadic and live in very low densities, but associating in packs they travel far and wide in remote wilderness areas. Originally they inhabited all of the northern hemisphere from the latitude of Mexico City and southern India northward to the Arctic Ocean, but have now been removed from most of western Europe, Mexico, United States and south-east Asia, and are at very low numbers in Arctic Russia.

Wolves are organised in a social system based on a double dominance order, within the pack; a separate male ("pecking") order and a female order, and it is assumed that it is the dominant male, who is the leader of the pack, and that he mates with the dominant female, while the others only work for the good of the "family", in a communal manner (Mech, 1970).

Wolves are territorial during the denning season, but the territories are quite huge, usually in the order of a thousand square miles, depending on the density of prey. There are few detailed studies of arctic wolves, but Mech (1988) studied one pack of seven wolves with 6 cubs in the Canadian high-Arctic. This den, which probably is the only inhabited high arctic den to be described, was dug out under an outcropping of rock and boulders with multiple openings on the wide open tundra. The breeding takes place in late March in the high Arctic and the cubs, on average 5, are born in a rather helpless state in the underground den in late May. In tundra regions dens are often dug into ridges of sand and gravel. Sometimes old fox dens are enlarged. If undisturbed the wolves will continue using the den year after year. Often the pack even move to a second den when the cubs are about five weeks old. During the denning season the pack spends its time on extensive periods of sleeping and intensive periods of hunting, while the cubs divide their time between sleeping and intensive playing. At times the whole pack join together in protracted group-howling for a variety of social reasons.

The wolves are said to be very friendly towards each other and amiable towards the pups, at least during the denning season, but any suggestion of amiability would probably call a lot of laughter from muskoxen and reindeer. Arctic wolves are specialized predators and prey consistently on reindeer or caribou, particularly calves, in the tundra region, where rather wild surplus killing of caribou calves have been observed (e.g. Miller, Gunn & Broughton, 1985). It is, on the other hand, also argued by scientists with long-term experience with wolves that it is primarily wounded, sick and aged animals which are most likely to succumb to wolf predation, but this remains to be documented. In the high Arctic some wolves specialize on muskoxen, which they kill, sometimes after terrific fighting, which might go on for several rounds over a period of more than an hour (Fig. 8.12). In the high north the wolves also take arctic hare, which in the Ellesmere Island area seems to be their most important prey. After a successful kill of, say, a muskoxen the wolves indulge in a feast, and thereafter go to feed the cubs at the den either by hauling chunks of meat, or by repeatedly regurgitation already eaten material, depending on the age of the cubs.

Arctic hare (*Lepus arcticus*)

Except for the lemming, no doubt the most numerous mammal in the high Arctic is the arctic hare. Weighing up to 5 kg, this animal is quite prominent, especially since it lives in herds of up to a hundred animals outside the breeding season, and because it has a snowy-white fur, with only the tips of the ears black, all year around (Fig. 8.28). The arctic hare is found in coastal western and northern Alaska, all over the North-west Territories and the northern coast of Labrador in Canada and in south-western, northern and north-eastern Greenland (Fig. 8.29). The reason for its absence in south-eastern Greenland is oceanic climate with lots of snow and over-icing of the ground.

The arctic hares feed primarily on arctic willow (*Salix arctica*) throughout winter, pawing sprigs out from under the snow and sometimes scratching out even the nutritious roots. In summer the diet is more diverse, but legumes dominate (Larter, 1999). The hares are primarily active during the night and spend the day in boulders or rock crevices where they are sheltered from the sun. This, of course, is no good during the high-arctic summer when the sun is up around the clock, but even then they are most active when the sun is at its lowest. Mating takes place at the end of April and at that time the winter herds break up

Figure 8.28 Three arctic hares sporting their year-round white fur, which make them rather visible during summer (Anon., 2001).

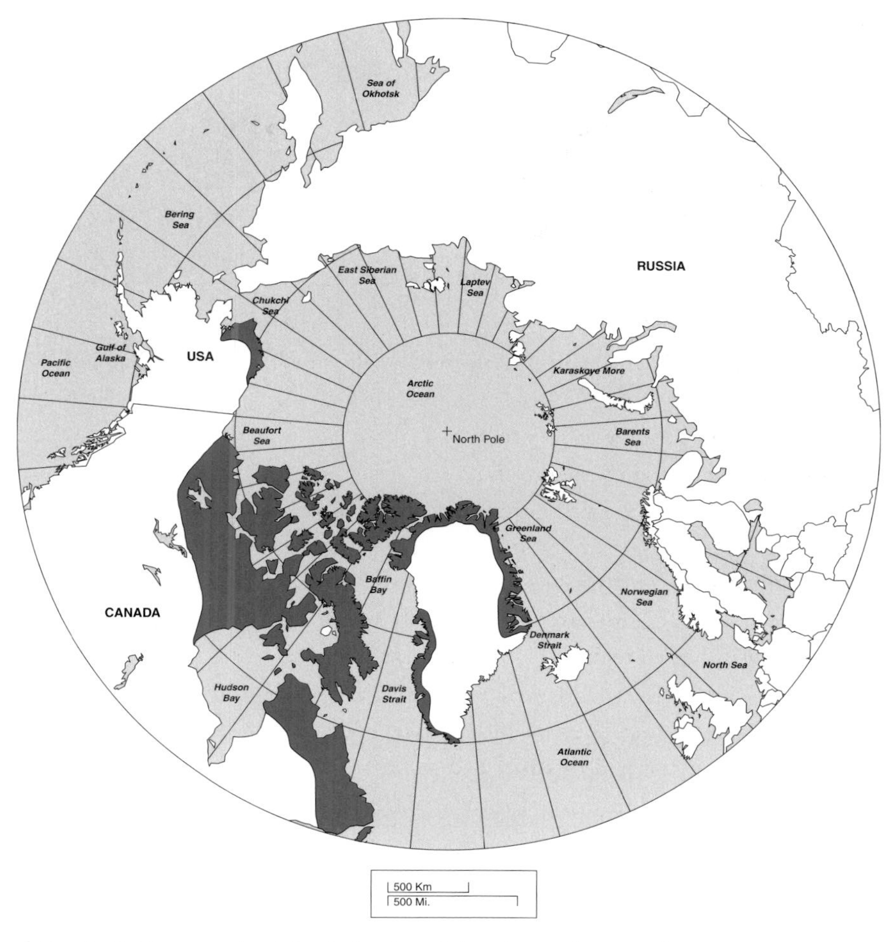

Figure 8.29 The geographical distribution (green) of arctic hare *(Lepus arcticus)*.

into couples which establish their own territories. Then, in June and July the hares produce 5-8 leverets, which they after about 3 days of continuous attention leave alone for long periods and nurse only once a day, but already after about two weeks the young have to cater for themselves. The leverets are a stony grey at birth, precisely matching most of their surroundings, and they seem to disappear the moment they stop moving (Fig. 8.30). This is obviously quite important for the preservation of the species, since the leverets are the favourite snack of the Arctic fox, ermine and ravens. In September, however, they are fully grown and have aquired the white pelts of the adults. The adults are very fast runners and have very few enemies in the high Arctic, except wolves and

Figure 8.30 Young arctic hares are cryptically coloured and tend to disappear completely when they lie still (Mech, 1988).

snowy owls in some areas. If the hares feel disturbed they have the peculiar habit of standing on their hind legs, even on their very toes, to survey their surroundings, and once they decide to take off, they hop like a kangaroo on their hind legs at a speed which make them hard to follow, even for a top trained wolf.

During winter these high-Arctic survival specialists often evade the full brunt of the storms by digging shelters in snowdrifts, or simply by balling up and letting themselves being drifted over by snow. In any case, Wang *et al.* (1973) have shown that their insulation is so superior that they may maintain body temperature with a metabolic rate which is only 60-80 % of the value predicted from their body mass. I am unaware of any data on fat deposition in arctic hares, but I am sure that they, like most other high arctic residents, stock up during the autumn for subsequent use when the weather turns bad during winter.

Ermine (*Mustela ermina*)

The ermine, which belongs to the *Mustelidae* family, is a small, predatory animal of some 200-300 mm length with a tail of 45-70 mm and a weight of only about 50-80 grams. During winter the animal is sporting a snowy-white fur with a

Figure 8.31 A bunch of nobility of different denominations contemplating the demise of an abundance of ermines (Photo: P. Lichfield).

prominent black pointed tail, which has adorned the ceremonial gowns of European sovereigns and noblemen probably for a millennium (8.31). During summer, from May till September, the ermine wears a less aristocratic dress of brown on the back, but still with white throat and belly and maintain the black pointed tail. However, in spite of its value to the aristocracy, the value of this fur to the animal is rather low! Its thermal conductance is higher than in other mammals of similar size, and although its colour changes seasonally, its thermal conductance does not (Casey & Casey, 1979).

The ermine has a very wide circumpolar distribution (Fig. 8.32), but is missing from south-east and all of western Greenland, where there are no lemmings, on which the ermine depend. Thus, in areas where the lemmings are at a low, the ermine is rare, while they seem to be everywhere when there is a lemming high. During summer they supplement their diet with eggs, ptarmigan chicks, passerine birds and leverets of the arctic hare. When ambient temperature approaches freezing in the autumn the ermine starts working overtime on collecting food for the winter, and hoards consisting of 150 lemmings and even 75 ptarmigan eggs, obviously collected already during summer, have been found in Norway.

The ermine spends most of its time in piles of rock debris, where it establishes its nest of grass and hairs and feathers from their prey, sometimes settling only for

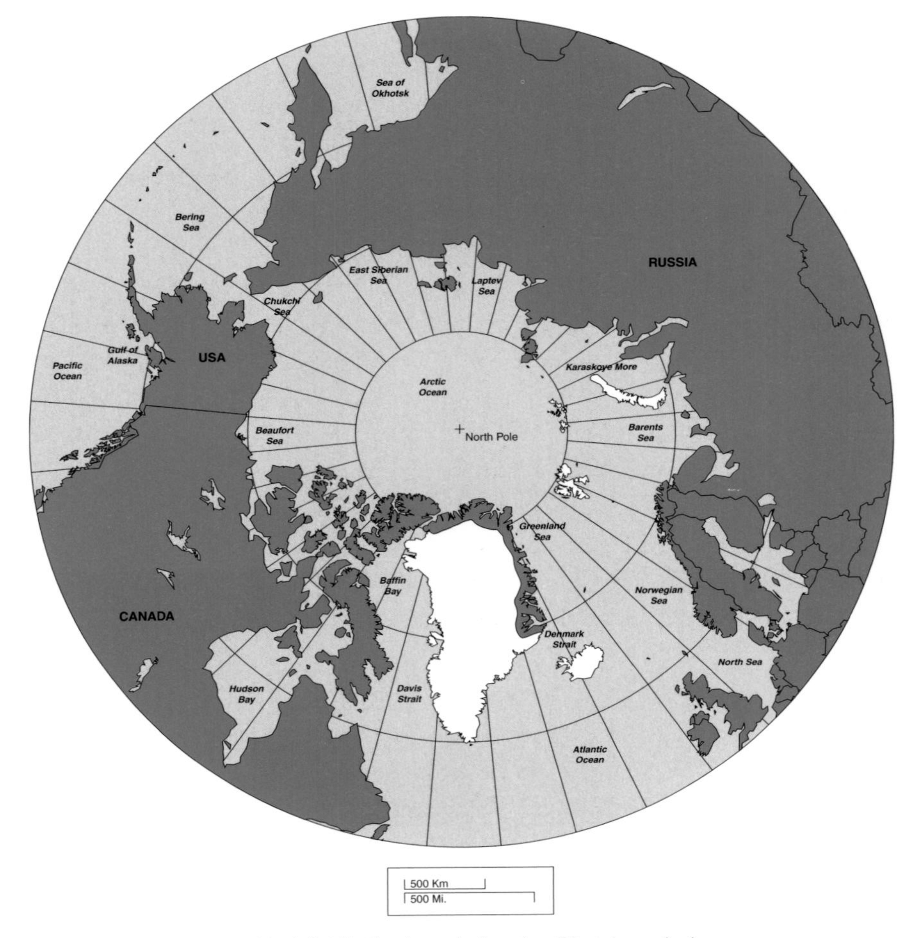

Figure 8.32 The geographical distribution (green) of ermine *(Mustela ermina)*.

an improvement of a lemmings nest after having killed and devoured its legal inhabitants. During winter the small ermine will not survive for very long periods above the snow, for lack of adequate insulation, and spends instead its time hunting lemmings in their burrows under the snow.

The ermine breeds during early summer, depending on latitude, but implantation is delayed until next spring, when the young are born one month later. The ermine has no less than 10 teats and may raise 7-9 young during a lemming high, while few, if any, are born when the lemmings are at a low.

The ermine is normally active only at night, but in the Arctic with continuous light around the clock it may be seen in broad daylight.

Polar bear (*Ursus maritimus*)

The polar bear has a circumpolar distribution, and is confined to the Arctic and sub-Arctic ice-covered sea areas. The bears are not evenly distributed, but are found in several more or less isolated populations (Fig. 8.33; Table 8.3). The polar bear is not really a marine mammal, and with its long beautiful fur, which offers prime insulation in air, it often walks long ways to avoid going into water, where the insulative value of the fur is lost. However, the structure of the fur is such that a couple of shakes after emergence onto the ice will effectively rid it of water and regain its insulative value. The Eskimos have for ages taken advantage of this particular property of polar bear fur and utilized it for trousers and mittens, which often get wet, while the more durable seal skin has been used for the rest of their outfit (Fig. 4.28).

Polar bears are long-lived, late maturing carnivores that have relatively low rates both of reproduction and natural mortality. The average weight of the males is 350-450 kg, but some bears may reach a weight of a thousand kilo. Females weigh on average 150-250 kg, but there are both regional and, not the least, seasonal differences in body weight of both sexes.

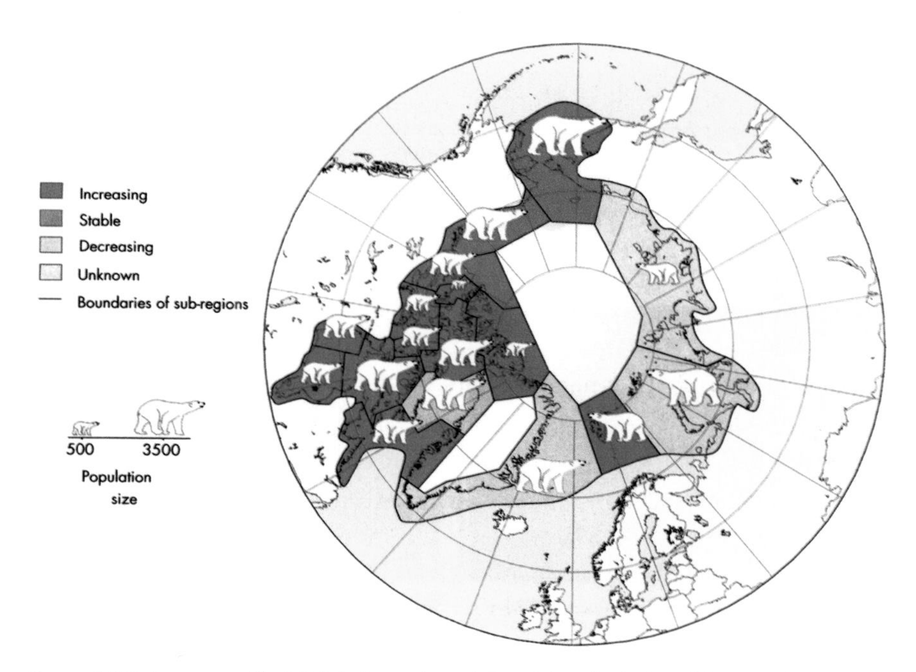

Figure 8.33 The geographical distribution and population status of polar bear (*Ursus maritimus)* in the Arctic (Redrawn from Anon., 1998).

Table 8.3
Polar bear population status

Population	Number	Current annual harvest
Western Hudson Bay	1,200	55
Eastern Hudson Bay	1,000	52
Foxe Basin	2,020	91
SE Baffin Island	950	52
Lancaster Sound/N Baffin Bay	2,470	200
Gulf of Boothia	900	41
M'Clintock Channel	700	32
Viscount Melville	230	4
North Beaufort	1,200	33
South Beaufort	1,800	64
Chukchi	2,000-5,000	87
Laptev	800-1,200	Incidental
Franz Josef Land/Novaya Zemlya	2,500-3,500	Incidental
Svalbard	1,700-2,200	Incidental
East Greenland	2,000-4,000	100

TOTAL POPULATION ESTIMATE = 21,470-28,370; most populations appearing to be stable.

Polar bears have been reported as far north as 88°N latitude. In winter and spring they are commonly found in three distinct types of ice: shorefast ice with deep snow drifts along pressure ridges, the floe edge, and areas of moving ice with 7/8 or more ice cover. In the western Canadian Arctic, subadult females and adult females with young cubs generally prefer shorefast ice, possibly because of the absence of large adult males in this type of habitat. When new ice forms in the fall, polar bears that have spent the summer on drifting ice north of Alaska move to the south. As ice breaks up and recedes north in the spring and early summer, the bears move north. There are no physical barriers to prevent polar bears from moving across international boundaries, but there is growing evidence for separation of different subpopulations (Fig. 8.34).

Seals are the primary prey of polar bears throughout their range, although other marine mammals, birds and vegetative food may also be consumed. In Alaska 94 % of the prey is ringed seals and 6 % bearded seals, while the percentage of bearded seals were twice as high in the western Canadian Arctic. Stirling & Oritsland (1995) found that 50 % of the ringed seals that were taken by polar bears in the Canadian Arctic during periods of high ringed seal productivity were young of the year seals. It has been estimated that an adult

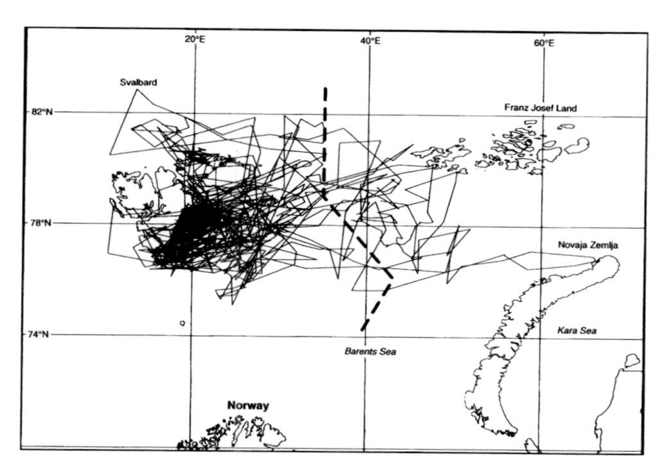

Figure 8.34 Movements of 36 female polar bears followed by satellite telemetry in the Svalbard area in the period from 1988 to 1993. The approximate border between Norwegian and Russian areas is indicated (Wiig, 1995).

polar bear would need to kill on average a ringed seal every 6.4 days to maintain body weight. This in spite of the fact that polar bears have a digestive efficiency of more than 90 % when feeding on ringed seal blubber (Best, 1988), which is the preferred part of the seal. The forest dwelling polar bears in the Hudson Bay often eat roots, grass, berries, rodents etc., but they also catch seals on the Bay when the sea freezes over. Also in the Hudson Bay some bears have specialized in leftovers from camps and settlements, and are frequent visitors to the garbage dump in Churchill, where some 50 bears may be seen at one time, something which has turned this dump into a major tourist attraction.

Polar bears have probably been hunted at all times and trade with polar bear hides has been documented since viking times. Through the 1950s, and particularly during the 1960s there was a rapid increase in the recorded number of polar bears killed. This was primarily due to a rapid increase in the price paid for polar bear hides and the introduction of effective methods for trophy hunting (Fig. 8.35). The polar bears were completely protected in the Soviet Union from 1956, because their populations were severely depleted due to intensive hunting over a long period of time. After much negotiation the polar bears were finally protected by an international (polar bear) agreement in 1973. But there are provisions in the agreement for subsistence hunt by aboriginal people and an annual average of 775 bears are still killed every year, total protection only being offered to the bears in the Svalbard region. The total world population of polar bears is supposed to be in the order of 25,000 individuals (Table 8.3).

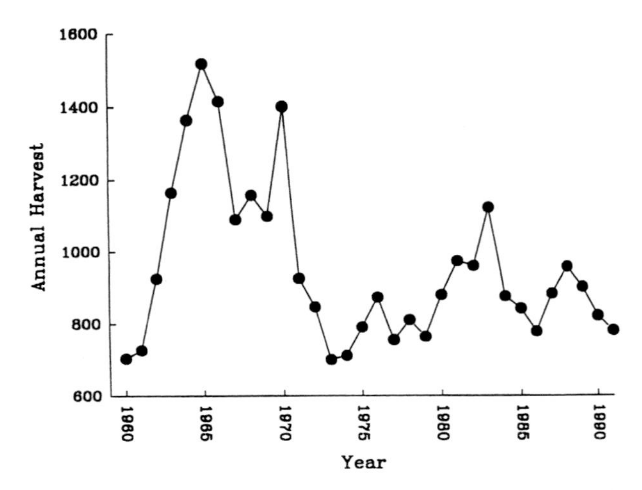

Figure 8.35 Total recorded harvest of polar bears from Norway, Greenland, Alaska and Canada from 1960 to 1991. The annual harvest during the 1990s has been fairly stable at an average of 775 bears, 75 % of which taken by Canadian Inuit (Prestrud & Sterling, 1994).

Female polar bears enter maternity dens which they excavate in the snow (Fig. 8.36), by late November; the very undeveloped young are born in late December or early January with a birth weight of only 600 grams! In most areas of the Arctic, family groups leave the dens in late March or early April, when the cub has reached a weight of about 10 kg (Fig. 8.37). The density of maternity dens varies geographically, the largest concentration of dens being found on Wrangel Island in Russia, on some of the islands in the Svalbard Archipelago, and near the west coast of Hudson Bay in Canada. Lentfer (1975) reported the first confirmed instance of polar bears denning on drifting sea ice of the Beaufort Sea in multiyear pack-ice, and recent studies indicate that this may be the rule and not the exception.

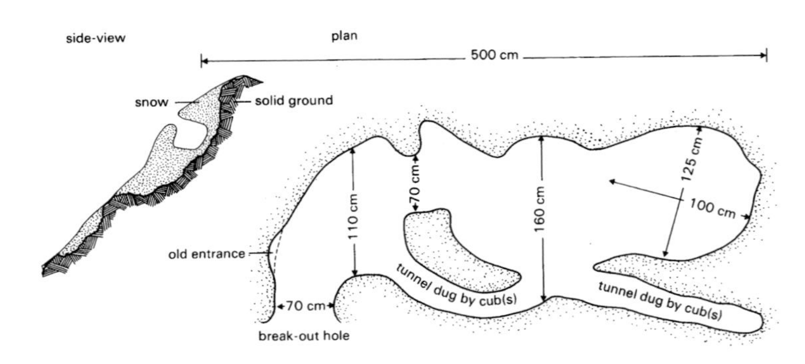

Figure 8.36 The plan of a polar bear den in Svalbard. Note the smaller tunnels dug around the main den by the cubs (Redrawn from : Larsen, 1978).

Figure 8.37 A polar bear cub at the time of emergence from the den at the end of March, at Point Barrow, Alaska (Photo: P. E. Reynolds).

The average litter size is probably very close to 2.0 at birth. Adult female polar bears in most areas are "available" for breeding in the spring if they are not accompanied by young of the year or yearlings. The average reproductive interval seems to be between 3.1 and 3.6 years, the average age of first reproduction being 6.4 years, with a range of 4-8 years. The maximum reported age of reproduction is 20 years. For further information, see: Stirling (1988).

CHAPTER 9

Summer Migrants to the Arctic

Birds

The spring in the Arctic brings migratory birds that make tundra, coasts and water lively with many summer species in numbers that far exceed the scant winter populations (Fig. 9.1). Migrants multiply the list of arctic winter species by more than ten and the numbers by many times tenfold. The major annual production of arctic birds results from the migrants that nest there in summer. The regular return of the birds occurs in spite of the vagaries of arctic weather which in spring can be extremely trying, particularly for small birds. However, migrants arriving in the Arctic are generally in good condition and fat enough to select and settle in their nesting places by the use of reserves adequate for some inclement weather. Though they are busily concerned with mating, nesting and attendant social activities, the food supply is evidently adequate for their support without undue competitive struggle. On the tundra, the major food supply of early season birds is derived from berries, seeds, buds and spouts of stems, and arthropods produced in the previous year and exposed in spring by melting snow. It is a fact that was pointed out by Irving (1972) that avian populations of the tundra decline markedly during August, at the time when annual production of food in the form of seeds, berries and invertebrates reaches its peak. Frost and snow then preserve a large part of the summers yield of food until the following spring, when melting exposes an unharvested supply for the returning migrants. When the young are

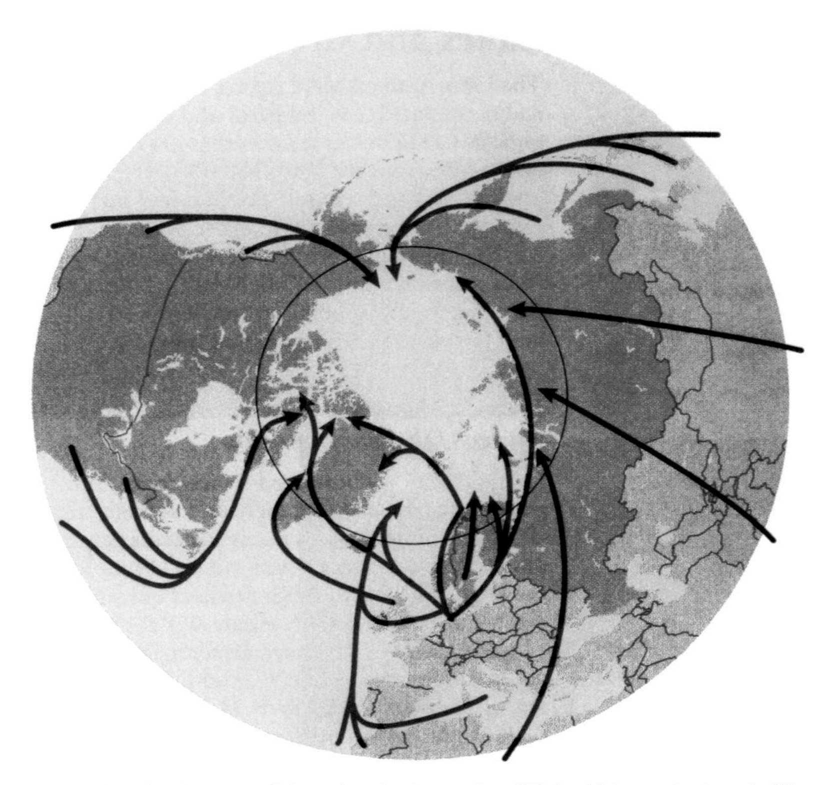

Figure 9.1 Map showing some of the major migratory routes of birds which spend only part of the year in the Arctic, migrating south to less hostile climates and better supplies of food during the winter (Anon., 1997).

hatched, green new growth is abundant and insects have emerged. The fresh production of summer provides ample nourishment that fattens the birds before their departure on migrations southward.

Willow ptarmigans (*Lagopus lagopus*) are common on the Alaskan tundra during summer, but migrate south in the fall. At Anaktuvuk Pass in the Brooks Range in Alaska, Irving (1972) recorded observations of the ptarmigan for many years, and their observations serve as a good example of the dynamics of migration: flocks of willow ptarmigan began to move through the Pass in a prevailing southward direction during October. The southward trend continued to be evident through November. In the dark and often stormy days of December and January flocks walked at rather orderly spaced intervals over the snow as they fed on the buds and twigs of willows, or rose syncronously in rapid and concerted flight. In their midwinter period, when human observers are often weatherbound, no directional trend was apparent. Late in January or early in February, however, the ptarmigan began to appear restless and talkative with evidence for beginning northward movements. These movements subsided during March but increased to a crescendo by the end of April, when several thousand birds were counted daily in obvious progress northward, with uncounted numbers heard passing in flight. During four years of observation they estimated that from 21,000 to 52,000 birds were seen between January and May in prevailing northward movement over a 2-kilometre wide area of the stream flat in the Pass.

Some 90 species migrate to nest on the interior tundra of Alaska, numerically submerging the few species and numbers of local residents. Thirty-four of these species are found wintering or migrating along western American coasts. For nesting in the interior of Alaska they resort to tundra and fresh water. The transition to the inland nesting habitat is most marked in the case of six species that breed mainly in the Arctic: two phalaropes (*Phalaropus fulicarius*) and (*Lobipes lobatus*) three jaegers *(Stercorarius spp.*), and the arctic tern (*Sterna paradisaea*). These six birds spend the non breeding season at sea and even over far southern oceans; or, in the case of the arctic tern, among antarctic ice floes. Thus, no other organism makes such a spectacular journey as the arctic tern. This bird has a circumpolar breeding distribution in the northern hemisphere, although 70 % of the population breed in Iceland.

The migration routes of the arctic tern are shown in Fig. 9.2. The absence of migrating terns on the eastern sides of the continents presumably reflects the availability of food in the surface waters of the oceans in which the terns must feed en route. Their return northward begins in March and by and large follows the same nutrient-rich routes north. Juvenile birds may spend more than a year in the southern hemisphere, making the migratory cycle of the arctic tern more complex than just a simple annual to-and-fro journey. Although many birds undertake migratory journeys almost as spectacular as that of the arctic tern, most do not, but make only short trips between breeding and non-breeding areas. Nevertheless, these journeys still function to place birds in the most favourable possible environments throughout the year.

In any part of Arctic Canada, Greenland and over Arctic Eurasia there are similar examples of regular migrations converging on arctic nesting-grounds from lands and waters far south, and from east and west. These migrations transport unbelievable numbers and masses of birds to produce and raise their young in the production of the arctic summer. In dealing with the myriads of birds which invade the Arctic during summer we will in the following first con-

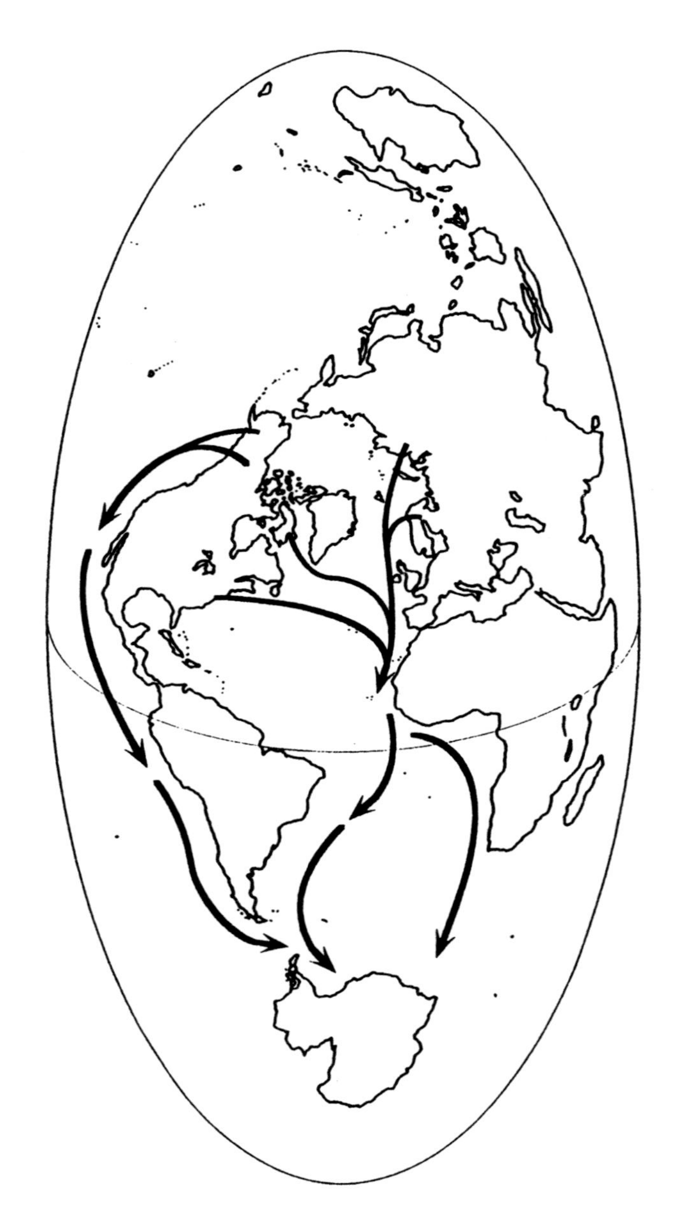

Figure 9.2 The southward migration of the arctic tern. Populations breeding in both eastern Nearctic (Greenland and North America) and western Palearctic (Scandinavia and western Siberia) converge in the eastern Atlantic. Off Africa the population divides to migrate down both sides of the Atlantic to antarctic regions. Birds from the western Nearctic (Alaska and western Arctic Canada) migrates south along the Pacific coast of the Americas (Dingle, 1996).

cern ourselves with those which are known to breed at the most high-Arctic locations (Table 9.1)

This table illustrates that a great variety of birds penetrate into the most hostile northern parts of the globe to breed in summer. Some of these have a circumpolar distribution, while some only occur in the northern Atlantic/Eurasian

Table 9.1

Migratory birds which are known to breed at the most high-Arctic locations
Ellesmere Island/Peary Land (E/P); Svalbard (S); and Frans Josef Land (FJL).

Red-throated loon *(Gavia stellata)*	E/P	S	FJL
Snow goose *(Chen hyperborea)*	E/P		
Pink-footed goose *(Anser brachyrhynchus)*		S	
Barnacle goose *(Branta leucopsis)*		S	
Brent goose *(B. bernicla)*	E/P	S	FJL
Long-tailed duck *(Clangula hyemalis)*	E/P	S	
Common eider *(Somateria mollissima)*		S	FJL
King eider *(S. spectabilis)*	E/P	S	
Northern fulmar *(Fulmarus glacialis)*		S	FJL
Glaucous gull *(Larus hyperboreus)*	E/P	S	FJL
Ivory gull *(Pagophila eburnea)*		S	FJL
Black-legged kittiwake *(Rissa tridactyla)*		S	FJL
Brunnich's guillemot *(Uria lomvia)*		S	FJL
Black guillemot *(Cepphus grylle)*		S	FJL
Little auk *(Alle alle)*		S	FJL
Pomarine skua *(Stercorarius pomarinus)*		S	
Arctic skua *(S. parasiticus)*		S	FJL
Great skua *(S. skua)*		S	
Longtailed skua *(S. longicaudus)*	E/P	S	
Arctic tern *(Sterna paradisaea)*	E/P	S	FJL
Ruddy turnstone *(Arenaria interpres)*	E/P	S	
Knot *(Calidris canutus)*	E/P		
Baird's sandpiper *(C. bairdii)*	E/P		
Sanderling *(C. alba)*	E/P	S	
Purple sandpiper *(C. maritima)*		S	FJL
Dunlin *(C. alpina)*		S	
Ringed plover *(Charadrius hiaticula)*		S	
Grey phalarope *(Phalaropus fulicarius)*	E/P	S	
Lapland longspur *(Calcarius lapponicus)*	E/P		
Snow bunting *(Plectrophenax nivalis)*	E/P	S	FJL
Gyrfalcon (*Falco rusticolus)*	E/P		

and Pacific/North American regions, respectively. It is also worth noticing that the Svalbard area is particularly rich in bird diversity in spite of its northern location. This is no doubt due to the unusually warm climate along the west coast of this archipelago.

Seabirds that nest in arctic colonies

Great numbers of birds assemble in localized colonies on certain cliffs, mountain slopes and shores of arctic seas (Tables 9.2, 9.3 & 9.4).

Table 9.2
Seabirds with circumpolar distribution within the Arctic region.

Northern fulmar *(Fulmarus glacialis)*
Arctic skua *(Sterocorarius parasiticus)*
Long-tailed skua *(S. longicaudus)*
Pomarine skua *(S. pomarinus)*
Glaucous gull *(Larus hyperboreus)*
Sabine's gull *(L. sabini)*
Black-legged kittiwake *(Rissa tridactyla)*
Arctic tern *(Sterna paradisaea)*
Common guillemot *(Uria aalge)*
Brünnich's guillemot *(U. lomvia)*
Black guillemot *(Cepphus grylle)*
Ivory gull *(Pagophila eburnea)*

Table 9.3
Seabirds with a North Atlantic distribution.

Great black-backed gull *(Larus marinus)*
Iceland gull *(L. glaucoides)*
Ross gull *(Rhodostethia rosea)*
Razorbill *(Alca torda)*
Little auk *(Alle alle)*
Atlantic puffin *(Fratercula arctica)*
Great skua *(Stercorarius skua)*

Table 9.4
Seabirds with a North Pacific distribution.

Mew gull *(Larus canus)*
Red-legged kittiwake *(Rissa brevirostris)*
Least auklet *(Aethia pusilla)*
Crested auklet *(A. cristatella)*
Parakleet auklet *(Cyclorrhynchus psittacula)*
Horned puffin *(Fratercula coniculata)*
Tufted puffin *(F. cirrhata)*

These aggregations form at sites appropriate for the nesting habits of each species, to which the populations return annually according to a tradition that is rather faithful to given localities and on a seasonal schedule that is likewise currently regular and which pursues a temporal tradition.

Several of these birds breed in colonies, some of which are enormous. The biggest known bird-cliff colony is in the Thule district in Greenland where a staggering 14-20 million little auks have been reported to aggregate to breed. The biggest known bird-cliff colonies in the Arctic are indicated in Fig. 9.3. This figure does not reflect a population of some 2 million common guillemots and some 2.5 million Brünnich's guillemots in the Bering Sea, since these are spread out in about 200 different colonies along the coasts and on a great number of islands in the area.

In addition to these birds there are several species of storm petrels, shearwaters, petrels, cormorants, shags, gulls and gannets that infringe on the arctic region, but normally breed outside the area. It has always been assumed that this multitude of birds consume enormous amounts of food and play a key factor in the transfer of energy in their ecosystems. The following example may indicate that this may not be so:

Assume a generous estimate of 100 million seabirds of an average body mass of 500 grams, which amounts to 50,000 tons of birds. Assume also that they eat 5 % of body mass per day, which amounts to 2,500 tons of food per day. Assume also that they all eat this amount for 300 days a year. Then the total amount of food eaten by the circumpolar populations of seabirds will amount to 750,000 tons per year. This amount is of course not trivial but it is still only half of what the minke whale population of some 110,000 animals in Norwegian waters alone consume during a 6 month sojourn in the area. Moreover, the minke whales consume primarily fish, while a large proportion of the feed taken by seabirds are small crustaceans and other invertebrates.

Ornithologists have also been promotors of yet another long-lived concept. That is that seabirds are the key components in transporting nutrients from the sea to the land. Sure enough, you can see that the vegetation below the bird-cliffs is much more lush than the surrounding areas, and Gabrielsen, Taylor, Konarzewski & Mehlum (1991) have even calculated that a colony of 70,000 pairs of the little auk at Svalbard fertilises the surrounding ecosystem with some 60 tons (dry weight) of faeces during the breeding season. Amazing as this seems, surely, birds that alternate between foraging trips to sea and the nest in the bird-cliffs are likely to limit most of their dumping to the very steep and limited area underneath the cliffs, where 60 tons of guano seems counterproductive from an agricultural point

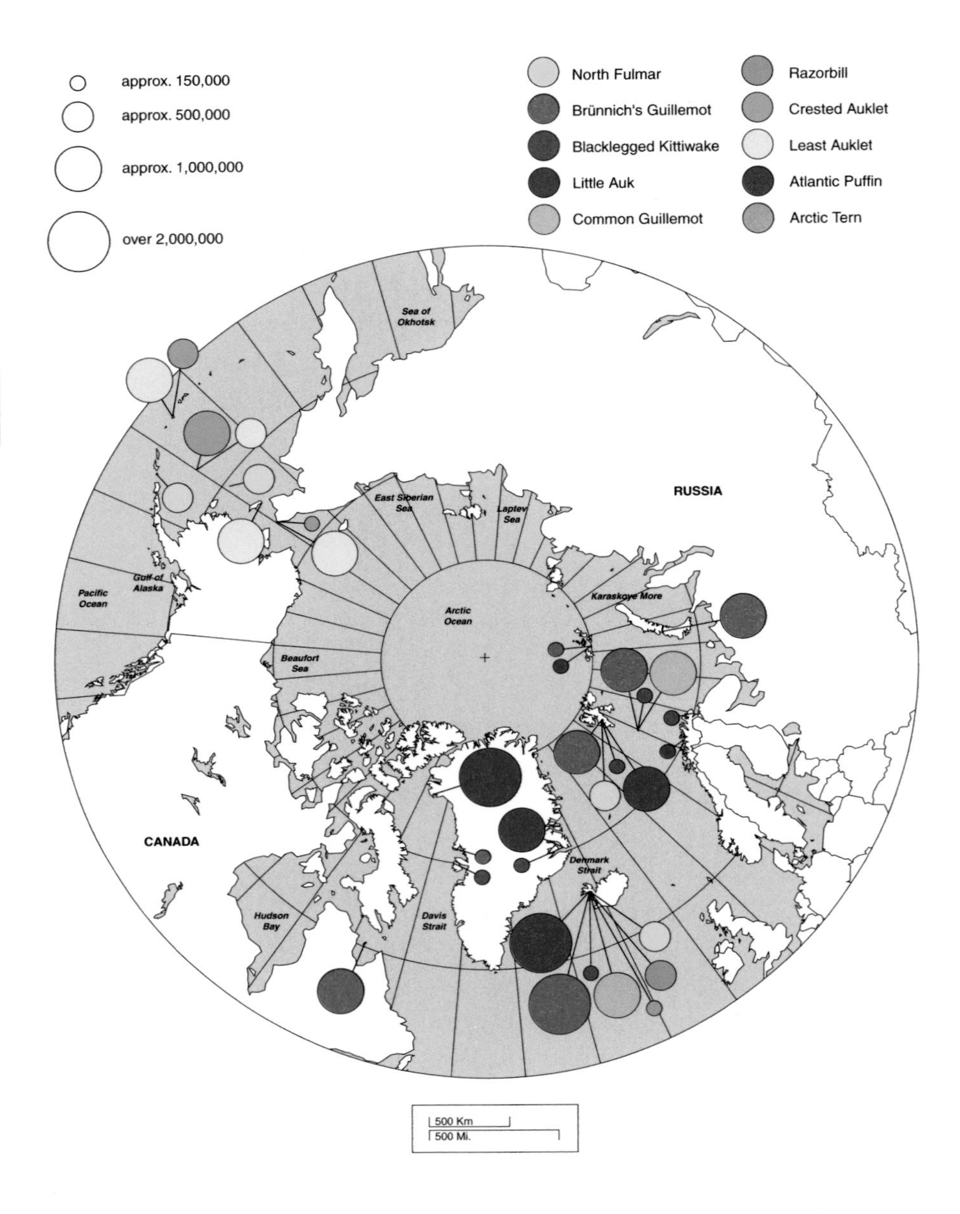

Figure 9.3 Major seabird colonies in the Arctic. The number of birds in each colony varies from year to year, probably due to changes in the availability of food. The numbers indicated are therefore approximate. The atlantic puffin colony in Iceland is supposed to number some 5 million birds, and the little auk colony at Thule, Greenland, numbered some 14 million birds, but has decreased in recent years. Note the lack of seabird colonies in Siberia and at the north coast of Canada-USA. This is mainly due to lack of cliffs and food at those locations.

of view. In any case, it is likely that the import of guano-based nutrients from the sea has a rather limited regional effect.

The Arctic is not only the venue for a multitude of seabirds during the hectic summer months. Also a great many sea ducks come to breed (Table 9.5). Several of these are only occasional visitors to the Arctic, with their main breeding area as well as wintering range further south. The Arctic tundra is also the summer home of a number of geese and fresh-water associated ducks (Table 9.6). The arctic tundra is also the breeding ground for Bonaparte's gull (*Larus philadelphia*) and herring gull (*L. argentatus*) and a number of loons (*Gavia spp.)*.

Table 9.5
Sea ducks with a circumpolar-distribution, which breed in the Arctic.

Common eider *(Somateria mollissima)*
King eider *(S. spectabilis)*
Spectacled eider *(S. fischeri)*
Steller's eider *(Polysticta stelleri)*
Long-tailed duck *(Clangula hyemalis)*

Table 9.6
Geese and fresh-water associated ducks which breed in the Arctic.

Whistling swan *(Olor columbianus)*
Snow goose *(Chen hyperborea)*
Blue goose *(C. caerulescens)*
Pink footed goose *(Anser brachyrhynchus)*
White fronted goose *(A. albifrons)*
Barnacle goose *(Branta leucopsis)*
Brent goose *(B. bernicula)*
Greater scaup *(Aythya marila)*
Pintal *(Anas acuta)*

Now, up and above this jamboree of avian sex and family activities the predatory birds are feasting on eggs and meat. In addition to the resident snowy owls and the ravens, Gyrfalcons (*Falco rusticolus*), peregrine falcons (*Falco peregrinus*), golden eagles (*Aquila chrysaëtos*), and a number of owls and hawks take part in the bonanza and invest in their own offspring. Finally, the arctic tundra is also the breeding ground for small birds of a multitude of species that it takes a sworn ornithologist to appreciate. Personally, I would have been content only with my personal favourite, the snowbunting, but such is not the case. The arctic tundra is also housing a variety of plovers, sandpipers, warblers, chickadees, sparrows, longspurs, redpolls and many others, but it would be far beyond the scope of this book to indulge in details over this overabundance of bird biodiversity. Besides, there are a number of excellent specialist textbooks available to the really consummate reader.

Mammals

Grey whale (*Eschrichtius robustus*)

The grey whale is a medium sized baleen whale without a dorsal fin. The body, which has a length of 13 m in the adult male, and 14 m in the female, is grey with white mottling and is colonized by barnacles (a host specific *Cryptolepas rhachianecti*) and three species of whale lice, which altogether give the animal a rather shaggy appearance. The upper jaw has 140-180 coarse yellowish baleen plates on each side.

There are two stocks of grey whales in the North Pacific – the Korean stock on the western side and the Californian stock on the eastern side (Fig. 9.4). The Korean stock is probably close to extinction. From the end of May through September, the Californian stock feeds mostly in the northern and western Bering Sea, the Chukchi Sea and the western Beaufort Sea, while some feed along the Kamchatka Peninsula, the Commander Islands, in the Gulf of Anadyr and around the Chukotski Peninsula as far as Wrangel Island, all the time being restricted to shallow waters of the continental shelf, generally in areas of high benthic biomass.

From November through December, the grey whales move south through Unimak Pass in the Aleautian chain, around the Gulf of Alaska to the lagoons of Baja California. The grey whale migration is thus up to 18,000 km, which is the longest migration of any mammal.

The grey whales have been hunted both by Americans, Russians and Eskimos for thousands of years. In 1948 the grey whale stock was depleted after a long period of intensive harvest, but since then a total of some 4,000 animals have still been taken by aboriginal whalers in Russia and America. The grey whales have now recovered and probably count in the order of 20,000 animals, which is probably the same as the original stock (prior to commercial exploitation, which started in 1846).

The grey whales characteristically raise their tail flukes into the air when they dive to the bottom, the animals being primarily, but not exclusively, bottom feed-

Figure 9.4 The migration routes (red arrows) and geographical *summer* distribution (blue) of grey whale (*Eschrichtius robustus*) in the Arctic.

ers. In faunal benthic species, especially of gammaridean amphipods such as *Anonyx nugax*, *Pontoporeia femorata*, *P. affins*, *Ampelisca macrocephala*, *A. eschrichti*, *Nototropis ekmani*, and *N. brueggeni* predominate. Polychaete worms and molluscs are poorly represented, suggesting that the whales are selective feeders, but of course, it may be that they select for area communities rather than being able to select for species. It is possible that the whales stir up the bottom sediments with their snouts, then filter the turbid water immediately above the bottom from which the heavier molluscs have settled out. The occurrence of sand, silt and gravel in the stomachs provides further evidence.

Sexual maturity is attained at a mean age of 8 years. Breeding takes place in early December. Pregnancy lasts about 13.5 months. The calves are born in December-January, when they weigh about 500 kg, and are nursed for about 7 months.

Minke whale (*Balaenoptera acutorostrata*)

The minke whale is the smallest of the rorquals and together with the poorly known pygmy right whale (*Caperea marginata*) the smallest of all the baleen whales. It grows to become 10 m in length and weighs up to about 10 tons. Coloration varies, but most whales are generally black or grey dorsally and rather white ventrally. In the northern hemisphere, the animals sport a wide white band across their front flippers, which together with their yellowish-white baleen plates are diagnostic of the species.

The minke whale is widespread in both hemispheres, but only those in the north (Fig. 9.5), will be considered here. In spring and summer the minke whales migrate northwards to arctic waters, along the coasts of Newfoundland to the south west of Greenland and Davis Strait and Baffin Bay in late summer. In the northeast Pacific, minke whales are found from the Kurile Islands to the Chuckchi Sea and Pt. Barrow, Alaska, and south to the Gulf of Alaska during summer. Few sightings exist for minke whales during winter, but it is assumed that the minke whales of the north Pacific are wintering in the Sea of Japan and the Yellow Sea and in the waters off Baja California, Mexico, while the animals of the northwest Atlantic Greenland/Davis Strait stock winter in the Gulf of Mexico and the West Indies. The wintering, and hence calving grounds, of the northeast Atlantic stock are not known, but these animals migrate into Norwegian and nearby Arctic waters in early April and have penetrated as far north as off West Spitsbergen in Svalbard and the Barents and Kara Seas in July/August.

The minke whales have been hunted by the Norwegians since the middle ages, while modern commercial whaling began in the 1930ies. After World War II such harvest also spread to Iceland, Greenland and Newfoundland. Historically, minke whales were also taken in very small numbers by natives of the Pacific Northwest of North America, while commercial whaling was started by Japan around 1920, by Russia in 1933 and by Korea in 1962. The reason for the late start in minke whale exploitation obviously is that these are small and very fast-swimming animals, which were not taken until the local stocks of bowhead, right whale, fin and blue whale were depleted.

The minke whale is presently (since 1987) protected under the International Whaling Commission, but based on sound scientific evidence of a population

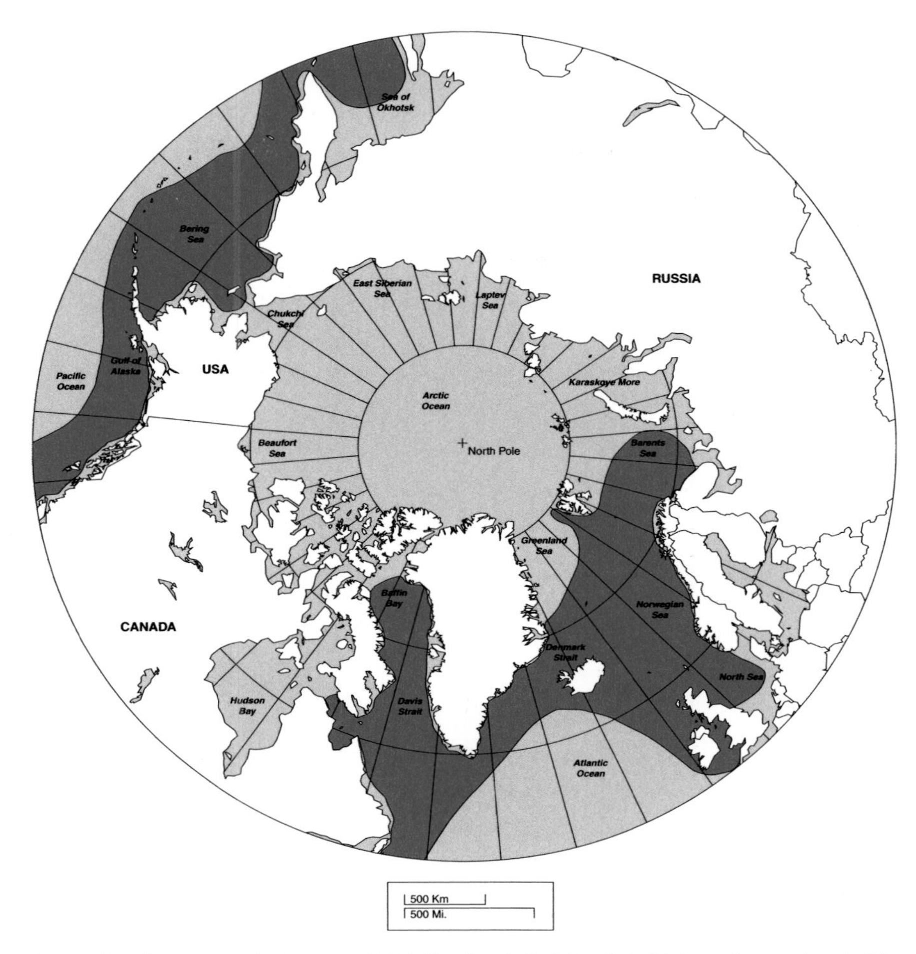

Figure 9.5 The geographical *summer* distribution (blue) of minke whale (*Balaenoptera acutorostrata*).

size of 110,000 animals in Norwegian waters, Norway resumed whaling at the level of 400-500 animals annually in 1994, after kilotons of political commotion. The catch figures for the Norwegian harvest is given in Fig. 9.6. In addition to a total of about 200,000 animals in the Atlantic, there are some 30,000 in the Pacific, while the Southern Ocean is believed to hold about one million minke whales.

In the Southern Ocean the minke whales feed almost exclusively on krill *(Euphausia sp.)*, as would be expected by a decent baleen whale. Recent research has shown, however, that the minke whales of the northeast Atlantic feed prima-

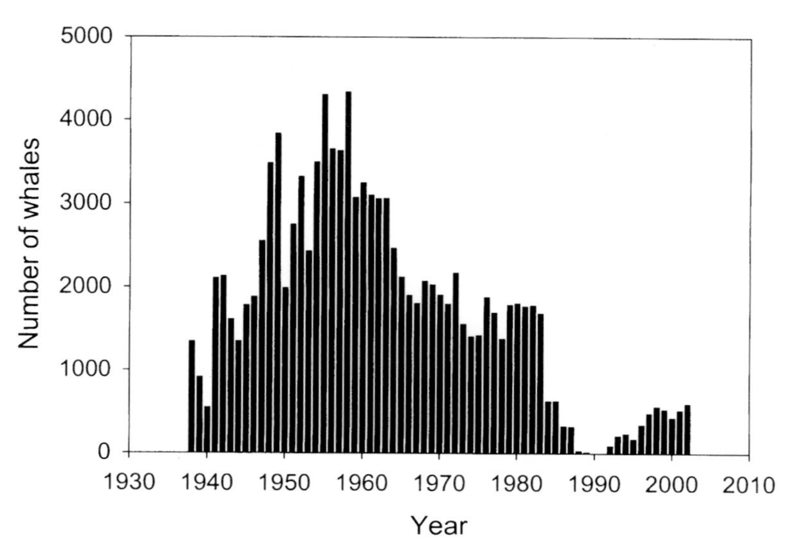

Figure 9.6 Annual catch of Minke whales in the North East Atlantic between 1938 and 2002 (N. Øien, HI-Bergen, pers.com.).

rily on fish, of which herring, capelin, cod and sand eel are most important, while krill (*Thysanoessa sp.*) are taken at the northern extension of their range.

In the North Atlantic mating is assumed to take place from October to March and gestation is approx. 10 months. Mature females give birth every year. Calving occurs from November to March, lactation lasts 4-5 months. Age determination is difficult in minke whales but age at sexual maturity is estimated to about 7 years.

Minke whales occur singly or sometimes in small groups of 2-3 animals. In the north, the blow of the minke whale is usually invisible, which makes the animals difficult to detect. The minke whales are reportedly able to produce sound in the form of low-frequency grunts (80-140 Hz, 165-320 msec duration) and a variety of other thumps, ratchets, clicks and pings, but in my own experience they are rather like cattle, who have the ability to vocalize, but seldom use it.

Together with Lars Folkow, I have studied the daily energy expenditure of minke whales (Blix & Folkow, 1995) and we found that cost of swimming is remarkably low in these large animals and that their estimated daily energy expenditure, on average, only amounts to 80 kJ/kg/day.

Olsen, Nordøy, Blix & Mathiesen (1994) also studied the digestive processes in these whales and discovered that they have a multichambered stomach system (like other whales investigated) and that the large forestomach contains huge numbers (10^{10} /ml forestomach fluid) of endemic bacteria, which like in ruminants, digest the food by microbial fermentation. As a result of this we have shown that the minke whale has a digestive efficiency of 93 %, when eating krill (*Thysanoessa sp.*). It is interesting that the antarctic crabeater seal (*Lobodon carcinophagus*), which has a single stomach and lack microbial fermentation only has a digestive efficiency of 84 %, when eating krill (*Euphausia superba*) with a similar chemical composition as the krill eaten by the minke whales (Mårtensson, Nordøy & Blix, 1994).

From these results, and knowledge of the chemical composition of the prey which are taken by minke whales throughout their six-month sojourn in Norwegian and Barents Seas, we have been able to calculate that the amount of food which is consumed by the north-east atlantic stock of minke whales is in the order of 1.7 million tons, of which some 80 % is fish and some 20 % krill (Nordøy, Mårtensson & Blix, 1995). It is also interesting to note that we have found that the blubber which is deposited during summer and autumn can only cover its energy expenditure for about 60 days, when in other waters during winter. It follows, that the minke whale must be feeding during winter.

Northern fur seal (*Callorhinus ursinus*)

The northern fur seal is not really an arctic seal (Fig. 9.7), but it breeds primarily on islands in the Bering Sea and has a very significant impact on the Bering Sea ecosystem. Besides, I have had the good fortune to study this animal on the breeding grounds on several occasions and I would like to revisit these friends of mine on this occasion. The northern fur seal breeds primarily on the Pribilof Islands in Alaska, some at the Commander and Kurile Islands in Russia, and a few on San Miguel Island on the west coast of America.

Like all otarid seals, the northern fur seal has external ears and very long flippers. When wet they all appear black, but when dry, they are silver-grey dorsally and reddish brown on the belly. Adult males are usually darker than the females and are sporting a rather impressive mane. The pups are born in a black lanugo pelt.

The northern fur seals that breed in Alaska appear to migrate south through the Aleutian chain and fan out during August-October into the North Pacific Ocean, where they tend to concentrate in areas of upwelling over sea-mounts and along the continental slopes. In May they start to return to the breeding rookeries where they are highly gregarious (Fig. 9.8). In this chaotic mess of bodies the big males aggressively defend well-defined territories through threats and ferocious fights, while the females give birth to a 5 kg pup. The males have

Figure 9.7 The geographical *summer* distribution (blue) of northern fur seal *(Callorhinus ursinus).*

sole access to the females within their territorial boundaries, where they may attend to a "harem" of about 50 females. The females start going off to sea to feed already a week *post partum*, usually a day after copulation, but keep returning to nurse the pup after progressively longer sea trips until the pup is about 3 months old. When the females are at sea the pups gather in large pods, and the females and the pups recognize each others calls when she returns to the rookery.

Fur seals are opportunistic feeders, taking a wide variety of prey (75 known species) in different parts of their range and in different seasons. In continental shelf areas of the Bering Sea fur seals eat mostly juvenile pollock, whereas in oce-

Figure 9.8 Northern fur seals on the shores of St. Paul Island, Alaska, during the breeding season. A bull "beach-master" in the centre, is guarding the mess of females and pups in his noisy harem of some 50 females (Photo: A.S. Blix).

anic areas of the Bering Sea they eat mostly squid. It is known that Steller sea lions, and killer whales in some areas, prey on northern fur seals.

The Pribilof population of northern fur seals has been exploited for skins since 1786, when this herd probably counted 2.5 million seals. The herd steadily declined to a low point in 1835 whereafter it increased, but unregulated harvesting drove the herd to another low in 1910, when fewer than 300,000 animals came on shore. This resulted in a treaty among the US, Canada, Russia and Japan for the management of the fur seals. The herd rallied again under new harvest restrictions and by 1950 was producing over 400,000 pups per year. From 1956 to 1963 the herd was purposely reduced by killing approximately 300,000 adult females. The total population of animals on the Pribilofs was about one million in 1990, but is down again to about 850,000 in 2002. The last commercial harvest on St. George Island was in 1972 and on St. Paul Island in 1984 (Table 9.7).

Steller sea lion (*Eumetopias jubatus*)

The Steller sea lion is a frequent visitor to the Bering Sea, but breeds primarily on the Aleutian Islands, on the Alaskan-Canadian coastline down to southern California, on the Kurile Islands, Kamchatka, and on the islands in the Okhotsk Sea.

Table 9.7
Estimates of population abundance and pup production for northern fur seals on the Pribilof Islands.

Year(s)	Stock size	Number of pups born
1949-1951	2,100,000	531,000
1970	1,200,000	306,000
1974	1,250,000	326,000
1983	877,000	198,000
1990	1,012,000	253,000
2002	850,000	172,000

There is an estimated world population of 250,000 Steller sea lions. The males of this species are formidable animals that weigh about 1,000 kg, while fully grown females weigh only 275 kg, pups born in mid-June weighing in the order of 17 kg. Like northern fur seals, the Steller sea lion is territorial and large bulls may attend to 6-13 females. When in the Bering Sea the animals are opportunistic feeders and may take capelin, pollock, halibut, flounder, squid, bivalves, shrimps and crabs.

CHAPTER 10

Physiologicial Adaptations

The Arctic is a hostile place in winter, yet, as we have seen, the cold, dark polar "wastes" sustain an abundant amount of life. Still, the environment is truly marginal and it is tempting to conclude that poikilothermic, or "cold-blooded", animals would freeze to death, and that homeothermic, or "warm-blooded" animals which spend the winter there must endure a truly marginal existence. Paradoxically, however, poikilothermic arctic animals survive in great numbers and the homeothermic animals usually neither freeze nor starve to death. They are, instead, as we shall see, well adapted to the several challenges of the environment, of which I will deal with temperature first.

Acclimation and adaptation

These two important adjustable processes occur simultaneously but within different time-scales. *Acclimation* is the non-heritable modification of characters caused by exposure of organisms to environmental change. It may occur several times within an organism's lifetime, sometimes within a span of hours, or even minutes. Individuals acclimatize to cold, for example, by adjusting physically or physiologically following cold exposure. Such changes are readily reversible and not known to effect the genome. *Adaptation* has two linked meanings, both implying longer-term hereditary changes. An adaptation is a modification of the genome resulting in structures, functions or behavioural patterns that increase the probability of an organism surviving and reproducing in a particular environment. The time-scale involved is seldom measurable, but spans many generations. Adaptation in the second sense is the accumulation, in a species, of enough of these changes to promote viability within the specified environment. Though individual acclimation is not inheritable, ability to acclimatize may well be inherited. Plants and animals that were incapable of acclimatizing to cold would not have survived, for example, in the cooling environment of the early Pleistocene (e.g. Stonehouse, 1989).

Organisms at low temperatures

Biological processes are chemically based and subject to the rules governing chemical reactions. All chemical reactions are temperature dependant, occurring more slowly at low temperature than at high. Within biological systems the rates of most reactions increase by a factor of two to three for each 10 °C rise in temperature. This is usually expressed as a Q_{10} relationship; the velocity constant K_1 of a reaction at temperature T_1 and the velocity constant K_2 of the same reaction at temperature T_2 are related by the expression:

$$Q_{10} = \left(\frac{K_2}{K_1}\right)^{\left[\frac{10}{(T_2 - T_1)}\right]}$$

This relationship for living systems holds good within narrow limits, but is severely upset by temperature extremes. Many intracellular reactions are enzyme catalysed and the enzymes that moderate the processes often work optimally only within a narrow range of temperatures. Thus, Q_{10} values vary considerably from one 10 °C range to another. It is therefore advantageous for most organisms, large or small, to work within a narrow range of body temperatures. Both plants and animals therefore have the means of optimizing and stabilizing their working temperatures.

Poikilotherms include most animals and all plants. With few exceptions the working temperatures within their bodies at any moment are dependent on the temperature of their environment.

Polar fish and marine invertebrates live at the near freezing temperatures of their environment, but most terrestrial polar poikilotherms are able to attain working temperatures that differ favourably from those of the environment – animals through behaviour and plants through small body movements. Solar radiation is an important factor in their lives; it may be the first factor that raises their temperatures from low ambient to effective working range. Hence, it does not generally pay poikilotherms to be insulated. In polar regions active life may therefore be restricted to slopes where summer sunshine allows sufficient warming for a few days or weeks, but overheating may occur in summer sunshine especially in dry areas.

Homeothermic animals (mammals and birds) are governed by the same laws of thermodynamics as the poikilotherms, but they have developed the ability to produce large amounts of body heat and the ability to regulate body temperature at a stable level, sometimes even at 100 °C above ambient temperature (e.g. Schmidt-Nielsen, 1990).

Cold defence in poikilothermic arctic animals

Poikilothermic animals exposed to low temperatures live in danger of suffering a lethal freezing of their body fluids, but in spite of this, there are many poikilothermic animals, and particularly insects, that live in Arctic regions and hibernate in sites where they may be exposed to the extremely low temperatures that prevail there in winter. Thus, studies by Zachariassen and Sømme of Norway, Miller and Baust of America, Block of Great Britain and others (e.g. Zachariassen (1985), Sømme (1995), Strathdee & Bale (1998)) have shown that poikilothermic animals may use two different strategies for survival under extreme temperature conditions. They may seek to *avoid freezing* by keeping their body fluids supercooled, or they may develop a *tolerance to freezing.*

But, before we go into just how this is done, we have again to agree on a few concepts and definitions:

The *melting point* (MP) of a fluid is the temperature at which the last tiny ice crystal in a frozen sample disappears during slow heating. The MP of pure water is 0 °C. The rate by which the MP drops with increasing solute concentration is given by the osmolal MP depression, which is 1.86 °C/Osmol. The *freezing point* is frequently used synonymously with the melting point. This is unfortunate, because the freezing point can easily by confused with other important concepts, such as the super cooling point. When a sample of water or an aquous solution is cooled, it will normally not freeze when the MP is reached, but will remain unfrozen even when cooled far below this temperature. A system that remains unfrozen at temperatures below its MP is said to be supercooled, and the temperature at which spontaneous freezing occurs in a supercooled system is termed the *super cooling point* (SCP) of the system. The SCP of pure water may be as low as -40 °C.

The difference between the SCP and the MP of a system is termed the *super cooling capacity* (SCC) of the system.

The body fluid of certain poikilothermic animals contains proteinaceous factors that prevent growth of ice crystals at temperatures below the melting point. The ability of proteinaceous antifreeze agents to prevent ice crystal growth is limited in that cooling to a temperature as much as 10 °C below the MP causes a sudden, rapid growth of the ice crystals. The phenomenon that MP is separate from the temperature where an ice crystal can start growing is termed *thermal hysteresis*, and the temperature at which the ice crystal growth begins in a system of this kind is termed the *hysteresis freezing point* of the system. *Freezing tolerance* is the ability to tolerate formation of ice in the body fluids at temperatures equal to or below the SCP, and *freezing sensitivity* is the lack of tolerance to formation of ice in the body fluid, while *freezing avoidance* is the survival strategy adopted by freeze-sensitive insects that are exposed to low temperatures (e.g. Zachariassen, 1985).

Active summer insects in the arctic regions, even those that are active on the snow in the winter, have SCPs in the range from -7 to -12 °C, which appears to be the general supercooling limit of active insects, regardless of season and geographical distribution, in spite of the fact that they should have had a SCP of about -20 °C on an osmolar basis. It follows, that active insects have components with the ability to induce freezing of water at high subzero temperatures. These components are called *ice nucleating agents* (INAs), of which food particles in the guts appear to be particularly important (Fig. 10.1).

Freeze avoidance

INAs of some kind are always present in the body fluid of insects prior to the winter hardening (i.e. physiological preparation for winter conditions), (Fig. 10.1), and for insects that base their winter survival on avoidance of freezing, the presence of these INAs represents a major problem. Thus, for the insects to obtain a high supercooling capacity, the INAs present in the gut or in intracellular compartments have to be removed or inactivated , leaving the insect in the state illustrated in Fig. 10.1B.

Normal summer insects have hemolymph osmolarities ranging from approximately 400 to approximately 600 mosmol/kg. By accumulation of polyhydric alcohols (polyols) and other low-molecular-weight solutes, the body fluid osmolality may increase to more than 3,000 mosmol/kg, and in some cases even far above this value, and thereby greatly increase their supercooling capacity (Fig. 10.2). Several different polyols and carbohydrates are accumulated by hibernating insects.

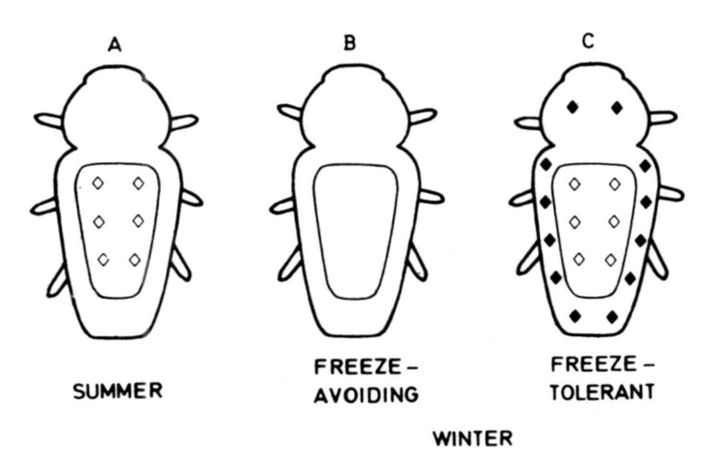

Figure 10.1 The distribution of ice nucleating agents (INAs) in body fluid of freeze-susceptible (avoiding) and freeze-tolerant species. During the summer, INAs are present in the gut or other internal compartments *(open symbols)* of both categories. In winter, freeze-susceptible insects are free of INAs, while many freeze-tolerant species have INAs in their hemolymph *(closed symbols)* (Zachariassen, 1985).

Most common in this respect is *glycerol*, which has been found in a wide variety of insect groups, but other polyols, such as sorbitol, mannitol and ethylene glycol also occur in concentrations high enough to offer cryoprotection. But, even insects without significant levels of polyols in their body fluid may be able to supercool to temperatures below -20 °C. In such animals the cold hardiness can instead be ascribed to the removal or inactivation of INAs from the body fluids.

Insects that hibernate in a supercooled state obviously risk lethal freezing due to several factors at all times, since the supercooled state is physically metastable, and a system can remain supercooled only for a limited time. After a period, the duration of which depends on the volume, the temperature and the solute concentration, the system will enter the thermodynamically stable frozen state. Thus, insects hibernating in a supercooled state should be expected sooner or later to undergo a spontaneous lethal freezing, and those that hibernate in frozen habitats, where they are in close contact with external ice, should, in addition, be expected to undergo freezing due to inoculation of external ice through the body wall. However, despite these expectations, many freeze-sensitive insects spend the winter in a highly supercooled state in habitats where they are in close contact with external ice without freezing. The physiological basis of this success in freeze avoidance is probably the presence of *thermal hysteresis factors* in their body fluids (e.g. Zachariassen, 1985).

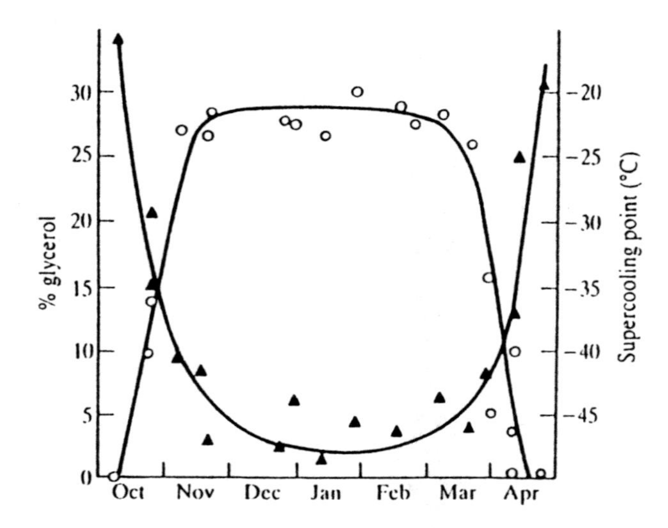

Figure 10.2 Seasonal changes in glycerol content (% of fresh weight; circles) and supercooling points (triangles) in larvae of the pine shoot tortricid *(Petrova recinella)* (Redrawn from: Hansen, 1973).

The phenomenon of thermal hysteresis has received considerable attention by De Vries who found it to be prominent in the blood plasma of certain polar teleost fishes, which may live their entire lives in a supercooled state. Like other teleost fishes , they have body fluid MPs ranging from about -0.6 to about -0.9 °C. The temperature of the ice-covered seawater in these regions is about -2 °C, and fish , being ectothermic animals, have the same body temperature. In spite of their supercooled state, the fish survive for years without undergoing spontaneous internal freezing or freezing due to seeding of external ice. The reason is that the blood plasma of these fish contains macromolecules, which counteract the growth of seeding ice crystals at temperatures below the plasma MP (e.g. De Vries, 1980; Zachariassen & Kristiansen, 2000).

The hysteresis factors found so far fall into two categories, glycopeptides and peptides, and in both cases they make up only 3-4 % of the body fluids, and can therefore make no significant colligative osmotic contribution to freezing point depression. Isolation and purification of the colloidal macromolecules have now been carried out for several species (Fig.10.3), and although the mode of action of all the hysteresis factors are somewhat obscure, it is suggested that glycopeptides adsorbed onto the surface of the ice crystal cause greatly increased curvature of the growing fronts of the crystal, and that this increased curvature causes high surface energy, with the result that growth can only continue if the temperature of the system is lowered. De Vries and associates demonstrated that a range of northern fish which possessed these factors in the winter but not in summer, lost them in response to a combination of heightened environmental temperature and long pho-

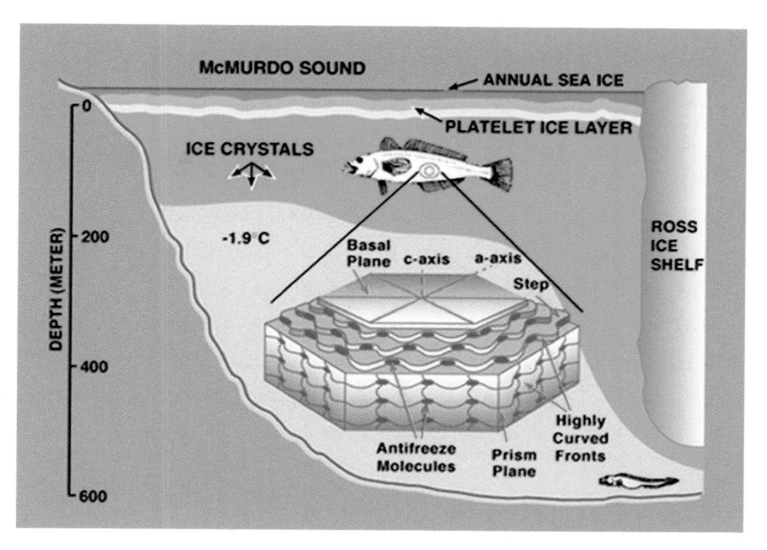

Figure 10.3 A schematic ice crystal is shown within a fish, in this case in antarctic waters, with antifreeze glycoprotein molecules adsorbed onto its prism planes that alter the crystal planes surfaces preventing growth of the crystal (DeVries & Cheng, 2004).

toperiod; neither factor operating alone was sufficient stimulus. These workers also found that complete loss of antifreeze took 3-5 weeks, and it is likely that pituitary hormones suppress the synthesis of antifreeze proteins by the liver during summer.

In many of the populations of capelin (*Mallotus villosus*) the eggs are laid subtidally, but in Iceland, Labrador, Greenland and a particular fjord, called Balsfjord, near Tromsø in Northern Norway, populations of this species have the peculiar habit to spawn between the tidemarks in early spring, but, not when the tide is in. Males and females, instead, leap out of the advancing waves on rising tides, onto the substratum where copulation takes place in air. Most eggs are laid on the middle shore, but substantial numbers are often found on the upper shore, and are therefore exposed for many hours each day at air temperatures which will often be well below the freezing point of sea water and fish plasma. Davenport (1992) established that these eggs could supercool, even in the presence of ice, down to -11 °C, and that an intact chorion (eggshell) was needed to prevent freezing of the embryo, since the cutting of even a minute hole in the chorion resulted in the freezing of the embryo as soon as the surrounding sea water contained ice (at around -2 °C). Other experiments clearly showed that there is no antifreeze involvement, and therefore chorion somehow prevents inoculation of the egg contents by ice crystals.

Another strategy adopted by a number of cold-water fish to avoid freezing is to migrate into deep, cold water during the winter months or to spend their whole life in this environment. Deep water in the Arctic has a temperature of about -1.8 °C, but, being slightly warmer than the freezing point of sea water, does not contain ice-crystals. Scholander *et al.* (1957) were the first to note that fish could exploit deep water as a refuge in which they avoided contact with ice and survived in a supercooled state. They found that fish, such as the tomcod *(Boreogadus kofoedi)* and snailfish (*Liparis turneri)* had slightly lower blood freezing points (-0.9 to -1.0 °C) than temperate species but not to an extent which significantly reduced the risk of freezing. They also demonstrated, however, that if ice was introduced into the environment of these fish, they instantly froze, presumably because of migration of seeding ice nuclei from ice to fish, even though the sea water itself did not freeze.

Freeze tolerance

Freezing of the body fluid involves several potentially lethal effects. First, freezing in the intracellular or intraintestinal compartments will cause an increased osmolality in the unfrozen fraction of these compartments, leading to an osmotic swelling of the compartments and eventually a rupture of the membranes that confine them. Second, extracellular freezing may cause cells to shrink to the extent that their external membrane rests on a "matrix" made up of intracellular membrane structures, making further shrinking impossible, and cooling and elevating the solute concentration of the system beyond this point may lead to intolerable osmotic stress across the cell membranes. Third, freezing will increase the concentration of inorganic salts to levels that may irreversibly change the structure of proteins and thereby the function of enzymes.

It has been known for some time that most insects tolerant to freezing have a conspicuously limited ability to supercool, and Zachariassen & Hammel (1976) were able to demonstrate that the hemolymph of freeze-tolerant beetles contains INAs , which are responsible for the high SCPs of these animals (Fig. 10.1). Later, INAs have been found in the hemolymph of several freeze-tolerant species, and with few exceptions, their presence in the hemolymph seems to be seasonal and limited to the cold part of the year (Zachariassen & Kristiansen, 2000).

The function of ice nucleating agents in the hemolymph of insects tolerant to freezing is probably to prevent *intracellular* freezing in the cells, but by initiating ice formation in the *extracellular* compartment. The extracellular fluid then becomes concentrated, and there is an osmotic efflux of water from the cells, which subsequently shrink. The general idea of extracellular freezing to prevent intracellular ice-formation was first conceived by Kanwisher (1959), who studied freeze-tolerant blue mussels.

Accumulation of polyols and carbohydrates, such as glycerol, threitol and erythritol and sucrose, commonly called cryoprotectants, is a common feature also among freeze-tolerant hibernating insects, but the accumulation of polyols has only a moderate effect on their SCP, while increasing concentrations of polyols strongly increase their cold hardiness. Miller (1969; 1982) found, for example, that several freeze-tolerant insects from interior Alaska tolerate freezing to temperatures ranging from -55 to -70 °C; individuals of one species even survived freezing to -85 °C. The lowest temperature reported to be tolerated by freeze-sensitive insects is about -50 °C, but tolerance to temperatures below -35 °C is uncommon, and will always require extremely high levels of polyols. Thus, the strategy of freeze tolerance appears to offer several advantages to that of freezing avoidance. This is supported by the fact that in the interior Alaska and Canada there are many freeze-tolerant species, some of which can tolerate temperatures as low as -85 °C. The freeze-avoiding insects hibernating in these areas all seem to belong to the group that displays extremely low SCPs and extremely high polyol concentrations.

Freezing tolerance may also offer an advantage with regard to water balance. Insects hibernating in a supercooled state may lose water during the winter, when the relative humidity in the air is frequently very low. In frozen insects, on the other hand, the body fluid stays in vapour pressure equilibrium with ice, and if they hibernate in a closed hibernaculum where ice is present, they neither loose nor gain water from the environment (Lundheim & Zachariassen, 1993).

Several review articles on adaptations in invertebrates to terrestrial Arctic environments are available (e.g. Bale *et al.*, 1977; Strathdee & Bale, 1998; Zachariassen & Kristiansen, 2000).

Cold defence in homeothermic arctic animals

For a homeothermic animal exposed to the full brunt of the arctic climate and with a thermal gradient between body core and the environment of up to 100 °C, the challenge is first and foremost to balance heat loss against the rate of metabolic heat production. In the Arctic where energy usually is in short supply, this is preferentially achieved by control of heat loss to the cold environment. There are three main avenues for heat loss from an animal : conduction/

convection, radiation and evaporation. Loss by conduction over the body surfaces, is determined chiefly by, and proportional to, the thermal gradient between the body surface temperature and that of the environment and the area over which the heat can be exchanged. The conductive heat loss can be increased substantially by convection, that in most cases means wind, which causes turbulence and breaks down the insulating layer of stagnant air between the hairs of the fur. An animal will also loose heat to the environment by long-wave (infra-red) radiation. The net heat transfer by radiation is described by Stefan-Boltzman's Law, which says that radiation heat transfer changes with the forth power of the absolute temperature, but if the temperature difference between the surfaces is not too great the rate of heat loss by radiation can also be considered proportional to the temperature difference. This implies that the rate of conductive and radiative heat loss from a warm-blooded animal in cold surroundings can be expressed as:

$$\dot{Q} = C\,(T_2 - T_1),$$

in which $\dot{Q}$ = rate of heat loss, C = a proportionality factor, and T_2 and T_1 = the temperatures of the body surface and the environment, respectively.

However, an Arctic animal can also receive heat by radiation from the environment (the sun), and we shall later see that such radiation may in some cases add a substantial amount to the energy budget of the animal.

When warmblooded animals breathe, cold air is inhaled into the lungs, where it is warmed to body core temperature, but it is also saturated by water vapour. Upon expiration, this heat and water are often lost. In an environment where this water has to be replenished either by cold water, which must be heated to body temperature, or, even worse, by snow, which first must be melted, with an additional loss of energy, this evaporative respiratory water loss may add a very significant amount to the energy expenditure of the animal in addition to the evaporative water loss which always to some extent will take place over the rest of the body surfaces. We shall see how Arctic mammals and birds have developed means to reduce and control also this form of heat loss.

Now, even in homeothermic animals, the body temperature does not always remain absolutely constant. Assume, for example, that heat loss is not quite equal to metabolic rate, but is slightly lower. The body temperature will then rise. This means that part of the metabolic heat remains in the body instead of being lost, and the increase in body temperature therefore represents a storage of heat. If, on the other hand, body temperature decreases, which happens when heat loss exceeds heat production, we can regard the excess heat loss as heat removed from storage. The amount of heat which is stored depends on the change in mean body temperature, the mass of the body, and the specific heat capacity of the tissues, which is usually assumed to be 0.8.

These factors can be expressed in a simple heat-balance equation:

$$H_{tot} = H_c \pm H_r \pm H_e \pm H_s,$$

in which H_{tot} is the rate of metabolic heat production (which is always positive), H_c = rate of conductive and convective heat exchange (+ for net loss), H_r = rate of net radiation heat exchange (+ for net loss), H_e = rate of evaporative heat loss (+ for net loss) and H_s = rate of storage of heat in the body (+ for net heat gain by the body) (e.g. Schmidt-Nielsen, 1990).

Now it is time to look into how arctic animals go about reducing heat loss to the environment, and we will start by dealing with conductive and radiative heat loss, keeping in mind that :

$$\dot{H} = \dot{Q} = C\,(T_b - T_a),$$

which says that the rate of metabolic heat production ($\dot{H}$) equals the rate of heat loss ($\dot{Q}$), which is proportional to the temperature difference between the body (T_b) and the environment (T_a); C being a conductance term that will be discussed later. In this equation the surface area of the animal over which heat is lost by conduction and radiation to the environment is included in the conductance term, which is considered constant. However, this is not always true. In fact, very many species often attempt to reduce their surface area by "balling up" into a more or less spherical posture (Fig. 10.4). This has the advantage that the surface area over which heat is lost is much reduced, while the heat producing mass of the animal is unchanged, and the sphere is not surprisingly the form for which surface area in relation to volume is the smallest. The less advantageous side of this behaviour is that you are rendered immobile and run the increased risk of being eaten.

Other behavioural means of reducing heat loss, with much the same risks, is to construct shelters in the ground and in the snow, or simply to "ball up" and let oneself snow down. This behaviour is, of course, driven to perfection by Arctic man, in the form of the igloos of the eskimos, while simple burrows in the snow are used temporarily by polar bears, wolves, foxes, hares and ptarmigans, as well. The obvious advantage of this arrangement during cold and windy nights is that

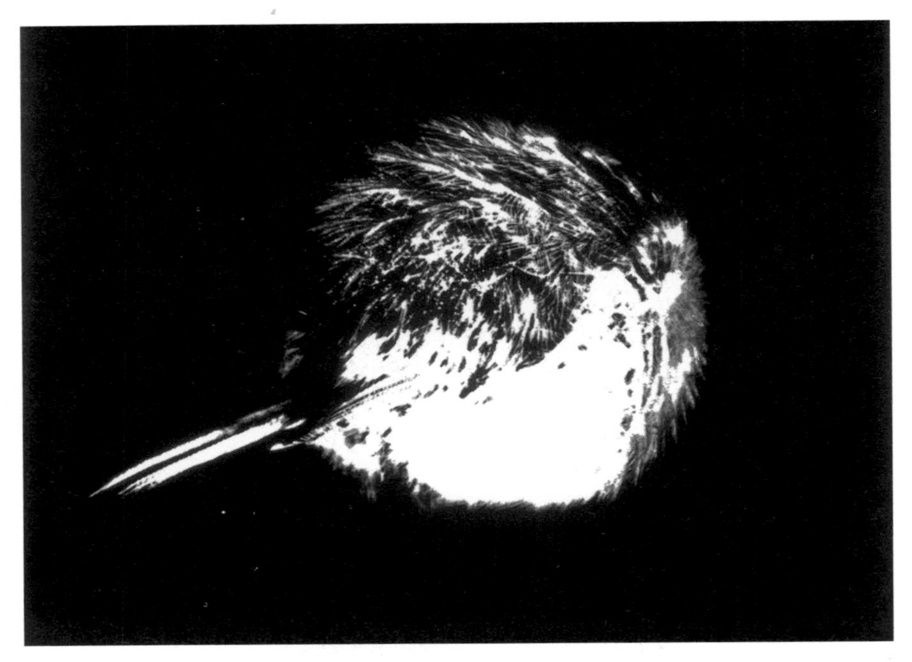

Figure 10.4 Grey-headed chickadee (*Parus cinctus*), roosting at an ambient temperature of -45 °C in Alaska, with its surface area reduced to the absolute minimum (Photo: A. Andreev).

it creates a microclimate where convection is almost eliminated and conductive and radiative heat loss is much reduced.

Another important aspect of this surface area/body mass relationship is that the surface area relative to body mass is reduced with increased body size. This implies that it should be advantageous for animals that inhabit cold habitats to be big! Hence, there should be a tendency for a given species (or taxonomic group) to evolve larger body size in the colder parts of its range. Bergmann, in 1847, offered this as an empirical rule derived from observations, and in fact many groups of mammals and birds obey it, but there are many exceptions, and this "rule" has gained much more importance in the ecological texts than it deserves! Scholander (1956), has drawn attention to its shortcomings: "The hopeless inadequacy of cold adaptation via Bergmann's rule may be seen by the following consideration. Take a body-to-air gradient in the tropics of 7° and in the Arctic of 70°, i.e. a tenfold increase. A tenfold greater cooling in the Arctic animal is prevented by covering the surface with fur a few centimeters thick. A relative surface reduction of ten times would require a weight increase of the animal of one thousand times". This view has later been reiterated by Geist (1987) and others.

Another long-lived "rule", which is almost as venerable as Bergman's is Allen's "rule" from 1877, which has the same empirical basis and the same limitations. It states that in cold environments there is a tendency for the development of shorter and more compact extremities, such as ears, limbs, tails and snouts. This "rule", in addition to all the exceptions, implies a reversal of the normal allometric consequences to be expected to follow from Bergman's "rule", and it is my suggestion that they should both be forgotten.

Chernov (1985), based on data collected by Syroyechkovskiy (1978), has pointed out a novel problem with increasing size among geese and swans which deserves attention in this context. That is that the duration of incubation and the duration of development of the chicks, from the moment of hatching until they can fly, increases with increasing body weight of the parents (Table 10.1), and it is the smallest species of the group which breed in the Arctic, while the largest are limited to the southern belt of the tundra, where they have sufficient time to raise their offspring.

Table 10.1

Indices of maximum weight and duration of incubation and development of nestlings from hatching to fledglings of some species of swans and geese (Syroyechkovskiy, 1978).

Species	**Weight (kg)**	**Duration of incubation (days)**	**Duration of development of nestlings (days)**	**Total days required for development**
Mute swan *(Cygnus olor)*	16.0	36	150	186
Whooper swan *(C. cygnus)*	13.0	40	120	160
Bewick's swan *(C. bewickii)*	6.0	30	45	75
Bean goose *(Anser fabalis)*	3.7	29	45	74
Blue (snow) goose *(Chen caerulescens)*	2.9	26	49	75
Lesser-white-fronted goose *(Anser erythropus)*	2.5	28	40	68
Red breasted goose *(Branta ruficollis)*	1.7	26	40	66
Brent goose *(B. bernicula)*	1.4	26	40	66

All Arctic mammals and birds also protect themselves physically against cold by growing a plumage of feathers in birds and a coat of fur in mammals, except whales. Most Arctic animals, notably the marine mammals, also improve their thermal protection by deposition of a layer of subcutaneous fat, called blubber.

The insulative properties of fur depend on the inherent thermal conductivity of the individual hairs themselves and their collective ability to trap and hold a layer of air next to the skin. This is important because the conductivity of still air is less than half that of most furs. The insulative value of the fur may be enhanced both by a high density of fibres on the skin and by their special structure. In reindeer, for instance, each of the guard hairs is hollow, containing thousands of air-filled cavities separated by thin septa (Fig. 8.16). Thus, the insulative value of fur from different animals differs a great deal. Some data from Scholander *et al.* (1950a) are shown in Fig. 10.5, and illustrate the fact that the insulation value increases with the thickness of the fur. This clearly puts the small animals at a disadvantage, since they need to have a relatively short fur in order to be able to move about, and the smallest mammals like lemming and ermine are conse-

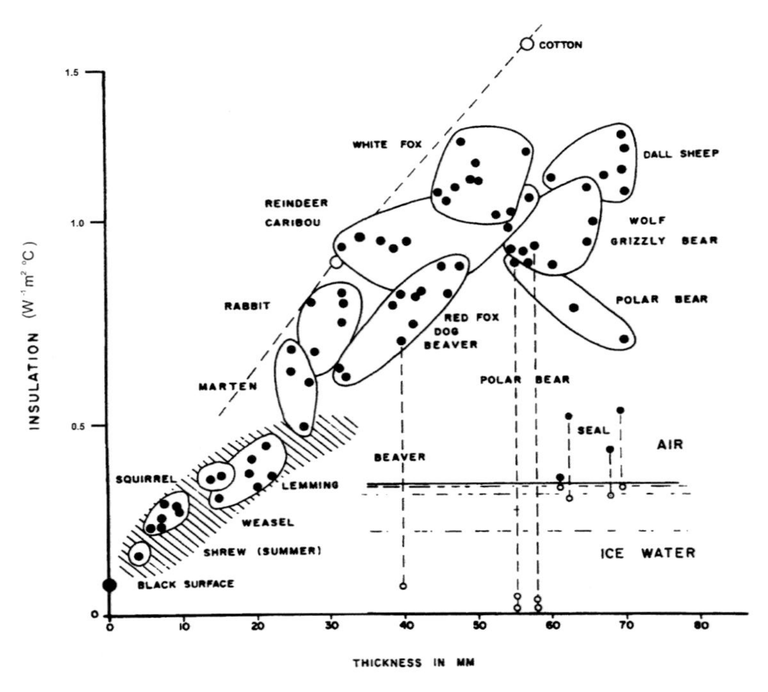

Figure 10.5 The insulation value of animal fur in relation to fur thickness. Insulation values for fur in air are indicated by closed circles; values for fur submerged in water, by open circles. The sloping broken line represents the insulation values of cotton of various thicknesses (Scholander *et al.*, 1950a).

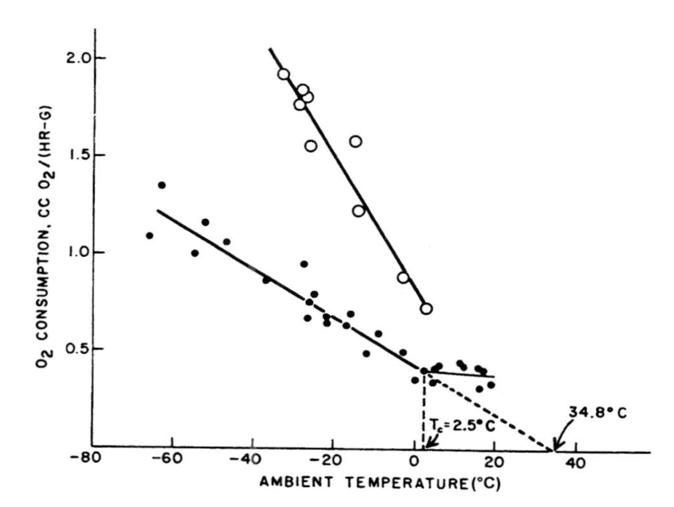

Figure 10.6 Energy expenditure (oxygen consumption) of four female snowy owls exposed to ambient temperatures from +18.5 to -66 °C at negligible air speed (filled circles) and three owls at an air speed of 7.47 m sec^{-1} (open circles) (Gessaman, 1972).

quently constrained to live their winter life in burrows under the snow, where they can take advantage of a less hostile microclimate. But, even in the largest of animals the insulative value of the fur is influenced by a number of factors. The fur of the polar bear is particularly interesting in this respect. The polar bear fur is very long and offers excellent insulation when it is dry in air, but loses, oddly enough, most of its insulative value when it is immersed in water, which conducts heat some 25 times faster than air. In the adult and well fed polar bear some insulation is provided by a subcutaneous layer of fat, but as we shall see, the young polar bear cub faces grave thermal problems when it is immersed in ice-water. The polar bear fur has one important advantage, however, and that is that it sheds the water very easily, when shook after emergence onto ice or land, due to the stiff and slippery nature of the hairs. The eskimos have always taken advantage of this and utilized polar bear skins for pants and mittens, which are frequently exposed to water, while the rest of their outfits have traditionally been made from seal skins (Fig. 4.28).

The insulative value of fur and plumage is also reduced by rain and sleet, and particularly by wind, which causes turbulence in the otherwise stagnant layer of air in between the feathers or hairs. Thus, Gessaman (1972) found that wind exerted a strong cooling effect on adult snowy owls at Barrow, causing their metabolic heat production to double to maintain body temperature in the face of even a moderate windspeed of some 7.5 m/sec at -30 °C (Fig. 10.6). The effect of wind has a relatively minor effect on large animals with a long and dense fur, such as muskoxen and Svalbard reindeer, in which even gale force wind seems to

Figure 10.7 A bowhead whale during flensing, showing parts of the enormous blubber deposits of this high-arctic plankton-feeding whale in the foreground (Photo: NARL).

cause little problem (Cuyler & Øritsland, 2002), while smaller animals, and as we shall see, most newborns are rather vulnerable to wind, particularly when it occurs in combination with rain and sleet.

In marine mammals, such as the seals, the fur is of little insulative value, even when dry in air, and is further much reduced when wet (Kvadsheim & Aarseth, 2002), or it is missing entirely such as in whales. In these animals insulation is provided by the subcutaneous layer of blubber, which is unaffected by water and sometimes reaches enormous proportions, notably in the Bowhead whale (Fig. 10.7). Thus, blubber generally provides much less insulation per unit depth than most furs, but has the great advantage of retaining its insulative value in water, and aside from some of the newborn forms the marine mammals have the advantage of a great body mass, which reduces their dependence on prime insulation.

In phocid seals, however, this insulative layer of blubber combined with the fact that these seals, unlike the otariid ones, do not have a scrotum and keep their testes in the abdomen would be expected to cause a problem: In most mammals the testicles lose their sexual potency unless they are kept at a temperature lower than that of body core, and to the best of my knowledge the purpose of the scrotum therefore is to keep the testes cool. Accordingly it has been long known that the testicles of northern fur seals are kept at least 6 °C lower than body core temperature in their pendulous scrotums, but one could argue, of course, that a scrotum hanging in ice-water may be a bit too much. Nobody had addressed this problem, of the utmost importance for the preservation of these species, until Bud Fay of Alaska and I (Blix, Fay & Ronald, 1983) looked into it and found that the testicles of phocid seals were indeed several degrees colder than body core, in spite of the fact that they were kept inside the insulating blubber. We also found that the way they manage this is by way of conspicuous venous connections between the hindflippers and extra-testicular venous plexuses, through which cold blood from the hind-flippers can flow and cool the para-abdominal testicles. So, that's the way it is.

The thickness, and hence, the insulative value, of the fur and plumage of most arctic animals changes throughout the year. These changes appear to be much less in small mammals than in the larger ones. In muskoxen, for instance, *all* the underwool, or qiviut, is shed in large sheets every spring and replenished in the autumn, while in others like the reindeer the changes are major, but less dramatic. Also the blubber layer of the marine arctic mammals undergoes very major seasonal changes, being at a low during summer (Fig. 4.1), but we shall see that this is more a consequence of prolonged fasting during the moult of their fur than of any thermal reasons.

The conductance of the animals may also be reduced by *physiological* means, such as to allow the peripheral tissues to cool, while the core temperature is maintained. This is achieved by reduction of peripheral and in particular cutaneous circulation and has the advantage of increasing the thickness of the insulating shell, while the core, which has to be kept warm, is reduced. This is illustrated by means of infrared photography, by which the surface temperatures of, in this case, a reindeer at different ambient temperatures are colour coded in Fig. 10.8.

Moreover, both in terrestrial and marine mammals and birds the appendages represent relatively large surface areas, that for several reasons are less well insulated than the rest of the body. Many mammals and birds, notably those that inhabit the northern regions are able to restrict heat loss from their legs and feet by special vascular arrangements that facilitate counter-current heat exchange. It had been known since the late 1920ies that a variety of aquatic and terrestrial mammals and birds are equipped with multichanneled arteriovenous blood vascular bundles at the base of the extremities, but the function of these bundles remained a mystery until Scholander and associates in the early 1950ies suggested that they were employed as organs for heat preservation. At that time Lawrence Irving of Alaska had already demonstrated that the feet of arctic dogs, reindeer, and sea gulls may

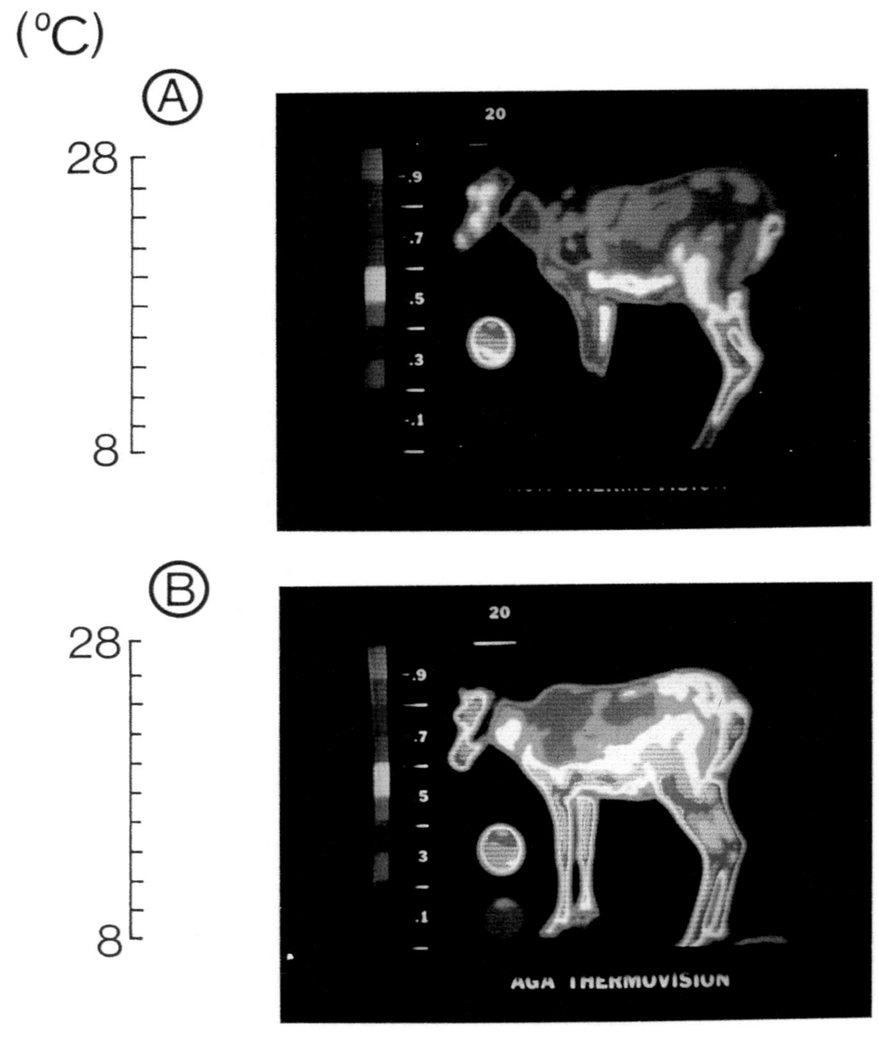

Figure 10.8 Typical radiative surface temperature as obtained from a reindeer by use of AGA Thermovision IR camera (A) before and (B) immediately after a 45-min period of running at 9.2 km.hr^{-1} on a level treadmill at a T_a *of + 2* °C in summer. The temperature scale is indicated to the left and the corresponding isotherms are shown in the pictures. Two black bodies, which were heated separately, were used as reference temperatures (shown in the lower left corner of the pictures) (Johnsen, Blix, Jørgensen & Mercer, 1985).

be near freezing temperature while the animals are at rest in cold air (Irving & Krog, 1955). The anatomical arrangement of the heat exchanger may vary from species to species, but the general structure consists of arteries running centrally

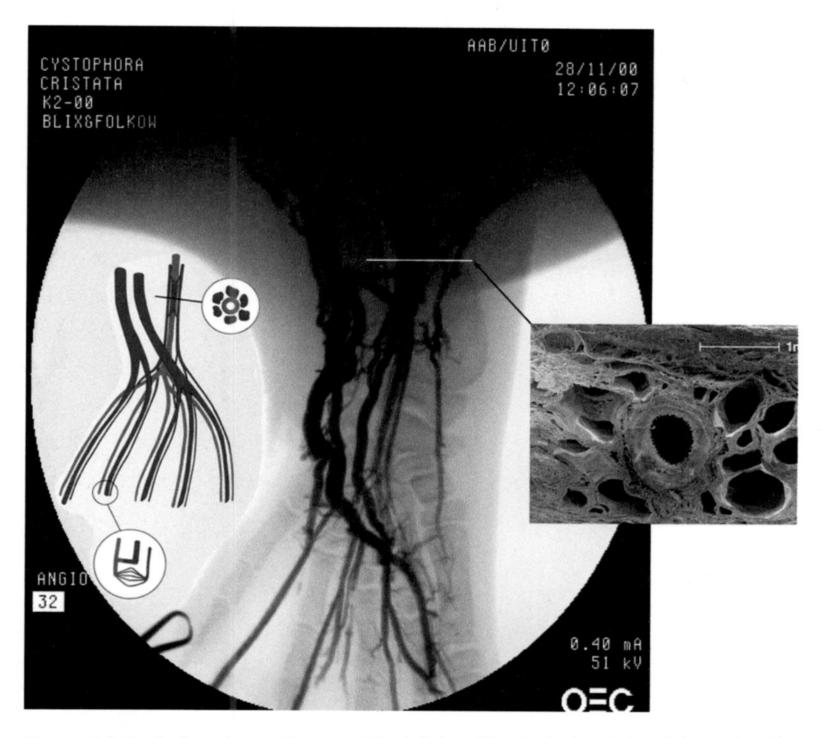

Figure 10.9 *Post mortem* angiogram of the left fore-flipper of a hooded seal (*centre*), with a scanning electron micrograph (*right*), showing how the deep internal veins are arranged around the artery for counter-current heat exchange (scale bar = 1 mm). *Left*: a schematic drawing of the flipper vasculature, indicating that the nutritive blood flow which reaches the flipper leaves by way of the deep internal veins and exchange heat with the incoming arterial blood, after having perfused the capillaries of the tissue, to save heat in the cold. When the animal, instead, wants to cool its brain during diving, or cool off in general, it may route the now much increased blood flow to the flippers through A-V shunts in the skin and return the cold blood by way of the large superficial veins, shown to the left in the angiogram, and thus, avoid the counter-current heat exchange in the periphery (Blix & Folkow, unpublished).

within a trabeculate venous channel. The result is two concentric conduits, where the warm arterial blood is cooled by the venous blood which has been chilled in the legs or flippers. But, the blood supply to the legs can also be routed through parallel veins bypassing the heat exchangers and returning to the trunk through superficial venous channels (Fig. 10.9). Effective vasomotor control allows variation in the distribution of blood flow between these pathways, so that the legs can function as heat dissipators or conservers, depending on the heat load on the animal. So, heat flow to the feet is minimized and results in a low foot temperature and reduced heat loss to the environment. Accordingly, Meng, West & Irving

(1969) have found that unsaturated fats of low melting points are selectively deposited in the distal parts of the extremities of a variety of arctic mammals, and they later demonstrated that the concentration of oleic acid ($C_{18=1}$) increased almost 50 % from the base to the tip of a caribou leg, while the concentration of palmic acid (C_{16}) dropped by about the same amount.

How the foot temperature is maintained near 0 °C even when animals are standing on substrates 50 °C or more colder, however, was not accounted for until a study by Henshaw, Underwood & Casey (1972) in arctic wolves and foxes at Barrow, Alaska. These workers discovered that these animals possess a cutaneous vascular plexus in their paw and toe pads which receives warm blood through four unbranched arteries, and by which foot skin temperature can effectively be regulated to within a degree of freezing. This, of course, begs the question of how the peripheral nerves can operate to attain this degree of sensory and vascular control at such low temperatures? Apparently, this works just fine, since Miller (1965), and others, have shown that the conduction velocity of peripheral nerves from a variety of arctic mammals is much reduced at very low temperatures, but still able to conduct, even when supercooled to -6 °C. In fact, it appears that this is also the case in temperate zone animals.

In addition to excellent insulation by fur and vascular control of skin surface temperature, arctic mammals and birds defend themselves against cold by restricting *evaporative* heat and water loss from the respiratory tract. This is very important when energy is a short commodity, since, for instance, in humans exposed to low ambient temperature , but warmly dressed and at rest, the heat lost in the exhaled air may account for more than 20 % of their energy expenditure. The reasons why this is not the case in arctic mammals and birds have been studied in detail in reindeer and seals in our laboratory over the last 20 years.

The basis for heat and water conservation in the nose is counter-current heat exchange in the nasal cavity, as pointed out by Jackson & Schmidt-Nielsen (1964), some 40 years ago. In reindeer this is organized as follows: Their nasal cavities are filled with an elaborate system of scrolled structures (conchae) (Fig. 10.10), which are coated with a richly vascularized mucosal layer. Cold air which the animal inhales passes over the warm mucosa and, as a result, is heated to body temperature and saturated with water vapour on its way to the lungs. The nasal mucosa is cooled as a result and stays cool while the air remains in the lungs. When the animal exhales, the warm, humid air passes down a temperature gradient as it flows over the cold mucosa and is therefore cooled, and the water vapour therefore condensed. The end result is that the animal expire cold and "dry" air and heat and water are saved for the animal.

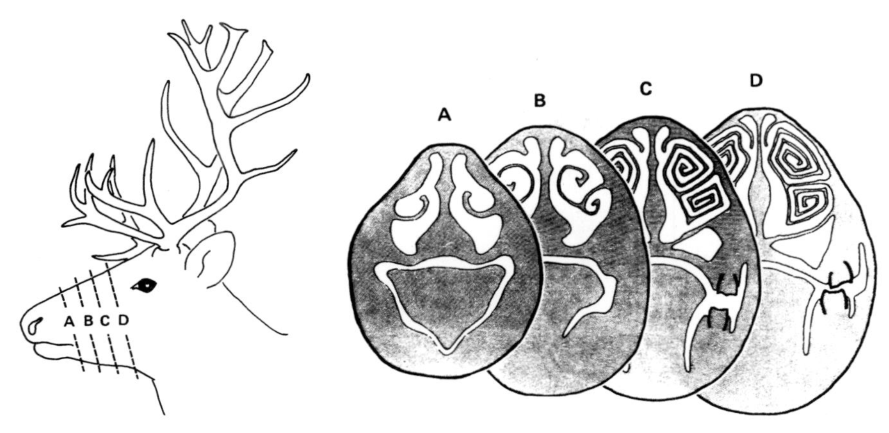

Figure 10.10 Cross sections of the reindeer nose obtained at four different levels (A, B, C, and D). Sections were made approximately at 1 cm intervals and illustrate the elaborate organization of the maxilloturbinates which project into each nasal cavity from the lateral wall (Johnsen, 1988).

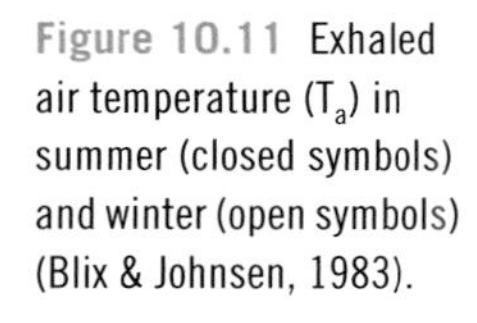

Figure 10.11 Exhaled air temperature (T_a) in summer (closed symbols) and winter (open symbols) (Blix & Johnsen, 1983).

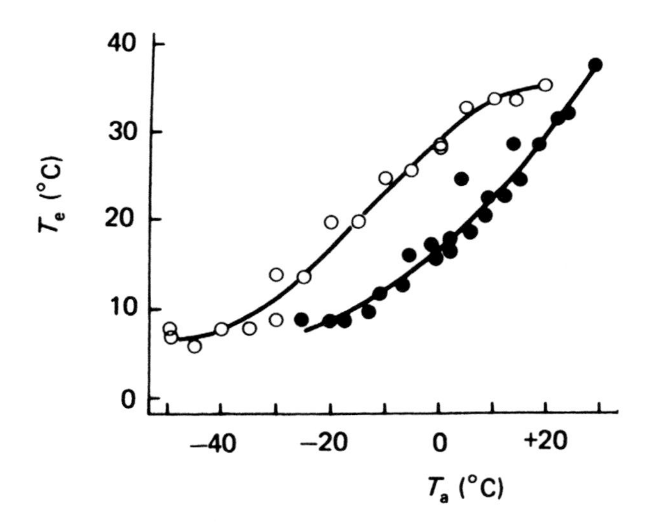

The concae of the nasal cavity both increase the surface area over which heat exchange can occur and divide the air flow into thin layers. Consequently, the distance between the centre of the air stream and the mucosal surface is reduced to a minimum. Both factors, large surface area and short distance between adjacent lamellae, promote rapid transfer of energy between the mucosal layer and the air stream. Nasal heat exchange is nevertheless not a passive process in reindeer. On the contrary, it is carefully regulated, enabling the animals to maintain thermal balance at widely varying ambient temperatures and work loads despite substantial seasonal changes in their fur insulation.

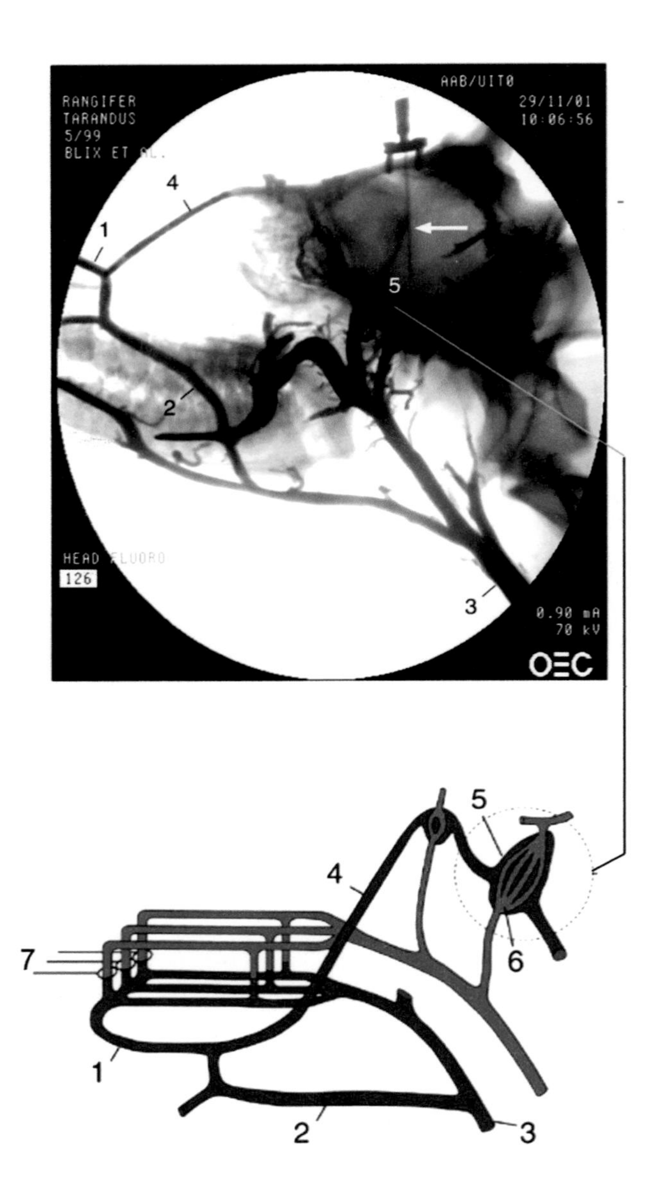

Figure 10.12 Heat and water conservation by nasal heat exchange in reindeer. *Top*: *Post mortem* angiogram of a reindeer head (snout; left) showing: 1: dorsal nasal vein; 2: facial vein; 3: jugular vein; 4: angular oculi vein; 5: cavernous sinus at the base of the brain; 6: carotid *rete*; 7: capillaries and A-V shunts in nasal mucosa. Arrow point to a re-entrant steel tube for measurements of brain temperature. *Bottom*: Simplified diagram of arterial and venous vasculature of reindeer nose. During heat conservation in the cold, blood runs counter-current in the *retia* of the nasal mucosa (left) and leaves through the sphenopalatine group of veins for the jugular vein (3), with the dorsal nasal vein (1) constricted. The efficiency of the nasal heat and water exchange will depend on blood flow to the nasal mucosa, the number of open A-V shunts, and respiratory rate (Redrawn from: Johnsen, Blix, Jørgensen & Mercer, 1985).

The first evidence for this came from my own laboratory (Blix & Johnsen, 1983), Fig. 10.11 showing that at any given ambient temperature (T_a) the expired air temperature of the reindeer is consistently lower in summer than in winter. The animals are evidently capable of adjusting the level of thermal exchange to conserve more heat at one and the same Ta in summer when they are relatively poorly insulated compared to in winter when their total body insulation

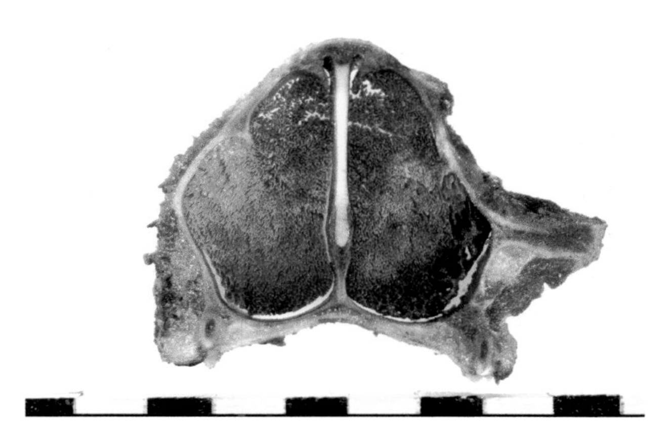

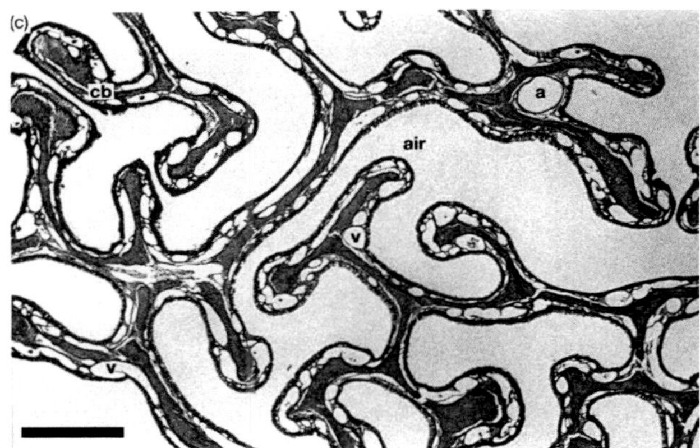

Figure 10.13 *Top*: Cross-section through the nose, just in front of the eyes, in a grey seal (*Halichoerus grypus*), illustrating the dense nasal conchae through which the air passes in respiration. (scale: cm intervals). *Bottom*: The nasal conchae at higher magnification (Scale bar, lower left; = 500μm); showing the conchal tissue (cb) in relation to the air spaces (air). An artery (a) and a large number of veins (v) are visible (Folkow, Blix & Eide, 1988).

is greatest. We subsequently showed that nasal heat exchange is regulated by vasomotor adjustments in the nasal mucosa and that those changes are under control by the thermoregulatory "center" in the mid-brain of the animal. The way this is done is explained in Fig. 10.12.

In seals, which we have also studied in detail, the same principles apply. In pinnipeds, however, the anatomical arrangement in the nasal passages are different from those of the ungulates, in that the conchae are more developed into a highly convoluted mass, reminiscent of the radiator of cars, usually with an air space of less than 1 mm between the lamellae (Fig. 10.13) (Folkow, Blix & Eide, 1988). This arrangement allows 66 % of the heat and no less than 80 % of the water added to the inspired air to be regained on expiration at an ambient temperaure of -20 °C, and we (Folkow & Blix, 1989) have shown that the water which is saved amounts to 10-30 % (depending on ambient temperature) of the total water flux of the animal. Keeping in mind that seals live in salt sea-water, and in spite of the fact that these animals do drink sea-water (Skalstad & Nordøy, 2002) they have little net gain of water from it, and it is quite likely that this mechanism has more value to marine animals in water conservation than in their management of the heat budget.

Now, with falling ambient temperature all the thermoregulatory means we have discussed above gradually become activated with the result that the heat loss from the animal is reduced and body temperature is maintained without any increase in heat production above what is termed the *resting metabolic rate*. This level of heat production, which is close to the lowest level needed to maintain cellular integrity in the resting animal, is characteristic of the species, but changes with season, age and other occurrences. The range of ambient temperatures over which the animal is able to maintain a constant body temperature without any increase in metabolic rate is termed the *thermoneutral zone*, and the ambient temperature at which this is no longer possible, the *lower critical temperature* (T_{lc}). Below the lower critical temperature (which is not all that critical, by the way), metabolic rate increases linearly with decreasing temperature, and the slope of the line will reflect the total conductance of the animal with all its heat conserving mechanisms at work. If we want to compare the thermal competence of a variety of animals, we can do so by assigning the value of 100 % to their different normal resting metabolic rates. This was done by Scholander *et al.* (1950b) and is presented in Fig. 10.14, which clearly shows that tropical animals have lower critical temperatures which are much higher than those of arctic animals, and their met-

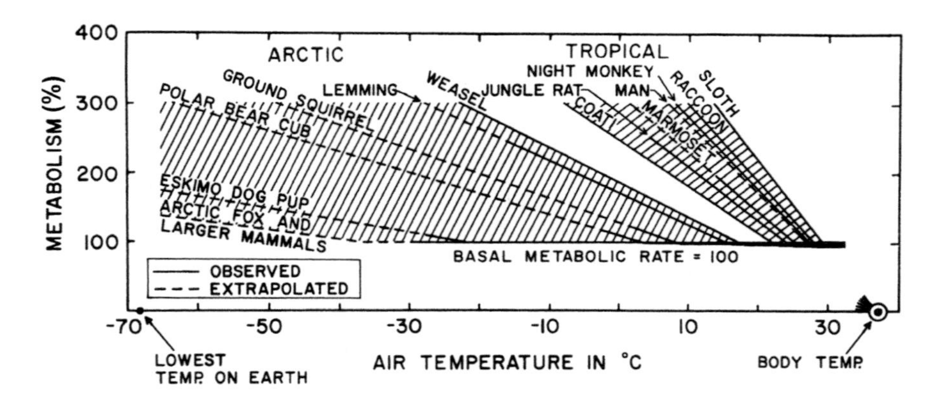

Figure 10.14 The metabolic rates of various mammals in relation to ambient air temperature. The normal resting metabolic rate for each animal, in the absence of cold stress (thermoneutral zone), is given the value 100 %. Any increase at lower temperatures is expressed in relation to this normalized value, making it possible to compare widely differing animals (Scholander *et al.* 1950b).

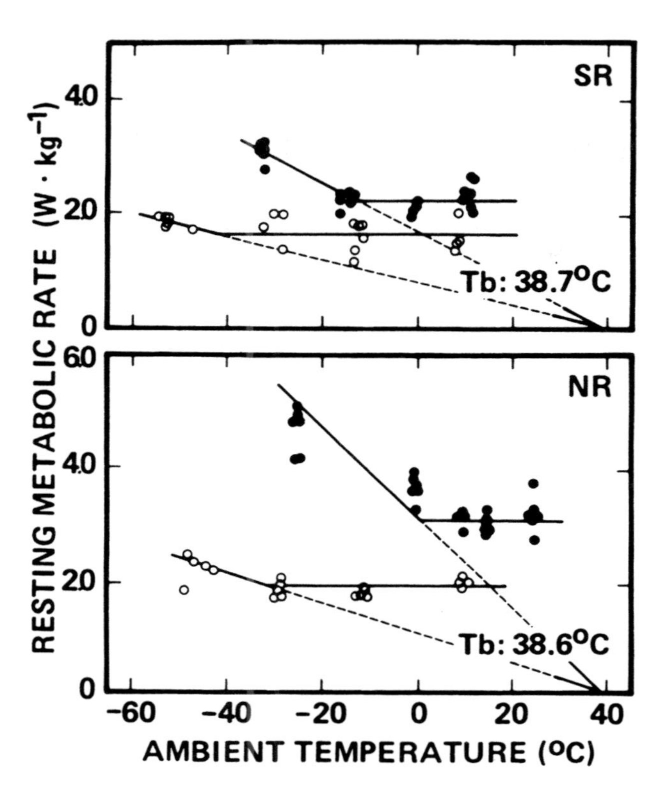

Figure 10.15 Resting metabolic rate at different ambient temperatures during summer (closed symbols) and winter (open symbols) in adult Svalbard (SR) and Norwegian (NR) reindeer fed *ad libitum*. Lines are fitted by eye to extrapolate through ambient temperature equal to deep body temperature (T_b) (Nilssen, Sundsfjord & Blix, 1984).

abolic rates increase much more rapidly below their lower critical temperatures, indicating that tropical animals, as a rule, have high conductances compared with the arctic forms.

For example, we have shown that the Svalbard reindeer is so well insulated that, while its lower critical temperature appears to be as low as -40 °C, metabolic heat production is still insignificantly increased at -50 °C (Fig. 10.15), which is the lowest ambient temperature this species has ever experienced in historic times. In summer, however, lower critical temperature of this animal increases to a little less than -20 °C, partly due to shedding of the winter fur, and partly due to an increased resting metabolic rate, which will cause the intersect with the line that describes its heat loss relative to changing air temperature to occur at a lower ambient temperature (Nilssen, Sundsfjord & Blix, 1984).

It follows, from all this that the lower critical temperature and the total conductance of an animal, as determined in the laboratory, is a good indication of the thermal adaptation of the species. But, the lower critical temperature so determined should by no means be taken as a measure of when an animal in the field must increase its heat production to keep warm! We have already learned that wind and precipitation may drastically increase the conductance of the animal, and Øritsland & Ronald (1973) have shown that solar radiation may very significantly contribute to the warming of the animal. These workers have shown that particularly white hairs transmit much of the visible and near-infra-red radiation from the sun down through the pelt, where it is absorbed by the dark skin. The transmittance through single hairs is reduced by pigmentation and air-filled pores, and the effect of solar radiation is negatively correlated with the coat reflectance. In any case, solar radiation may cause skin temperatures well above the deep body temperature in animals such as seals and polar bears, and the heat that is emitted from this warm skin then becomes trapped by the white hairs, warming the layer of insulating air in the pelt and thereby contributing to a reduced conductance of the animal.

As a side-show to these studies Lavigne & Øritsland (1974) discovered that when polar bears and some seals were photographed through quartz lenses (which transmits near-UV light) the animals appeared black on the white backdrop of the snow, since the UV light is reflected from the snow, but absorbed in the pelts of these animals. This technique, which vastly improved the chances of discovering polar bears during population censuses made it into "Nature" and naturally made quite a splash. Unfortunately, this very nice story has later been perverted by others into the myth that polar bears are efficient absorbers of the UV component of sunlight because they have hairs reminicent of quartz fiber and that UV light very significantly contributes to their energy budgets. This version, which has even made it into "Scientific American" (Mirsky, 1988), is all wrong. In fact, UV contributes little if anything to thermoregulation in polar bears or any other arctic mammals, while the visible part of the solar spectrum does, as outlined above.

Besides minimizing heat loss in the cold by means of reduced thermal conductance, arctic animals can reduce energy expenditure by reducing their total daily locomotor activity. This has been examined in detail in reindeer in our laboratory at Tromsø and in caribou at Fairbanks by White and associates (for references, see: Fancy & White, 1985). The daily energy cost of locomotion is influenced by three variables: the total distance which the animals travel per day, the overall distance which they climb and the nature of the surface over which they travel. Two other variables, running speed and the angle of ascent, have less significance, first, because the weight-specific net cost of climbing in ungulates appears to be almost independent of the angel of ascent, at least on moderate slopes, and, second, because in reindeer and caribou, like other terrestrial mammals, the relationship between running speed and the rate of energy expenditure is effectively linear (Fig. 10.16). Consequently, the net cost of travelling a given distance is, for an animal of a given body weight, largely independent of the speed at which it moves (Nilssen, Johnsen, Rognmo & Blix, 1984).

The nature of the surface over which the animal travels, on the other hand, has a very important influence on the cost of locomotion. In reindeer, for instance, 30 % more energy is expended when walking on wet tundra compared to on hard-packed roads (White & Yousef, 1978). The costs of walking across soft or crusted snow are even higher and rise exponentially as the animals sink deeper (Fig. 10.17). In one case, in which a caribou sank to 60 % of brisket height

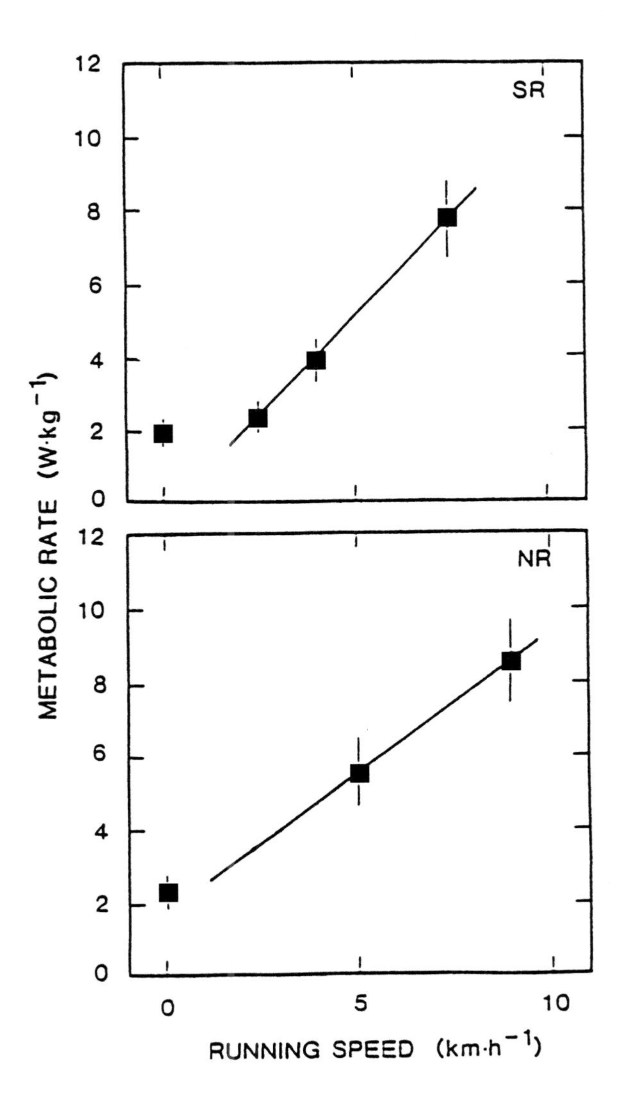

Figure 10.16
Relationship between metabolic rate and running speed in Svalbard reindeer (SR, *top*) and Norwegian reindeer (NR, *bottom*) (Redrawn from: Nilssen, Johnsen, Rognmo & Blix, 1984).

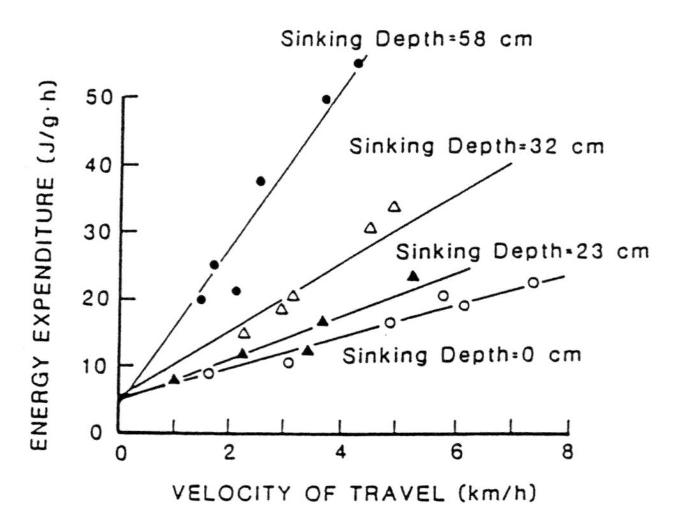

Figure 10.17 Effect of sinking depth on the energy cost of walking in snow for a 100-kg elk calf (Parker, 1983).

at each step, the relative net cost of locomotion increased almost six times (Fancy & White, 1985a). The capacity of snow to support an animal depends on the hardness of the snow and the pressure (foot load) which the animal exerts on it. Thus, if snow hardness consistently exceeds foot loads, animals can walk on top of the snow or will sink to only a fraction of its total depth. The broad, spreading feet of reindeer and caribou, which produces a noticeable click when they are put down, a well-known characteristic of this species, is clearly an adaptation to walking on snow, through minimising the extent to which they break through the crust and sink in.

The Svalbard reindeer appear to be able to reduce their activity to a level at which the total daily energy cost of locomotion becomes almost negligible at the level of only about 2 % of their daily energy expenditure. These animals are typically sedentary, walking, on average, less than 700 m per day in winter and rarely run at all unless provoked. We have also found that they maintain very low levels of thyroid hormones during winter, but this does not affect their metabolic rate and although both the reindeer and its compatriot, the Svalbard ptarmigan, show a rather lethargic behaviour, they are always alert and maintain a constant body temperature throughout the year. I have therefore suggested that the term "Arctic resignation" is best in describing the state of mind of these intriguing creatures (Blix, 1989). We will later discuss the physiological back-up behind this behaviour.

I have already mentioned that the net energy cost for a reindeer of a given body weight moving from one place to another is by and large independent of the speed at which it travels. This begs the question: Why don't reindeer always run?

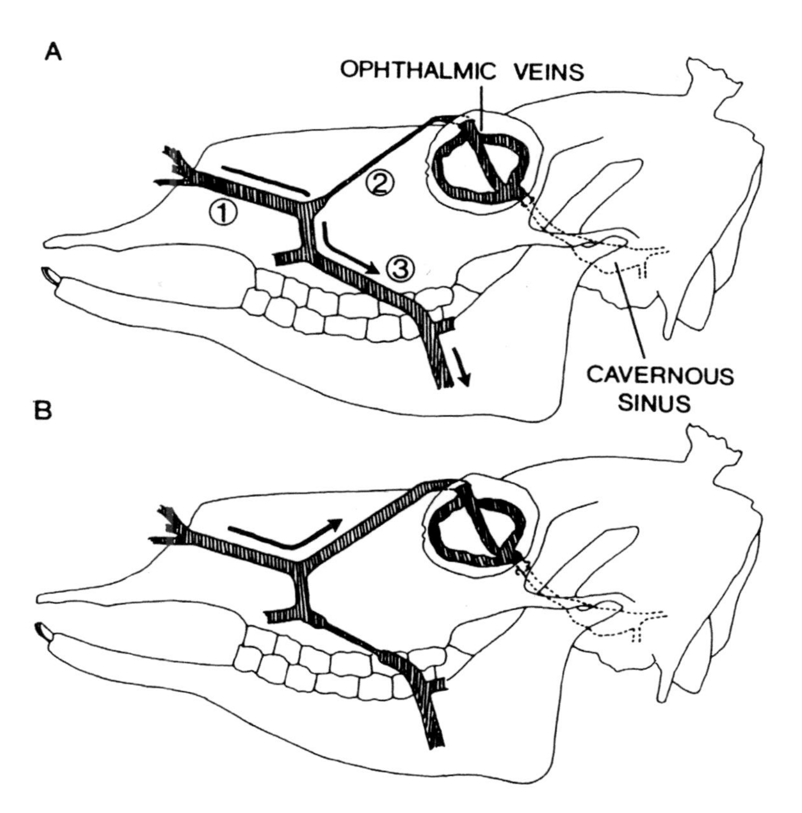

Figure 10.18 Schematic diagram of superficial veins of nose and their connections with the cavernous sinus at the base of the brain in relation to the outline of the skull in reindeer. Figures illustrate the proposed mechanism for distribution of cold blood from nose of hyperthermic reindeer. A: operation of mechanism during moderate heat stress, where cold venous effluent from the mucosal lining of the nasal surfaces returns to the caval veins via the facial veins (arrow), thus by-passing the cavernous sinus. It is suggested that the direction of venous return through this pathway is due to sympathetic stimulation of the angular oculi and facial veins, which results in release of pressure-induced inherent myogenic tone in the facial veins and simulataneous constriction of angular oculi veins. In contrast, when body temperature exceeds the threshold value for onset of brain cooling, sympathetic activity to the veins in question is reduced. This will result in dilatation of the angular oculi veins and activation of inherent myogenic tone in the facial veins (B). In this situation cold blood returning from the nose is directed mainly to the cavernous sinus for selective cooling of the brain (arrow); 1: dorsal nasal vein; 2: angular oculi vein; 3: facial vein (Johnsen & Folkow, 1988).

The answer is that the insulation which enables them to survive under extremely cold conditions predisposes them to heat stress if they exercise. In fact, heat production increases very rapidly with increasing running speed, and we have shown that the rate of heat production in Svalbard reindeer trotting at 8 km/h, for example, already is 4 times higher than the rate of heat production when standing

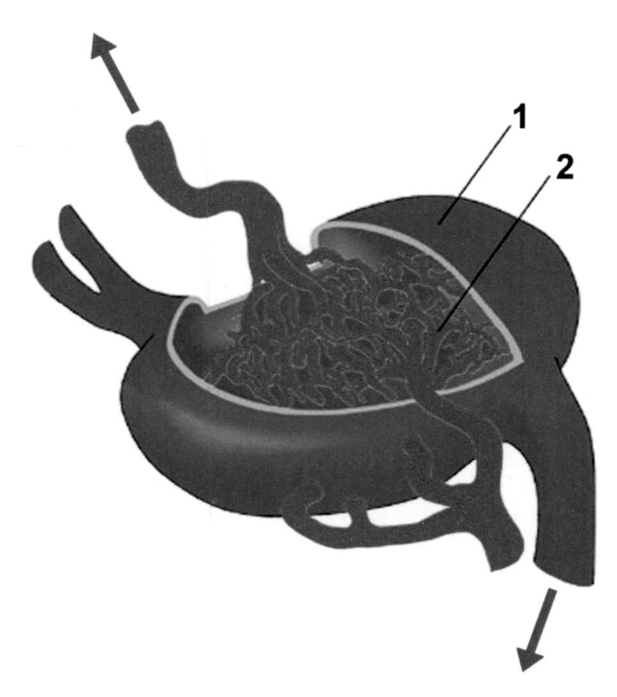

Figure 10.19 Carotid *rete* of a sheep exposed in a cutaway drawing. This structure lies at the base of the brain. Cool venous blood, draining from the nose, enters an enlargement of the venous tributaries called the cavernous sinus (1). Here it bathes a network of small arteries called the carotid *rete* (2), which is inserted into the carotid artery, and leaves via veins that enter the internal jugular vein at the right. At the same time warmer arterial blood arriving from the heart enters the *rete* via branches of the external carotid artery and is subsequently cooled by the venous effluent before continuing up to the brain (Redrawn from Baker, 1979).

(Nilssen *et al.*, 1984). Hot reindeer can, as we already know, increase heat loss both by peripheral vasodilatation and by panting, much the same way as seen in dogs. Nevertheless, if an animal is chased by a predator, or a snow-scooter, and runs hard during winter, it can soon reach a stage at which it produces heat more rapidly than it can lose it, with the consequence that body temperature rises. It would, of course, be advantageous, in terms of survival, to be able to prolong the length of time it can maintain a high speed, but to do so requires that thermally sensitive tissues, such as the brain, are protected when heat is stored and body temperature increases.

Several species in the order *Artiodactyla*, including reindeer, have developed a system which enables them selectively to cool their brain tissue independently of the rest of the body core. The principle of selective brain cooling, as we have discovered it, is illustrated in Figs. 10.12 & 10.18. During mild heat-stress, venous blood, cooled at the evaporative surfaces of the nasal *conchae*, flows via the facial veins directly to the caval veins and is used for general body cooling. If heat stress becomes more severe, however, the facial vein is closed off and the cooled blood is directed, instead, through the *angularis oculi* veins to a venous sinus at the base of the brain (Fig. 10.19). Here heat is exchanged with warm blood in the carotid artery, into which a *rete* is inserted to facilitate heat transfer. The result of this is that the brain is cooled selectively while heat is stored (as an increase in temperature) in the rest of the body – to be dissipated subsequently when the stress is past.

The polar bear is not fortunate enough to have any brain cooling mechanisms, but then again until man rather recently appeared with helicopters and snowscooters this animal did not have any enemies that could cause it to run, and, in any case, polar bears can normally remedy the situation by jumping into ice-water, which normally is readily at hand. Much worse is it that Hurst *et al.* (1982) have demonstrated that the metabolic cost of locomotion in this species is unusually high, in fact, twice that predicted by the general equation for quadruped locomotion by Taylor, Schmidt-Nielsen & Raab (1970). They also found that equilibrium deep body temperature increased exponentially with speed of locomotion and that the animals, as we would expect, were relatively unable to

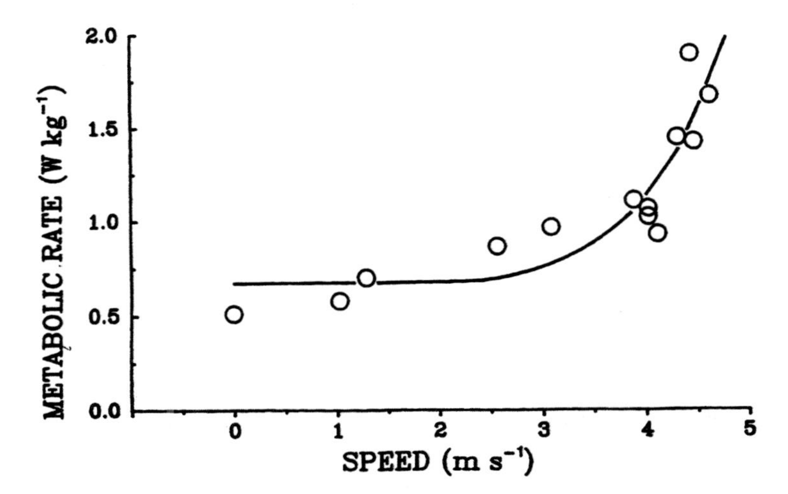

Figure 10.20 Energy expenditure at different swimming speeds in four ~ 4,000 kg minke whales which were instrumented with sonic swim-speed and dive-depth transmitters. The zero value is the calculated minimum average total rate of heat loss of minke whales of similar size in cold water (Folkow and Blix, 1992). Optimum speed was calculated to be 3.25 m s^{-1} (Blix & Folkow, 1995).

dissipate heat at high walking speeds. This inefficiency of polar bear locomotion is probably due to the animal's heavy legs and paws and the sideways motion which is typical of their gait, and make these animals prone to overheating when chased by man.

Whales are much more fortunate when it comes to the costs and thermal consequences of locomotion. First, we can safely assume that any additional heat load that is caused by swimming can effectively be dumped into the cold oceans, in which the arctic whales live, by cutaneous vasodilatation (Folkow & Blix, 1992). Second, we found in the minke whale that the cost of swimming, and hence, daily energy expenditure is remarkably low (Blix & Folkow, 1995). In fact, metabolic rate only doubled from the value at rest to a swimming speed of about 4 m/sec, which as it happens is the commonly observed (upper) cruising speed of these whales. At swimming speeds above this, however, metabolic rate increased exponentially (Fig. 10.20).

Birth in the arctic – a chilling experience

Although different polar species give birth at different times under various thermal conditions, their offspring still have some features in common: they are much smaller and have a far larger exposed surface area relative to body weight than the adults. The newborns also have less insulation, especially at the wetted moments of birth, or hatching, and the lower critical temperature therefore is far higher than in the well-insulated grown animals. By what mechanisms, physiologically or otherwise, are these thermally under-equipped young forms able to survive their debut to life in polar regions?

The offspring of both mammals and birds can roughly be classified as either precocious or altricial (e.g. Blix & Steen, 1979).

Altricial mammals

Altricial offspring are born in a very immature and helpless condition, they are very small relative to their parents, blind and fairly naked and are thermally dependent on their parents for a variable period of time. Mammals that deliver altricial offspring usually produce a litter of young in a den, cave or nest.

The polar bears may be used as an example of animals which produce altricial young : They give birth to one to three cubs in late December or early January, when the temperature at the high arctic coasts frequently drops to -40 °C. At birth the cubs weigh only 600-800 grams (!). They are blind, lack both fur and blubber insulation and any shivering capability is unlikely (Fig. 10.21). Despite the rela-

Figure 10.21 Polar bear cub, born 27 December, aged only 2 hours. Nose to tail length is 34 cm and weight is 637 grams. At birth these cubs have hardly any fur, and they are blind and helpless. An American quarter dollar coin is shown for scale (Photo: A.S. Blix).

tively mild microclimate of the den, which Blix & Lentfer (1979) have shown to have a temperature close to freezing, survival of the cubs depends on protection provided by maternal sheltering. According to Kost'yan (1954), the female curls up, grasps her hindpaws with her forepaws, and presses the cubs to her nipples with her heavily furred legs. Under normal conditions the cubs grow rapidly; when the den is deserted after 3 months of constant occupation, they have aquired a weight of 10 kg and relatively thick fur insulation (Fig. 8.37). Resting metabolic rate is unusually high and lower critical temperature is -30 °C (Fig. 10.22). At this stage in life very little subcutaneous fat is present and no brown fat (see below) is found. When the cub is immersed in ice water, we found that the wettable pelt loses most of its insulative value and that the animal rapidly becomes hypothermic and dies if it does not get out of the water (Blix & Lentfer, 1979).

Arctic wolf and fox, which, likewise give birth to litters of undeveloped young in late May and June, use a cave in the ground to much the same effect as the polar bear den, and ermine, which give birth in summer, and lemmings, which give birth both in summer and winter, use a nest made out of straw under the snow, where ground heat creates a fairly comfortable microclimate.

At birth, which in the lemming might take place even in midwinter, these animals, which weigh only about 5 grams, are essentially poikilothermic and completely dependent on maternal body heat. In her temporal absence, however, the supreme survival factor in these, and many other altricial young , is profound tolerance to hypothermia. In fact, Østbye (1965) cooled a number of newborn

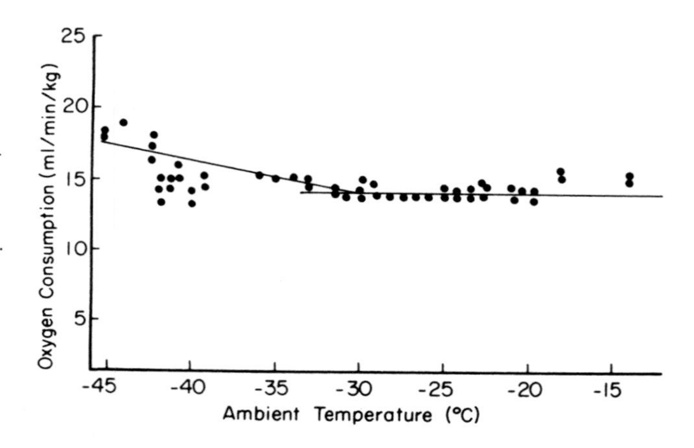

Figure 10.22 Oxygen consumption of resting polar bear cub, aged approximately 3 mo, at different ambient temperatures. Weight of the cub was 12.5 kg (Redrawn from Blix & Lentfer, 1979).

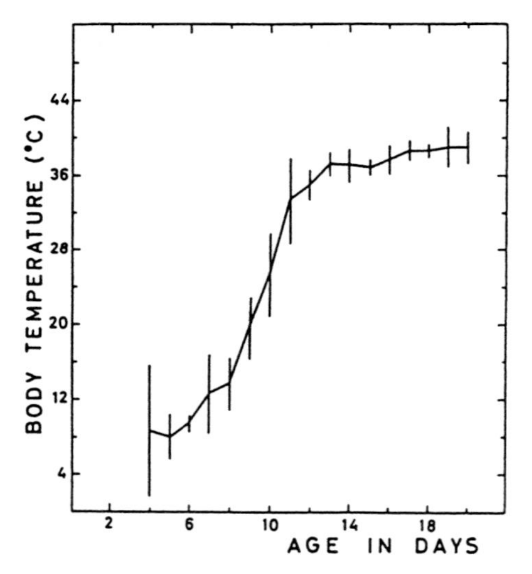

Figure 10.23 Body temperatures of young Norwegian lemmings after 30 minutes exposure to air temperatures of 2 °C (90 measurements of 18 animals; means and SEM) (Østby, 1965).

lemmings *(Lemmus lemmus)* to deep-body temperatures of 2-5 °C (Fig. 10.23) without observing any mortality or ill effects if the young were subsequently brought back to the mother in the nest. The progressively increasing ability to thermoregulate in this species is reflected in increased ability to shiver, improved insulation, growing size, and development of brown (thermogenic) fat.

Young and more or less thermally dependant littermates usually make use of their numbers and huddle when they at intervals are left alone for one reason or the other. The advantage of huddling, beside its cozy aspects, is that the common surface area of the group is much reduced, while the common heat producing mass is the same, and the heat loss from each and every individual is therefore much reduced and the drop in body temperature much retarded.

The arctic hare, which gives birth in June and July produces a litter of, so called, leverets, which are somewhere between altricial and precoccial (see below). The newborn have a good fur insulation and the female does not invest in a nest. After three days the leverets are thermally independent and are said by Pedersen (1934) to huddle in a rosette pattern when exposed to cold or danger. The physiological basis for their apparent tolerance to cold is not known, but fur insulation, brown (thermogenic) fat and huddling are probably the three most important factors.

Altricial birds

Among arctic birds the distinction between altricial and precocious hatchlings is often difficult to make. The typical altricial bird has virtually no power of thermoregulation at hatching and relies instead on brooding by their parents and huddling with nest mates. By such measures their body temperatures are usually maintained above 34 °C, but hardly at the adult level for the first few days after hatching.

Ravens are the earliest breeders of all truly arctic birds. In northern Greenland they have been observed incubating eggs as early as the 6th of April, when ambient temperatures down towards -30 °C may occur and temperatures far below freezing are common. The eggs are continuously brooded by the female, which is fed by her attentive spouse. No information about the thermal tolerance of these originally naked and helpless chicks is available.

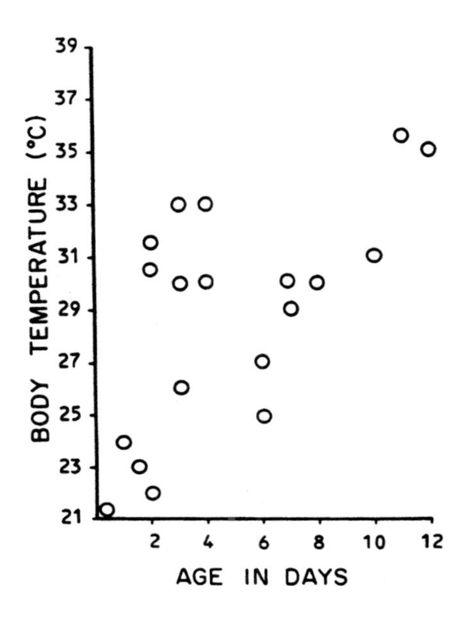

Figure 10.24 Effect of brooding on body temperature in snowy owl chicks. Field observations at ambient temperatures ranging from 5.5. to 10.5 °C. Body temperature after 5-15 min exposure without brooding (Redrawn from: Barth, 1949).

The snowy owls are also early breeders and their chicks better known. The eggs hatch in June and the chicks look like balls of grey down. At first they huddle when exposed to cold, but they appear to attain thermal independence at the age of 3-4 weeks. The body temperature of these chicks has been recorded daily by Barth (1949) from hatching to the age of 12 days (Fig. 10.24). Body temperature in these chicks clearly depends on brooding, and recently hatched birds have little capacity for thermoregulation. Deep and prolonged hypothermia, however, seems to be tolerated without ill effects, just like in lemmings.

Precocious birds

When it comes to birds that produce precocious (i.e. those that are hatched in a far developed state) young, they all hatch their young during the hectic arctic summer and the eggs and chicks are, of course, exposed to the same elements and hardships, whether their parents are residents or only summer visitors to the region.

Thus, precocious avian offspring are found among waterfowl, gulls, waders, and terrestrial birds. Newly hatched eider (*Somateria mollissima*) ducklings are able to maintain homeothermy at an ambient temperature of -7 °C, whereas 7-day old birds can tolerate -16 °C. This rapid development of cold resistance is related to increased metabolic capacity and insulation. One day old herring gulls *(Larus argentatus)*, on the other hand, are unable, even for short periods, to maintain normal body temperature when exposed to 10 °C, but exhibit high tolerance to hypothermia. A high tolerance to hypothermia is likewise reported for other species of gulls. Newly hatched murres (*Uria aalge inornata* and *Uria lomvia arra)* in Alaska are virtually poikilothermic, but after 9 days, when they still weigh less than 20 % of the adult weight, they can tolerate 2 °C for 1 h without a noticeable reduction of body temperature. Hatchlings of ducks, gulls, and murres apparently share the features of prime down insulation and high metabolic capacities. Consequently they develop homeothermy at an early age; ducklings almost immediately, murres and gulls after 1 and 2 weeks, respectively (e.g. Blix & Steen, 1979).

Among terrestrial precocious birds, rock ptarmigan and willow ptarmigan have been studied most. The hatchlings of both are typically precocious but more dependent on intermittent brooding than eider ducklings. We have studied the willow ptarmigan chick extensively at Tromsø, and we have found that they usually will return to their mothers when their body temperature approaches 35 °C, and even though they have a substantial thermogenic capacity and may be able to

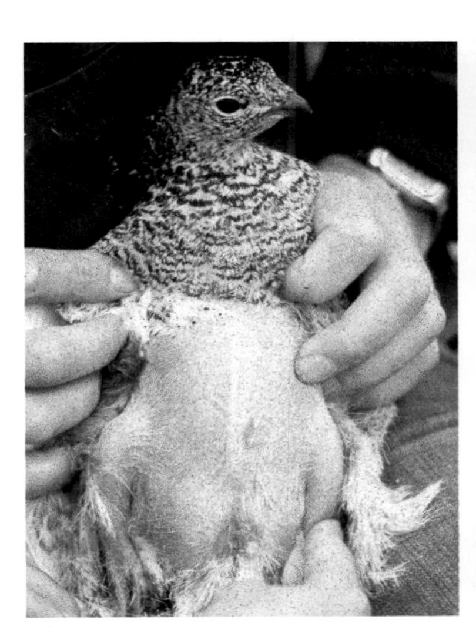

Figure 10.25 Incubation of the eggs in willow ptarmigan. *Left*: The bird plucks the feathers from her breast to create a brood patch during the incubation period. *Right*: Huge blood vessels lead warm blood to the skin of the brood patch to warm the eggs (Photos: G.W. Gabrielsen).

maintain body temperature for a while, they do not utilize their full capacity to maintain body temperature. Apparently, they do instead use it to restrict body cooling, the rate of which depends on the ambient temperature, whereby feeding time will be less dependent on ambient temperature. Indeed, the time spent on feeding excursions by 8 day old chicks under controlled conditions in the laboratory was the same at ambient temperatures of 2 and 12 °C, and the drop in body temperature was not significantly different at the two ambient temperatures (Jørgensen & Blix, 1988). This strategy is, of course, only possible when there is heat by brooding to be had at the end of the excursions, and the ptarmigan hen is, indeed, well endowed to provide just that (Fig. 10.25)!

We have also found that the chicks had relatively good tolerance to changes in the availability of food, which is the case, for instance, in bad weather. In such cases they simply returned immediately to the hen, while they were able to compensate with intense feeding activity when food was available. If the quality of the food was reduced, however, they were unable to compensate. It is likely that in such birds, which grow at an enormously rapid rate, the crop is filled at all times and in cases when the quality of the food is reduced compensation by overeating is impossible (Jørgensen & Blix, 1985). Therefore the availability of high quality food seems to be crucial for their survival, while low ambient temperature *per se* seems to be of little consequence. Willow ptarmigan hens are consequently known to lead their brood to areas where, due to differences in plant phenology, high quality food is easily available. It is more than likely that these aspects of willow ptarmigan chick lifestyle also apply to the young of the high Arctic rock ptarmigan chick.

Thus, young precocious birds show different mechanisms of thermoregulation. Ducks and other aquatic birds rely on good insulation and high thermogenic capacity and maintain themselves homeothermic from an early age. Gulls and gallinaceous birds have dispensed with strict homeothermy and rely instead on intermittent maternal brooding and high tolerance to hypothermia.

Precocious mammals

Terrestrial forms

There are two terrestrial arctic mammals which give birth to typically precocious young : the reindeer / caribou and the muskoxen. The muskoxen gives birth in late April on the snow of the open barren tundras of Greenland and the Canadian North West Territories, when ambient temperature frequently drops to -35 °C, while reindeer and caribou give birth to their calves all over the Arctic tundra in May and June, when the pastures are normally still covered with snow.

Figure 10.26 The newborn muskox is blessed with an outstanding fur and can withstand birth at -40 °C (Photo: S.D. Mathiesen).

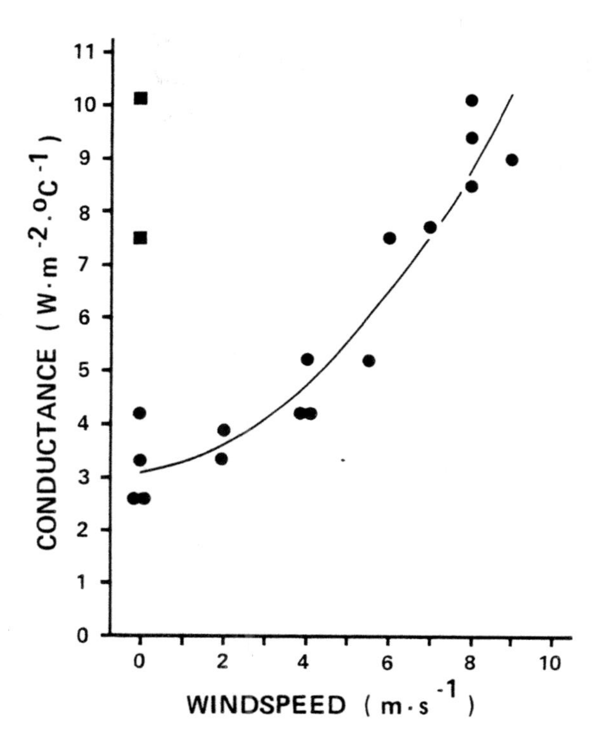

Figure 10.27 The conductance values (filled circles) for newborn muskoxen (back) pelt samples measured *in vitro* at different wind speeds. Included are also two values (filled rectangles) obtained for pelt samples wetted by ice-water. A conductance value of 31.3 $W \cdot m^{-2} \cdot °C^{-1}$ (not included in the figure) was obtained when the sample was permanently immersed in ice-water (Blix, Grav, Markussen & White, 1984).

At birth the reindeer calves weigh about 4-5 kg and are insulated in a pelt of air-filled hairs that offers prime insulation, and gives the animals a lower critical temperature of 10 °C, as long as it is dry. The newborn muskox calves weigh about 8 kg and are equipped with a qiviut pelt of probably unsurpassed insulative value (Fig. 10.26), resulting in a lower critical temperature of -10 °C, and less than a 30 % increase of metabolic rate at -30 °C. We have shown, however, that both these pelts lose much of their insulative power when exposed to wind (Fig. 10.27), and even worse, wind in combination with rain or sleet, which increases the conductance of the pelts some three times in the newborn muskoxen (Blix, Grav, Markussen & White, 1984), and some five times in the reindeer. The legs of newborn muskoxen are heavily furred and counter-current heat exchange is not in operation, subcutaneous temperature just above the hooves being almost as high as the subcutaneous temperatures on the back of the animals. Whether, or not, counter-current heat exchange occurs in the legs of newborn reindeer is presently unknown, but to be expected.

I have seen both reindeer and muskoxen calves shiver visibly just after birth, when they are still wet, but they are normally not seen shivering thereafter. In

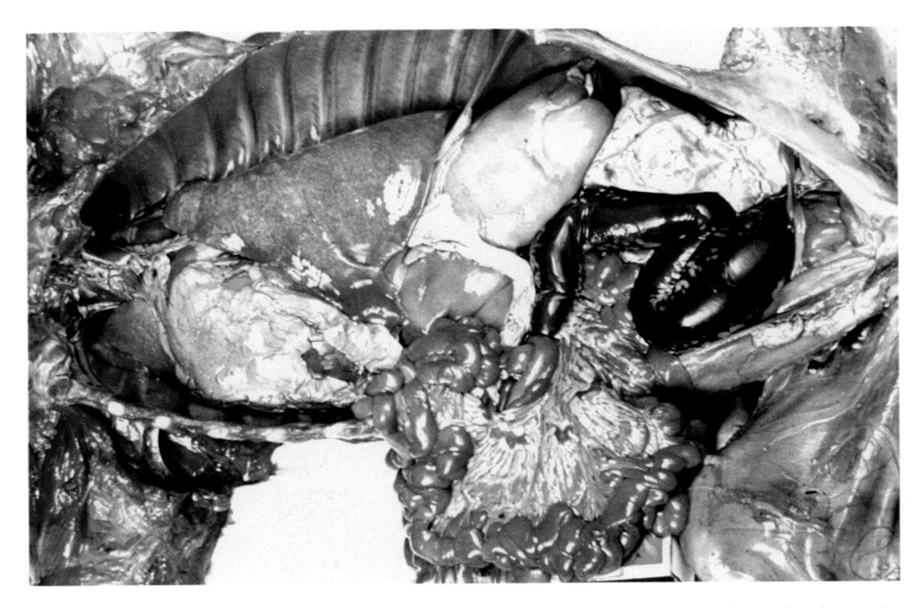

Figure 10.28 The thoracic (*left*) and abdominal (*right*) cavities of a newborn muskoxen showing vast amounts of highly thermogenic brown fat on and among internal organs (Blix, Grav, Markussen & White, 1984).

these animals deposits of thermogenic brown fat, which are large in reindeer calves and huge in the newborn muskoxen (Fig. 10.28) are instead the major sources of heat (Blix *et al.*, 1984).

Brown adipose tissue (BAT) is a tissue that produces heat, by so called non-shivering thermogenesis, when called upon to do so by thermal stresses imposed on the animal from outside. Heat production is aerobic and localized in numerous large mitochondria (Fig. 10.29). Its onset is elicited by noradrenaline, either released from sympathetic nerve endings in the tissue itself, or reaching the beta-adrenergic receptors in the tissue via the blood. BAT is also characterized by a multilocular distribution of its triglycerides, instead of the one big drop of fat in white adipose tissue. The reddish colour of BAT is obtained from a very rich vascularization and the high concentration of cytocromes in the numerous mitochondria.

In all tissues other than BAT, the rate of metabolism is determined by the rate of utilization of ATP, hence, by the rate of dissipation of the mitochondrial proton electrochemical gradient. BAT differs from other tissues in that it possesses an alternative mechanism for dissipating the proton electrochemical gradient. This is the proton conductance pathway, mediated by an "uncoupling protein" (UCP-1)

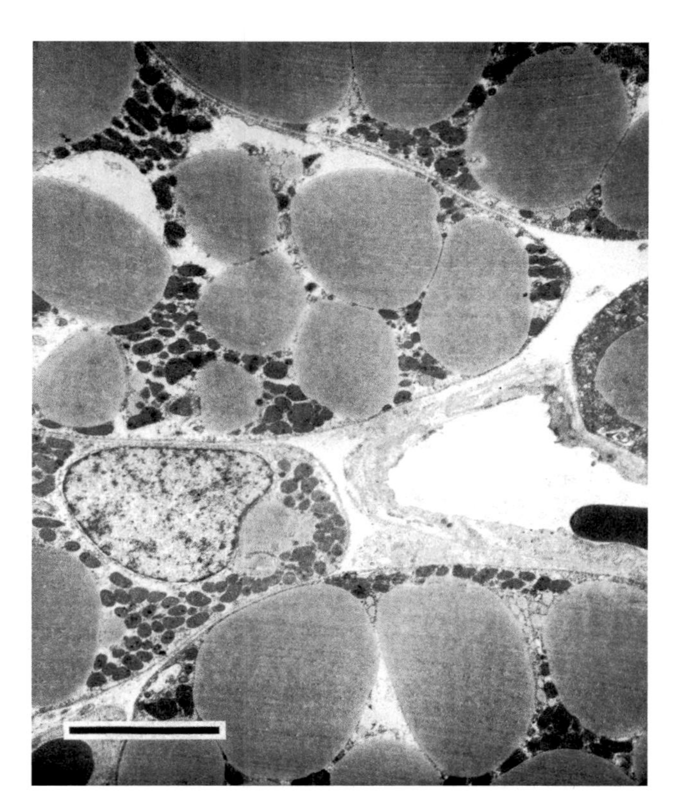

Figure 10.29 Electron micrograph of thermogenic adipose tissue that embeds the venous plexuses in a newborn harp seal, showing multilocular fat and an abundance of dense mitochondria. Scale (lower left) $3 \cdot m^{-6}$ (Blix & Steen, 1979).

that is unique to BAT mitochondria. When the function of the UCP is stimulated by way of the beta-adrenergic sympathetic nervous system, BAT mitochondria are effectively uncoupled and metabolism proceeds at a maximum rate, which is only limited by substrate (Free fatty acids; FFA) availability. The UCP acts as a proton-translocator that is regulated by FFA, which activate it, and by purine nucleotides, which inhibit it. When BAT is stimulated, the functioning of UCP as a proton translocator substitutes for the functioning of the proton-translocating ATP synthetase (oxydative phosphorylation) (Fig. 10.30) (Horwitz, 1989).

Newborn reindeer and muskoxen are obviously heavily dependant on this non-shivering thermogenesis in brown fat for their cold defence, since we have shown that in newborn reindeer calves exposed to low ambient temperature and wind the rate of fall of their deep body temperature increased some 5 times when the activity of the BAT was blocked by injection of beta-adrenergic blocking agents (Fig. 10.31). In fact, we were able to calculate that BAT might contribute as much as 60-70 % to the total metabolic heat produced in reindeer calves when summit metabolism is required (Markussen, Rognmo & Blix, 1985).

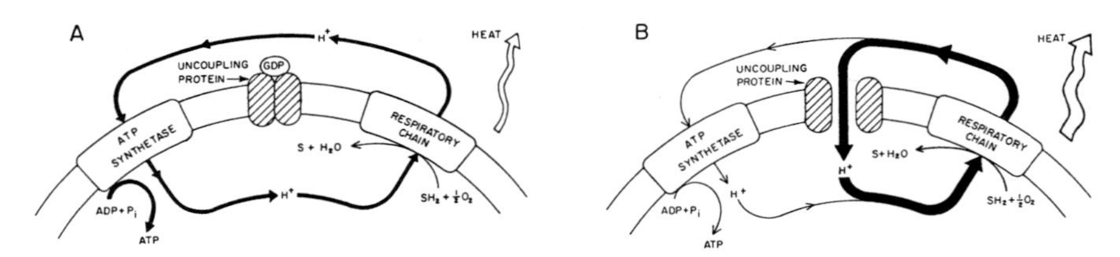

Figure 10.30 Proposed mechanism for the increased rate of mitochondrial fatty acid oxidation in stimulated brown adipocytes. **A**: Under "resting" conditions (i.e., when brown adipocytes are not being stimulated), the protons extruded from the mitochondrial matrix during the operation of the respiratory chain [as reduced cofactor (SH_2) is reoxidized (S)] reenter the matrix coupled to ATP synthesis. That is, the proton conductance pathway regulated by the uncoupling protein is essentially closed; and the respiratory rate is limited by the rate of oxidative phosphorylation (ADP+P_i to ATP). **B**: When the cell is activated by NE, the conformation of the uncoupling protein is altered, thereby allowing protons to reenter the mictochondrial matrix via a pathway independent of the ATP-synthesizing pathway. The rate of substrate oxidation is faster during this uncoupling because the rate-limiting step of ATP synthesis is bypassed. [B depicts one way whereby the conformation of the uncoupling protein could be altered, namely, as a result of displacement of bound purine nucleotide (GDP). However, the precise mechanism (or signal) responsible for the conformational change is not yet known] (Horwitz, 1989).

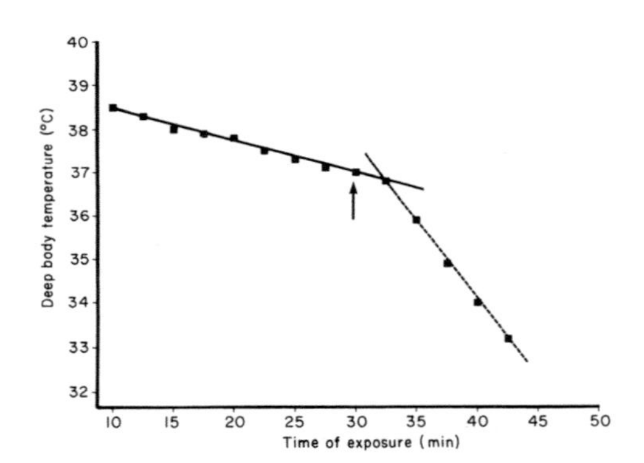

Figure 10.31 Deep body temperature of a newborn reindeer calf exposed to an ambient temperature of -25 °C in combination with a 10 m · s^{-1} wind. The beta-adrenergic blocking agent, propranolol (1 mg · kg body weight^{-1}), was infused (i.v.) after 29 min exposure (arrow), illustrating that the thermogenic activity of brown fat is under beta-adrenergic sympathetic nervous control (Markussen, Rognmo & Blix, 1985).

Marine forms

Most of the arctic marine mammals give birth on open ice-floes during winter or early spring. At the moment of birth the pups have to take the full brunt of the arctic winter and survive exposure to a thermogradient which may exceed 60 °C. The one exception to this rule is the ringed seal, which unlike all other pinnipeds give birth in a snow lair on the sea ice, with one opening into the sea through a hole in the ice (Fig. 4.4). The temperature inside this den is 0 to +2 °C (T.G. Smith, unpublished data).

The size and insulation of the newborn arctic seals show an amazing variation: The ringed seal pup is 5 kg and has a white lanugo fur, the spotted seal pup is 7 kg and has a white lanugo fur, the harp seal pup is 10 kg and has a white lanugo fur (Fig. 4.16), the bearded seal pup is 35 kg and has a dark brown lanugo fur (Fig. 4.8), the hooded seal is 35 kg, but has shed its grey lanugo fur before birth and appears in the normal short and stiff seal fur (Fig. 4.10), while the walrus calf is 45 kg with a short and thin light chocolate brown lanugo fur. The insulative value of these pelts, with the exception of that of the walrus, are rather good as long as they stay dry, but most of their value is lost when exposed to wind and wetness. When exposed to solar radiation, on the other hand, they may contribute to the heating of the animal, as outlined above. The newborn hooded seal is the only one of the club which is born with a 2 cm thick layer of blubber, which, of course, offers some insulation in addition to its sealskin coat. Finally, we have shown that even newborn harp seals have fully patent counter-current vascular heat exchange in their flippers, and it is most likely that all newborn arctic pinnipeds are able to restrict body cooling by such means (Blix, Grav & Ronald, 1979).

These pups go through a "terrestrial" stage of widely varying duration. This stage is terminated at the time of weaning, when moulting, in all but the hoods, occurs, and transition to aquatic life takes place. But, it has been known for long that the walrus and bearded seals venture into water shortly after birth, and recent studies by Lydersen & Hammill (1993) have revealed that ringed seal pups are quite active in the water even during the nursing period, when the pups are supposed to stay in their dens. It follows that only harp, hooded, spotted and possibly ribbon seal pups seem to adhere to the rule and stay put on the ice.

The nursing period also differs greatly among these species from a record short of 2-4 days in the hooded seal, through 10-14 days in the harp seal and several weeks in the ringed seal, to more than a year in the walrus. During the nursing period the pups gain a lot of weight, primarily as subcutaneous deposits of fat. In fact, even the hooded seal pup doubles its weight during the 2-4 days of nursing. During the nursing period the walrus is the only example of a pinniped which makes extensive use of brooding of the calf, while the others, with the notable exception of the ringed seals, are fully exposed to the elements at all times.

Working in the Magdalene Islands in eastern Canada in the 1970ies with Hans Grav, we (Grav, Blix & Påsche, 1974) discovered that the newborn harp seal possessed huge amounts of highly thermogenic brown adipose tissue (BAT). In this particular species BAT is found as a thin subcutaneous layer at birth, but this layer is transformed into blubber in a few days, as suckling commences. We also found that the vast venous plexuses, which are found in the thoracic region of most pinnipeds, and with a functional significance which until then had been obscure, were embedded in BAT in the newborn (Blix, Grav & Ronald, 1975). It was therefore suggested that the assembly of venous plexuses and thermogenic fat function as heat exchangers under sympathetic nervous control. These heat exchangers are, moreover, in series with the peripheral counter-current circulation and ensure that the venous return from the periphery is heated to normal deep-body temperature before entering the central circulation. As the harp seal grows larger the adipocytes of these internal deposits become unilocular and their mitochondrial concentrations decrease until adult life, when they appear as well vascularized white adipose tissue (Blix, Grav & Ronald, 1979). We have later found large amounts of thermogenic BAT in newborn ringed seals, which appears to be the only species which maintains functional BAT as adults, and I am certain that functional BAT is present also in newborn spotted seals and ribbon seals, which I have never been able to lay my hands on.

The newborn hooded seal, bearded seal and walrus do not have functional BAT. These newborn forms have one thing in common, and that is that they are all big. It is therefore possible that their surface area in relation to body mass is so favourable that thermogenesis in BAT is simply not needed. This view is further supported by the fact that newborn whales, as we shall see, also lack brown adipose tissue.

We have already agreed that the northern fur seal is not a truly arctic animal, but that it should be included on occasions because I myself have such good memories from working with them (Blix *et al.*, 1979). And, it is, indeed, interesting to compare the mechanisms of thermoregulation in this otariid species with those of the true phocid seals of the north. The weather at the Pribilofs, where most of them are born, is usually cold, wet, and windy during the breeding season. At birth the pups weigh a tiny 5 kg and are insulated only in a partly wettable pelt and a thin (2-3 mm) layer of blubber. About 40 % of their surface area is made up of their oversized naked flippers, but they have an amazing ability to control their peripheral blood flow from the very moment of birth and heat loss from the extremities is probably minimal. The pups have a high resting metabolic rate and a lower critical temperature slightly below the record low ambient temperature during the breeding period. During rainy weather, however, much of the insulative value of the pelt is lost. Internal deposits of BAT are of little significance at birth and the pups respond instead to cold with visible shivering, but, we discovered that these pups make heavy use of non-shivering thermogenesis in loose-coupled mitochondria in skeletal muscle (Grav & Blix, 1979). In any case, the fur seal pups apparently are brought close to their limit of tolerance during

Figure 10.32 The newborn beluga calf (*front*) is usually born in estuaries and river mouths, where the water is warmer than at sea.

wet and windy days and only escape hypothermia and death by vigorous periodic shuddering which temporarily improves the insulative value of the fur.

Although ringed seals, bearded seals and walruses are sort of semi-aquatic at birth, arctic whales are, of course, truly aquatic mammals from the very chilling moment of birth. In fact, their constant exposure to the arctic seas must represent the ultimate challenge to a newborn homeotherm!

The newborn arctic whales are equipped with a relatively thin layer of blubber (Fig.10.32) and, like some of the seals, benefit from their large body mass. The newborn narwhal and beluga, for instance, are both about 80 kg, whereas the newborn bowhead weighs more than a ton. Thus, in spite of the presence of spectacular venous plexuses in the thoracic cavity, I have found that no internal deposits of any type of fat are evident in belugas of any age. However, the skeletal muscles are rich in fat, glycogen, and mitochondria at birth and probably are employed in non-shivering thermogenesis in addition to the heat some of the muscles produce as a result of swimming.

Unlike the bowhead and the narwhal, which give birth in the open arctic seas, the pregnant beluga actively seeks the usually warmer water of estuaries and lagoons for the delivery and nursing of her calf, and there are indications of poor calf survival when the spring break-up in the lagoons occurs late. This might indicate that belugas do not seek the lagoons, where they have been harassed by the Eskimos for centuries, just for the fun of the hunt.

The substantial amount of energy required to support the high level of metabolism in newborn polar mammals during their first critical period of life is obtained from the milk. The milk of several species has been analyzed and, although few samples and odd conservation methods occasionally have been employed, the milk of polar species obviously is very rich in energy (Table 10.2). Thus, marine mammals produce milk that ranges in fat content from 40-50 %, whereas terrestrial polar species have milk that ranges in fat content from 10-20 %. One peculiarity of seal milk is its lack, or very low levels, of carbohydrates. The seal pup nevertheless has a normal blood glucose level.

Table 10.2

Gross composition of milk from a number of arctic and domestic mammals (from numerous sources, e.g. Oftedal, 1984)

Gross Composition of Milk, % of total

Species	Fat	Protein	Lactose
Beluga	27	11	0.7
Harp seal	52	6	0.9
Hooded seal	40	7	0.0
Fur seal	53	9	0.1
Polar bear	33	11	0.3
Reindeer	16	10	3.1
Muskox	11	12	2.1
Arctic fox	12	12	5.4
Wolf	10	9	3.4
Sheep	7	6	4.8
Cow	4	3	4.8
Dog	13	8	3.1
Man	4	1	6.0

Food and digestion

Aside from such peculiarities that polar bears feed primarily on ringed seal, and that ringed seals are only found in the Arctic, there are hardly any such thing as an arctic diet. Here, as everywhere else, everything and everybody gets eaten by somebody, except possibly the vastly abundant plant *Cassiope*, which is avoided by herbivores. It follows that it is equally unlikely to find a typically arctic digestive system or typically arctic digestive adaptations, aside perhaps from adjustments of the general carnivorous and herbivorous *modus operandi* in both the mammals and birds of the region.

Thus, polar bear, wolf, fox and ermine feed on a variety of meats, and have the typical carnivorous single stomach and short small intestinal digestive system. Best (1977) has reported that the polar bear is able to assimilate 92 % of the energy in a ringed seal diet, and that the apparent digestibility of ringed seal blubber is 98 % in these animals.

The different species of seals of the arctic region are also all carnivorous animals, feeding on a variety of fish, squid and invertebrates, and again have a typical carnivorous single stomach and short small intestine, caecum and colon. Recently, we have learnt, however, that even though the small intestine of seals is relatively short, it differs in length from 5 times body length in the antarctic Ross seal (*Ommatophoca rossi*), via 14 times body length in the harp seal to 25 times body length in the elephant seal (*Mirounga leonia*) (Mårtensson, Nordøy, Messelt & Blix, 1998). One would assume that this in some way is related to the diet of the animals, but although the diets cover everything from penguins to fish, squid, krill and other invertebrates intestinal length does not seem to correlate with the preferred diet of the animals. It has been suggested that intestinal length is correlated with diving capacity, in such a way that long duration divers (like elephant seals) need a long intestine to compensate for the diving induced intestinal vasoconstriction and ischemia in order to effectively absorb nutrients from the intestines, but we have recently also shown that this is not the case. So, the reasons for this striking variability in seal intestinal length is presently unknown.

The bowhead whales feed, as we have already learned, almost exclusively on small crustaceans, while narwhals and belugas feed on a great variety of fish, squid and invertebrates. Still, they share a compartmentalized stomach-system which appears to be common to all whales. (Fig. 10.33). This includes an initial non-glandular forestomach with high concentrations of indigenous anaerobic bacteria that ferment the prey, while the glandular part of the stomach is divided into a main gastric chamber, the fundic chamber, which possesses gastric glands, a connecting channel which appears to prevent the passage of large food particles, and the pyloric chamber, lined with mucous cells, and separated from the intestine by a pyloric sphincter. The multi-chambered stomach system is followed by a duodenal ampulla which marks the beginning of the small intestine, which in the minke whale is only four times body length (Olsen, Nordøy, Blix & Mathiesen, 1994). The bacteria in the forestomach of this species have been investigated in some detail by Monica Olsen and associates at our laboratory, where they have found that some of the species of bacteria even have the unusual ability to

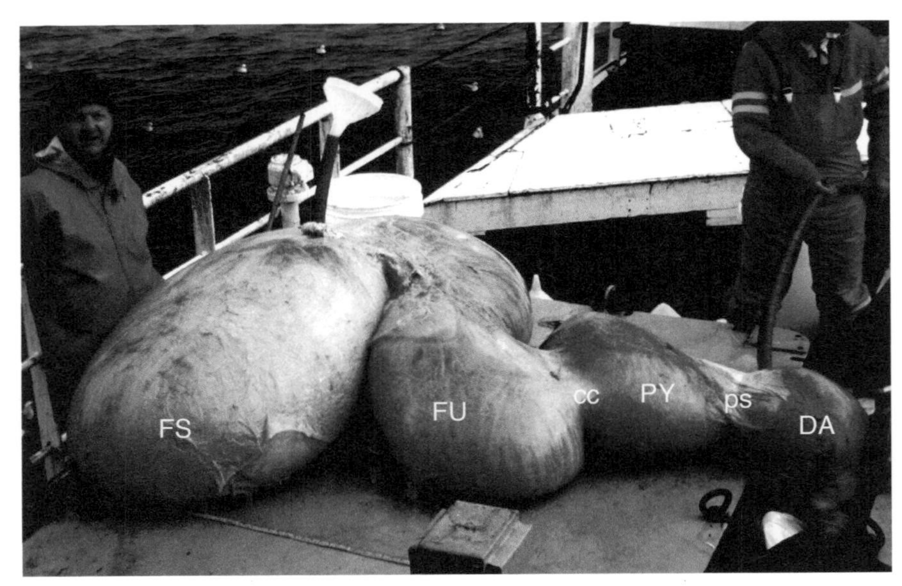

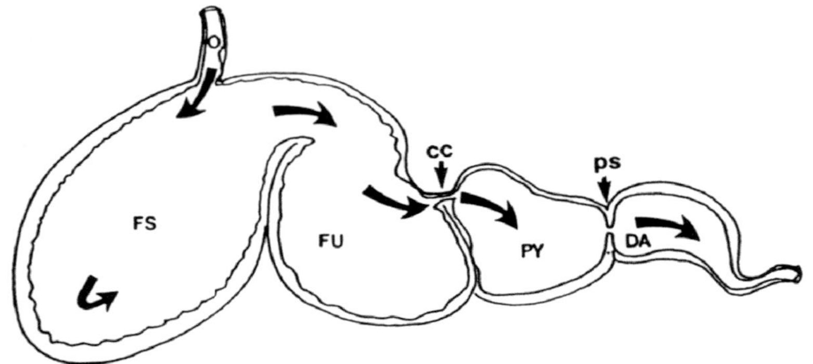

Figure 10.33 *Top*: Photograph of the intestinal system of a minke whale of about 4 tons. (Olsen, Nordøy, Blix & Mathiesen, 1994). *Bottom*: Schematic drawing of the same structures, showing forestomach (FS); fundic chamber (FU), connecting channel (cc), pyloric chamber (PY), pyloric sphincter (ps) and duodenal ampulla (DA).

digest chitin (Olsen *et al.*, 2000), which is a major constituent of the shell of crustaceans. Krill is also very rich in wax-esters, which are normally poorly digested in terrestrial mammals, but Nordøy (1995) has found that minke whales are able to digest as much as 94 % of the wax-ester dry matter, probably again due to the action of the fore-stomach bacteria. The fermentation products of all the different varieties of bacteria are fatty alcohols and volatile fatty acids, which are taken up in the blood and subsequently used as energy for the animal.

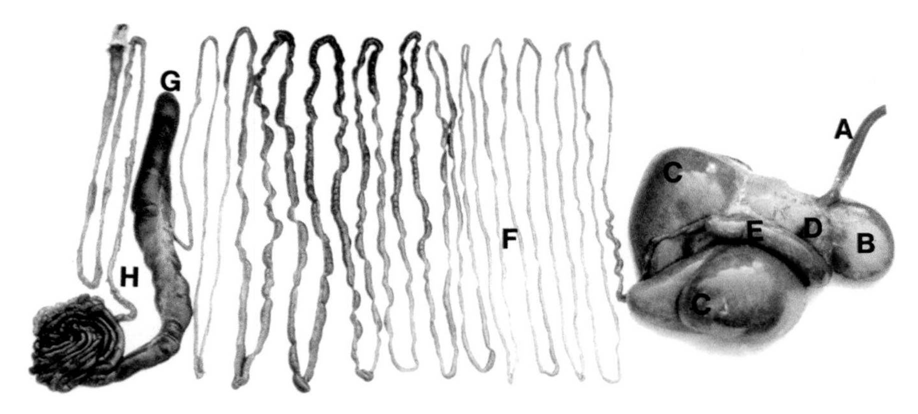

Figure 10.34 The intestinal system of reindeer, showing: A: Esophagus, B: Reticulum, C: Rumen, D: Omasum, E: Abomasum, F: Small – intestine, G: Caecum, H: Colon (Photo: S. D. Mathiesen, T.H. Aagnes & W. Sørmo).

Since minke whales and crabeater seals in the Southern Ocean both feed almost exclusively on krill *(Euphausia superba)* we found it of interest to compare the digestive efficiency in these species. We found that the minke whale digested as much as 93 % of the energy in the krill, while the value for the crabeater seal was 84 % (Mårtensson, Nordøy & Blix, 1994). When the ability of minke whales and harp seals to digest fish, (capelin), were compared we found a value of 95 % in the whale and 94 % in the seal. This suggests that the complex multi-stomach system of whales holds no advantage over the single-stomach system of seals when it comes to digestion of fish, while it was certainly superior with regard to digestion of krill, most likely due to the presence of the chitinase-producing bacteria in their fore-stomachs (Olsen *et al.*, 2000). It should be noted, however, that even without such bacteria the crabeater seals are still amazingly good in utilizing its krill diet.

Muskoxen and reindeer or caribou are herbivorous ruminants also with a multistomach-system (Fig. 10.34), of which the forestomach, or the rumen, like the forestomach of whales, contains myriads of anaerobic bacteria, but unlike whales, also contains protozoa, mainly ciliates, and fungi. The bacteria ferment the plant material and again produce volatile fatty acids, which may contribute as much as 80 % of the energy budget of the animal, in some cases. The function of the protozoa and fungi is not understood. Unlike whales, the arctic ruminants also have long small intestines, some 10 times body length (Fig. 10.34), and a large caecum, which also contains high numbers of anaerobic bacteria that ferment the intestinal contents into volatile fatty acids, and hence, in some cases, very significantly contribute to the energy budget of the animals.

The microbiology and digestive processes in both mainland Norwegian and Svalbard reindeer have been investigated in detail by Mathiesen and associates in our laboratory in cooperation with Orpin of Great Britain. We found that the rumen microflora is highly specialized and that, particularly in the Svalbard reindeer, unusually high numbers of fibre-digesting and ureolytic bacteria which permit optimal use of poor quality forage occur (Orpin, Mathiesen, Greenwood & Blix, 1985) (Fig. 10.35). We also found that most naturally occurring plant material in their diet is digested some 50-80 %, depending on species and season, and that in the high-arctic Svalbard reindeer even mosses, which are normally not eaten, are digested some 11-36 % in the rumen and some 11-27 % in the caecum during winter, when other plant material, as we shall see, usually is in short supply (Mathiesen, Orpin, Greenwood & Blix, 1987). Interestingly, Thomas & Kroeger, 1980) found that the Peary carbou, which live under pretty much the same austere nutritional conditions as its Svalbard relative, only was able to digest mosses some 3-11 %. A reason for the apparent importance of mosses in the diet of the Svalbard reindeer may be that lichens, which elsewhere are important food items for reindeer during winter, are not present in the area where the reindeer occur. In fact, the old "truth" that reindeer pretty much depend on lichens during winter is sim-

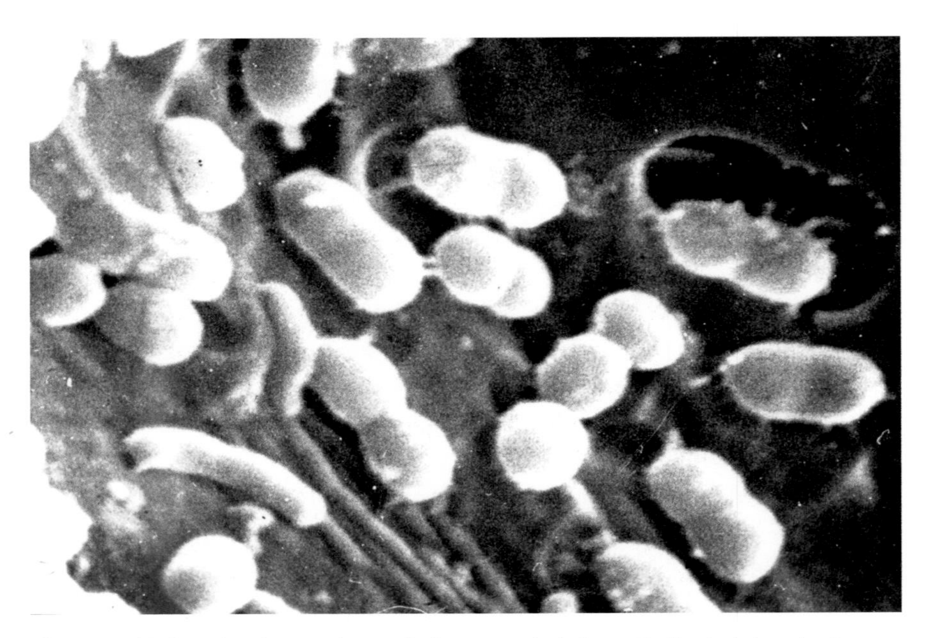

Figure 10.35 Scanning electron micrograph of rumen contents from a Svalbard reindeer showing numerous bacteria "chewing" on the plant material (Cheng *et al.*, 1993).

ply not true. The Norwegian reindeer that were introduced to South Georgia in 1911-1912 and again in 1925 (e.g. Leader-Williams, 1988) and have thrived there ever since on a pure grass diet is proof enough of that (Fig. 8.18).

The continental reindeer and caribou are particularly interesting animals from a nutritional point of view in that vascular plants make up the bulk of their diet during summer, while they feed to a large extent on lichens during winter. Lichen is composed of an algae and a fungus growing in intimate symbiontic association and is therefore mainly composed of hemicellulose and some lichenin, but very little cellulose and almost no protein. This results in a negative energy balance and loss of muscle mass in animals which live on lichens alone, in spite of the fact that it has been shown that reindeer are able to recycle urea to some extent, in order to regain nitrogen. Staaland *et al.* (1983) have moreover shown that reindeer become sodium deficient on a lichen diet and it is known that reindeer occasionally eat seaweed when they arrive at the coasts of northern Norway during their summer migrations.

Muskoxen have also been shown to be pretty efficient in digesting their normally rather poor winter diet. In fact, Adamczewski *et al.* (1994) have demonstrated that muskoxen were able to maintain body mass on a low quality diet, with a daily dry matter intake which was only 1/3 of that of cattle on the same diet.

The digestive system of the arctic hare has not been described, but the odd habit of recycling the faeces of other species of hares and rabbits, has received quite a lot of attention. In fact, in 1939/40, when England was fighting for her life at the beginning of World War II, no less than 5 articles on rabbit coprophagy, which is the term for this activity, was published in "Nature" (e.g. Eden, 1940), over a period of less than a year! From these and later studies (e.g. Kenagy & Hoyt, 1980) we now know that rabbits, and most likely also the arctic hare, produce two kinds of faeces. One consists of the familiar almost spherical dry pellets which are high in fiber and low in crude protein and are the result of a direct passage through the gut, while the other is soft and consists of mucous covered agglomerates of much smaller pellets with a high concentration of crude protein and a low content of fiber. In spite of the fact that the former appears to be less digested, it is some 50 to 80 % of the latter, which, according to eye-witnesses, is eaten directly from the anus. It has been suggested that coprophagy in hares primarily occurs in response to nutritional constraints, but this does not appear to be the case. Although much more remain to be known about this phenomenon it is quite clear that the soft faeces is produced in the caecum and that its high concentration of protein is of microbial origin. It is quite likely that this protein is of nutritional importance in providing, for instance, essential amino acids and it is also possible that other nutritional elements, such as vitamins, are obtained by this rather disagreeable behaviour.

Lemmings are also interesting from a food processing point of view, in that they are among the few that to a large extent depend upon the utilization of mosses, which have a low standing on the menu of most animals because of low digestibility. Lemmings, like hares, will also occasionally indulge in coprophagy, but this activity is apparently much less important in lemmings. Instead, it appears that the lemmings meet the problem of low digestibility values, which have been reported to be as low as 25 % for mosses, by processing large amounts of food by high rates of passage through the gut. However, the low retention times, which in the lemming may be little more than an hour, is likely to create difficulties when food of low digestibility, at least in part, is processed by micro-organisms. An equally low retention time for caecal bacteria would lead to a rate of loss of protein and other organic matter impossible for the animal to sustain.

It has been known since the turn of the century that lemmings and some other rodents are equipped with a peculiar proximal colon, which is shaped as a spiral (Figs. 10.36, 10.37). This structure has been studied in detail by Serber *et al.* (1983) who found that the first two windings of this part of the colon are functionally, but not physically, separated by an intricate system of internal folds into two different channels, the narrow channel and the main channel.

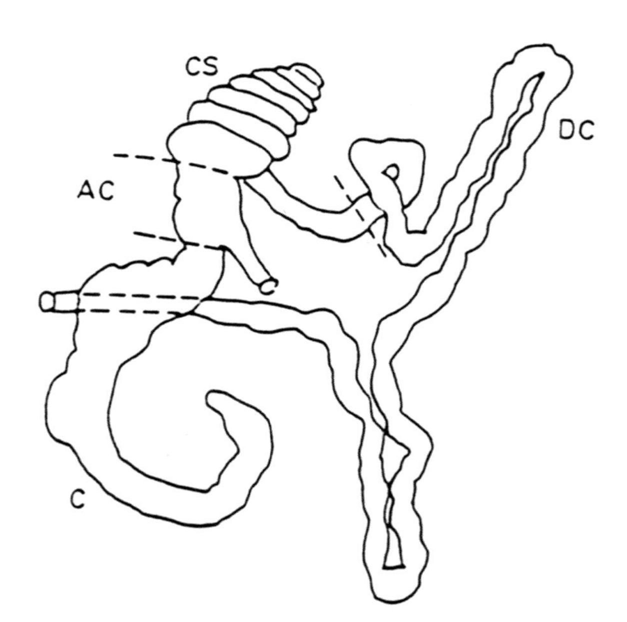

Figure 10.36 Caecum and colon of Scandinavian lemming. AC: ampullae coli; C: caecum; CS: colonic spiral; DC: distal colon (Tullberg, 1899).

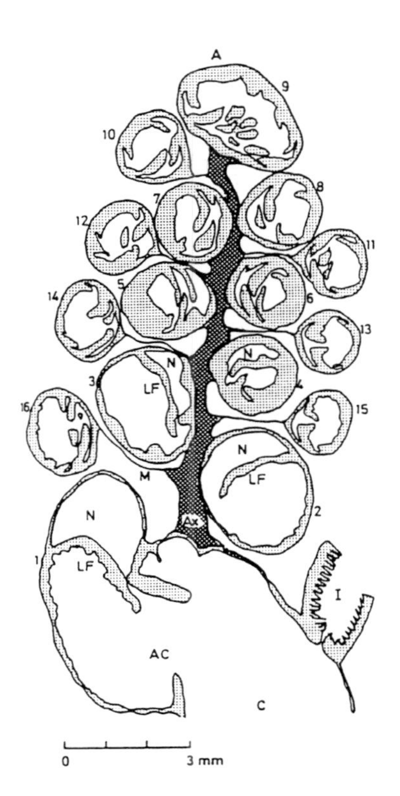

Figure 10.37 Section through the colonic spiral of Scandinavian lemming. Composite from sections of 21-day-old animal. *A:* apex; Ax: axial structure; *I:* ileum; *LF:* longitudinal fold; *M:* Mesentery; *N:* narrow channel. Numbers 1-8 refer to sections through inner spiral, 10-16 to sections through outer spiral (Serber, Bjørnhaug & Ridderstråle, 1983).

The contents of the main channel appears to be similar to that of the caecum, while in the narrow channel food residues are rare and the contents are instead made up of a dense mass of bacteria mixed with mucus, which becomes more scarce towards the end of the spiral. Thus, the pellets that are voided by lemmings on natural food hardly contain any bacteria. Lange & Staaland (1970) have shown that the post-caecal spiral is also important for sodium and potassium retention, but it appears that the real value of this structure lies in that it somehow is able to extract the microorganisms and to transport them by antiperistaltic movements by way of the narrow channel back to the caecum, while the food residues are passed on through the main channel to the distal colon and are voided through the anus.

Separation mechanisms as part of colonic function also exist in lagomorphs, but are much less developed than in lemmings, and hence, the former appears to be more dependent on the more vulgar form of coprophagy.

The digestive system of birds shows great variation among the different groups and species. Thus, in spite of the fact that the food items are swallowed entire in most birds, there is great variation in both beak, tongue, salivary glands and the esophagus. The latter may have a great diameter and extensive foldings to accommodate a lot of food for temporary storage, for example in sea-birds and snowy owls, or have a clearly differenciated diverticulum, called the crop, in herbivorous birds, like the ptarmigan, where it provides storage and a constant supply of food to the gastric apparatus, but also moisture and hence softening and swelling of the food. In some seed-eating birds, like the arctic redpoll, which relative to its own small size has a huge crop that extends laterally up and around the neck (Fisher & Dater, 1961), it also has the additional function of a storage chamber for food for the young.

The esophagus transports the food to the gastric apparatus, usually with two distinct chambers: First, the anterior glandular stomach, or proventriculus, where the food is stored, particularly in fish-eating birds, but where also secretion of gastric juices takes place, and second, the posterior muscular stomach, or gizzard, where acid and some enzymes are secreted and the mechanical digestion occurs, in many ways being analogue of the teeth of mammals. In herbivorous birds the gizzard often contains small stones, called grit, to help in the grinding of the food material (Norris, Norris & Steen, 1975).

The small intestine, which follows the gastric apparatus, is the principal site of chemical digestion, also in birds, being long in herbivorous birds and relatively short in carnivorous ones. At the junction of the small and large intestines most birds have a pair of caeca, which are small in fish-eating birds, but well developed in herbivorous birds and owls. It has been known for long that herbivorous birds are able, to some variable extent, to decompose cellulose by microbial activity in their intestines, and we shall see that the caeca are supposed to be important in this and other respects.

The relatively short large intestine of birds, which is employed in water resorption, ends in the cloaca, which serves as a storage chamber for both feces and urine and is also the pathway for the products of the sexual glands of both sexes.

The digestive system of the ptarmigan has been investigated in some detail by Hanssen (1979a,b) in our laboratory. Like other galliforme birds the ptarmigan has a large crop and a gizzard which is very muscular and contains grit, usually of quartz, and two caeca, protruding, as usual, from the ileo-colonic junction. The caeca are especially well developed in these birds and contains about 10^{11} bacteria per g (wet weight), as well as protozoa, while the small intestine of ptarmigan has very few microorganisms. The size of the caeca is usually strongly related to the normal diet of the bird. Accordingly, during winter, the gizzard of the ptarmigan is some 25 % heavier and the caeca are some 25 % longer than in summer, presumably to improve digestion of the particularly rough (fiber-rich) winter diet, while the length of the small intestine remains unchanged.

The caecal microorganisms have been found to be able to utilize the major nitrogen compounds which enter the caeca as a source of nitrogen for *de novo* synthesis of amino acids. It has previously been assumed that the caeca participate in recycling of excretory nitrogen, but this does not seem to be the case. In fact, it seems that the caecal microflora plays no role in making nitrogen more available for the bird (Mortensen & Tindall, 1981). What is made available for the bird are the waste products from the energy metabolism of the microorganisms, which again are volatile fatty acids, that can be utilized as substrates for the birds' own energy metabolism, while at the same time avoiding end-product inhibition of the microbial fermentation. According to Mortensen (1984), the caecal microorganisms most likely also participate in the detoxification of the bird's nitrogenous waste products, such as urea, and may be also ammonia, and although the significance of the detoxification is still unknown, it may be that the presence of the large caeca of the ptarmigan is related to just this. Finally, Gasaway (1976) has presented evidence for cellulose digestion in captive rock ptarmigans, and he claims that the degradation occurs as a result of microbial activity in the crop and small intestine, but probably also to a significant extent in the caeca. It follows, that this important aspect of bird biology deserves much more attention.

The digestive system of owls, including the snowy owl, is rather different from the more general picture painted above. These birds usually swallow their prey entire, with fur, feather, bones and the lot. The food may be stored temporarily in the large and distensible esophagus, but it soon passes into the gizzard, which in these birds is large and thin-walled with little musculature and obviously not adapted to any great degree of motility. The pyloric opening into the small intestine is moreover only about 1 mm in diameter and at the same level as the entrance of the esophagus, which makes something like a container in which the food can be digested. Interestingly, there is no free acid produced, but the peptic activity is about three times higher than for instance in the dog, and therefore the prey becomes completely dissolved in 12 to 24 hours, while the bones are unaffected. The bones and other undigestible materials are subsequently expelled by regurgitation as a typical pellet, while the digested material passes into the small intestine as a liquid (Reed & Reed, 1925).

Energy balance and growth

We have learnt by now that energy balance in an animal is a matter of balancing the energy intake in the form of food against the energy expenditure of, mainly, basal metabolic rate, thermoregulation, reproduction and locomotory activity. We also know that animals in the cold strive to reduce their cost of thermoregulation, by such means as good insulation and nasal heat exchange and by reducing their activity. Another potential way of reducing energy expenditure is to reduce the basal level of heat production itself. In winter, basal metabolic rate (BMR) accounts for between 50 and 70 % of the daily energy expenditure of adult reindeer and consequently, even small reductions in metabolic rate would potentially contribute relatively large overall savings to the animal's energy budget. Rather surprisingly, however, and despite an earlier speculation that the metabolic rate of Svalbard reindeer, which have extremely low thyroxine levels, might fall in winter, this does not seem to be the case. It is true that resting metabolic rate of reindeer fed *ad libitum* shows a marked reduction from summer to winter (Fig. 10.15), but Nilssen, Sundsfjord & Blix (1984) have shown that that is caused by changes in the animals' voluntary food intake and does not represent a physiological adaptation to energy conservation by a reduction in fasting metabolic rate. This conclusion was further supported by measurements of resting metabolic rate in reindeer which received only 15 % of their *ad libitum* food intake over a period of several days. As expected the animals' metabolism fell rapidly, but reached a stable minimum level after 5 to 10 days. There were no differences between summer and winter minima in either Svalbard or Norwegian reindeer. The other arctic ruminant, the muskox, provides an interesting contrast in this respect. These animals appear to have developed two metabolic adaptations to winter. First, their metabolic rate shows a pronounced seasonal variation. Fasting metabolic rate actually fell approximately 20 % from summer to winter. Second, their fasting metabolic rate in winter appears to be almost 40 % lower than in the Svalbard reindeer (Nilssen, Mathiesen & Blix, 1994). Aside from the muskoxen, however, there is little to indicate that other, non-hibernating, animals make use of reduced basal metabolic rate as a means to combat energy expenditure in winter.

Arctic animals not only need to be able to withstand the seasonal changes in ambient temperature by minimizing their heat loss to the environment. They also have to cope with the even more remarkable seasonal changes in both *availability* and, not the least, *energy density*, of their food. This problem probably is at the top of the agenda all over the Arctic, but probably not of such extreme importance as on the high-arctic archipelago of Svalbard. It is evident from Fig. 10.38

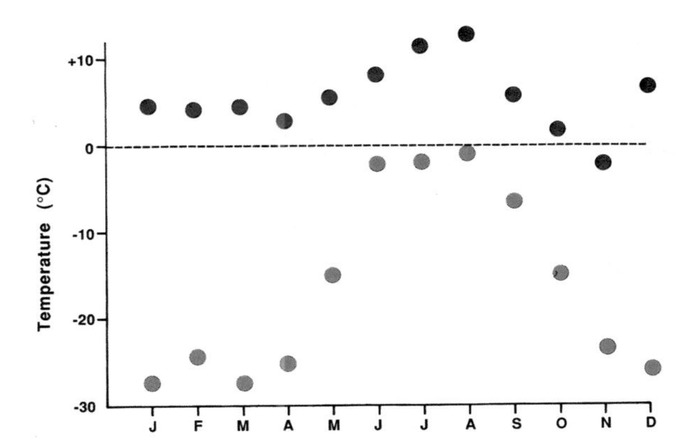

Figure 10.38 Ambient temperatures at Isfjord Radio on the west coast of Svalbard (78°N) throughout one year. *Top*: highest temperature every month. *Bottom*: lowest temperature every month. Notice the occurrence of above freezing temperature throughout winter (Source; Norwegian Meteorological Institute).

that the winter, particularly on the western coasts with its relatively warm maritime climate, is characterized by periods of warm, often rainy weather. Such episodes of mid-winter rain (at 80°N !) followed by periods of low temperatures result in over-icing of the range and a marked reduction in the availability of the plant material which in the first place has a rather low energy density during winter. And on top of this there is a significant additional energy cost of digging through the snow and ice (e.g. Fancy & White, 1985b). We shall see that most arctic mammals, and at least some birds, prepare themselves for this ordeal by deposition of body fat during times of plenty in the summer and autumn, albeit none as huge as those found in all the animals that winter in Svalbard.

Already the Norwegian naturalist Keilhau (1831) noted that the Svalbard ptarmigan is very fat in the autumn, while its closely related Norwegian rock ptarmigan which live under seemingly, similar climatic conditions, only deposits insignificant amounts of fat in the same period. At the Department of Arctic Biology at Tromsø we caught an early interest in the Svalbard ptarmigan and based on examination of 224 birds that were shot in the wild, some in each and every month, we were able to describe the seasonal variations in the amount of dissectable fat in these birds (Fig. 10.39). We found that the build-up of fat started in August and that the fat stores were at their richest in late October to the beginning of November, but that they rather surprisingly were nearly depleted at the beginning of March. The highest value recorded was 350 g in an adult male, being 32 % of total body weight in a flying bird (!), but there were no consistent differences in the amount of body fat among the sexes, except that females deposited some fat prior to egg laying in June, while the males were almost devoid of fat from mid-February till August (Mortensen, Unander, Kolstad & Blix, 1983). It is worth noticing that the build-up of fat

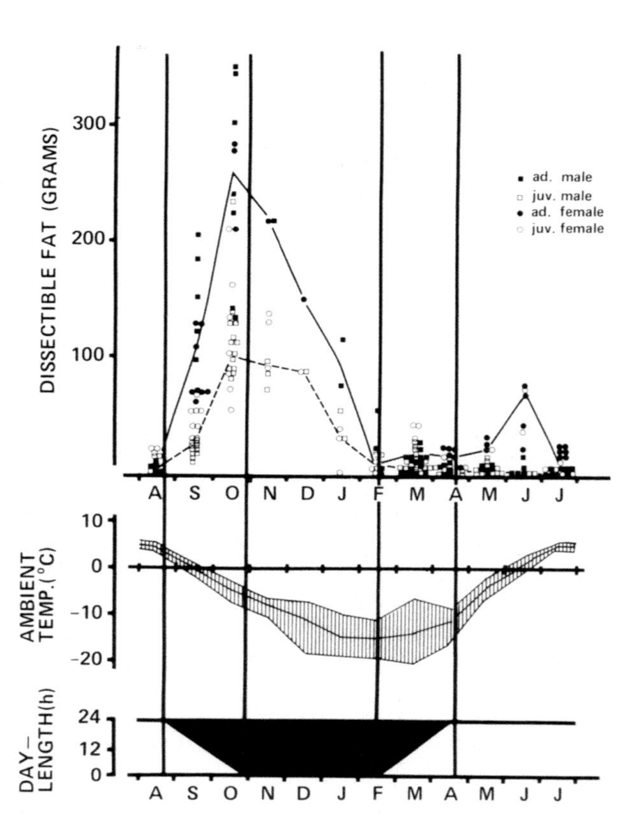

Figure 10.39 *Top*: Seasonal changes in dissectible fat in Svalbard ptarmigan. Each point represents one bird and lines are drawn through the median values. From August to February: Solid line represents adult birds of both sexes whereas dotted line represents juveniles of both sexes. From February to July: Solid line represents females and dotted line males, of all ages, respectively. *Middle*: Monthly mean temperature at Longyearbyen. Solid line: mean, and hatched area: range. *Bottom*: Seasonal changes in photoperiod at 78°N (Mortensen, Unander, Kolstad & Blix, 1983).

in these birds started when the period of continuous day-light ended, and that the fat stores reached their maximum at the time when the continuous winter night began. Moreover, the fat stores were depleted the day the sun returned, while the females started a build-up of fat at the time when the midnight-sun again shone. It is also worth noticing that neither of these changes in body composition of the birds could be related to changes in ambient temperature, and we have later learnt that the autumnal fattening is a delayed consequence of the spring increase in daylength, but with the prerequisite that the birds have become photorefractary (i.e. do no longer respond to stimulation by light) due to long day stimulation in summer.

Later on we were able to measure the feeding activity continuously throughout a 13-month period in captive birds under natural conditions of light and ambient temperature at 79°N in Svalbard. We then found that food intake was persistently high from March until August, including the period when daylight is continuous, whereas it was low from November until January, when it is perma-

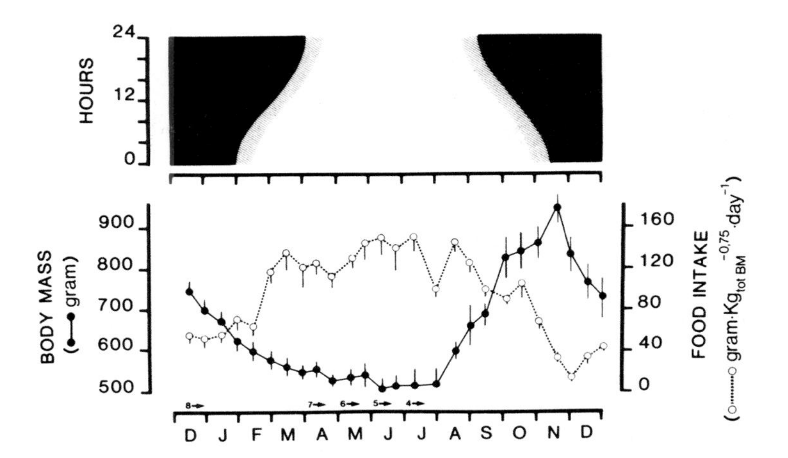

Figure 10.40 *Bottom*: Seasonal changes in body mass (BM) *(closed circles)* and food intake expressed as gram · kg total BM$^{-0.75}$ · day^{-1} *(open circles)* in captive Svalbard rock ptarmigan exposed to natural temperature and light conditions *(top)* for 13 mo at Svalbard (79°N). During this period, birds were given standardized high-quality feed and snow and water *ad libitum*. Period when sun is above horizon is shown in white, night in black, and civil twilight by hatched area. Number of birds is indicated on abscissa, and vertical bars indicate SEM (Stokkan, Mortensen & Blix, 1986).

nently completely dark (Stokkan, Mortensen & Blix, 1986). Much to our surprise, we moreover found that body mass doubled from August till November, while food intake dropped to one-third, and body mass fell rapidly from mid-November until April despite a doubling of food intake from February until March (Fig. 10.40). It follows from these observations that body mass and composition is not only regulated by appetite but depend heavily on seasonal changes in energy expenditure associated with locomotor activity.

We shall later discuss the significance and control of these fat stores, but first, we shall have a look at the reindeer both from Svalbard and Scandinavia, which we studied at the same time at Tromsø. We offered these deer *ad libitum* access to high quality food throughout the year and found to our surprise that voluntary food intake changed dramatically throughout the year, with very low values from January through March, increasing to values almost four times higher from May through August (Fig. 10.41). We also found that the volume, as well as the lipogenic capacity, of isolated fat cells changed in parallel with the changes in food intake (Larsen, Nilsson & Blix, 1985). These results are well supported by Tyler (1987) who studied reindeer in Svalbard for many years, and who found that the mean live weight of adult female Svalbard reindeer fell by almost 30 % , from a seasonal maximum in autumn to a seasonal minimum in late winter (Fig. 10.42).

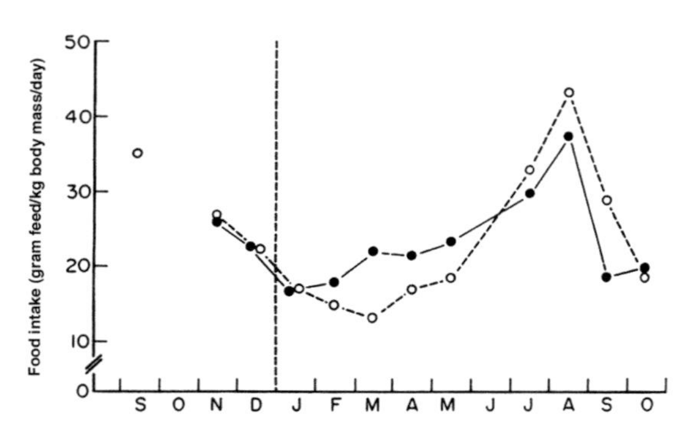

Figure 10.41 Seasonal changes in voluntary feed intake of Norwegian (filled symbols) and Svalbard (open symbols) reindeer fed *ad libitum* throughout the year (Larsen, Nilssen & Blix, 1985).

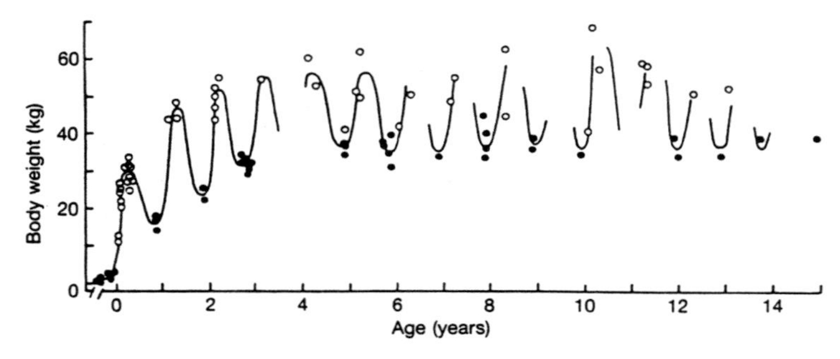

Figure 10.42 Growth curve for female Svalbard reindeer. Total body weight (kg), less the weight of the reticulo-rumen (all specimens) and the weight of the uterus and its contents (winter specimens). Each point represents one reindeer. Summer and autumn (July to October) (open symbols); late winter (April to May) (closed symbols) (Tyler, 1987).

Large seasonal changes in body weight are not unusual among free-living cervids. Patterns of growth similar to that of the Svalbard reindeer have been found both in barren-ground caribou and wild reindeer. The amplitude of the cycle, however, is much greater in Svalbard reindeer than in other reindeer and caribou. The principal cause of the seasonal changes in body weight of this animal is, as you by now should have guessed, deposition of fat, and the Svalbard reindeer are really fat in autumn (Fig. 10.43). On average, about 17 % of their body mass is fat, but one case has been recorded in which the value was almost 30 %. It is interesting that while almost 90 % of this fat is deposited under the skin over the back, flanks and rump and between the superficial skeletal muscles, it is unlikely that it contributes significantly to the insulation of the animal, since Ringberg, Nilssen & Strøm (1980) found that it consists mainly of fully saturated fatty acids and would actually go solid if it isn't maintained at close to deep body temperature. So, it is quite clear that the fat deposits serve mainly as an energy reserve, and usually, like in ptarmigan, most of it is used up by the end of winter (Fig. 10.44). The protein in the skeletal muscles is also mobilized, however, and Svalbard reindeer apparently lose almost half their skeletal muscle mass during winter, but this still represents only a minor energy reserve compared to that of fat. It should, nevertheless, not be disregarded, since the amino acids are the principal precursors for glucose synthesis in fasting animals.

Now, what are the fat reserves of the Svalbard reindeer used for? We have calculated that the Svalbard reindeer carry sufficient reserves of energy stored as fat and muscle protein to contribute, on average, approximately 25 % of their energy requirements during winter, cost of gestation not included. It follows, that these animals will normally have to meet no less than 75 % of their daily energy requirements by feeding. Evidently, therefore the fat of the Svalbard reindeer, just as in the Svalbard ptarmigan, is used as insurance against death by acute starvation in periods when the food is unavailable to them. But, there is also evidence from Tyler's studies that pregnant females delay mobilization of their remaining fat reserves during the last two months of gestation apparently in anticipation of the energy demands of lactation (Tyler, 1987). This implies that pregnant

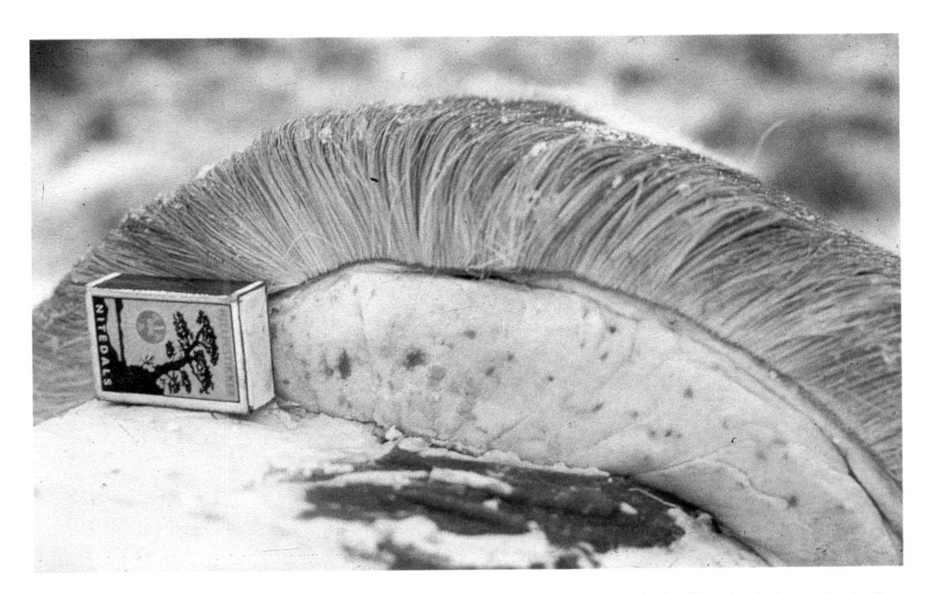

Figure 10.43 A cut through the huge fat deposits on the rump of the male Svalbard reindeer prior to the rut in the autumn. Note also the long and very dense fur (Photo: K. Bye).

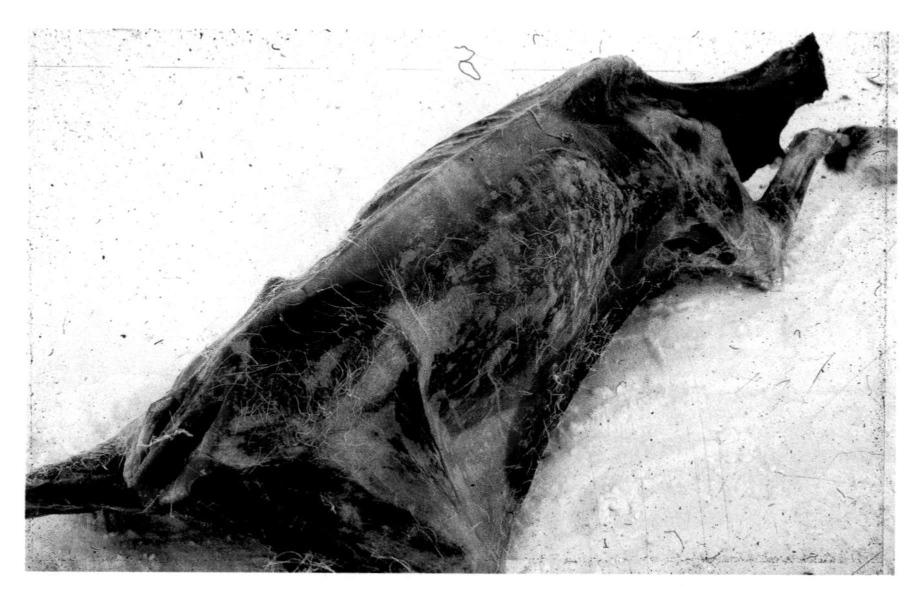

Figure 10.44 A typical skinned carcass of an adult and apparently healthy female Svalbard reindeer which was shot late in winter (April), when muscle mass has atrophied and the fat deposits (seen in Fig. 10.43) have been mobilized and used in anticipation of spring and a new growth season (Tyler, 1987).

females normally are able to find, eat and digest almost enough food to meet their energy requirements in April and May even though the range usually at that time still is covered by hard snow. How can pregnant females, with their high energy requirements, manage almost to maintain body mass late in winter while barren females, which presumably require less food, continue to lose weight? A probable explanation is that pregnant females gain preferential access to the feeding craters and the best of what little food is available, due to the dominant social status that follows from the fact that pregnant, unlike barren (non-pregnant) females and males, usually retain their antlers until calving time. Unlike the females, the mature bulls use up most of their fat reserves during the rut, before the start of winter, and this combined with the fat management outlined above for the females may indicate that the role of the fat reserves of reindeer is to enhance reproductive success, while it in Svalbard serves the additional purpose of insurance against death by acute starvation during the dark winter months.

The importance of the fat reserves in the promotion of reproduction at the conception end of this endeavour has been highlighted by Thomas (1982) in the Peary caribou. He found that fertility and conception in this animal of the Canadian Arctic is dependent on a certain body weight and fat content to the extent that a difference in body weight of 4 kg in March and April increased pregnancy rate from 8 % to 75 % (Fig. 10.45). The causes of infertility in malnourished female reindeer and caribou are poorly understood, but may include : failure to come into estrus, failure to ovulate, resorption of embryos and abortion, while survival in malnourished males would increase if they failed to rut entirely.

We are now ready to discuss the concepts of "maintenance energy requirements" and "energy balance", which to most people have become so familiar that there is sometimes more than a tendency to consider maintenance of energy balance almost as an end in itself, if not even synonomous with survival. However, we have now learnt that wild reindeer and caribou normally lose weight throughout winter. So, evidently they can survive perfectly well despite being in negative energy balance for months at a time. Clearly, therefore, maintenance of energy balance *per se* is not a prerequisite for survival. Moreover, an important point which often is ignored, is that slowed growth and even weight loss are not necessarily consequences of undernutrition, since we have seen that even captive reindeer fed high-quality food *ad libitum* show pronounced seasonal changes in fattening and body weight. As we shall see, later on, intrinsic cycles of growth and fattening, modified by photo-period, appear to be adaptations for survival in seasonal environments in which animals are confronted with long, predictable periods of potential undernutrition. Slowed rate of growth, and, to an even greater extent, actual loss of weight, have the effect of reducing an animal's daily energy requirements. This may be literally vitally important in winter when food is not only scarce and of poor quality but is also energetically expensive to aquire. In any case, it is misleading to regard the body weight of reindeer and caribou simply as

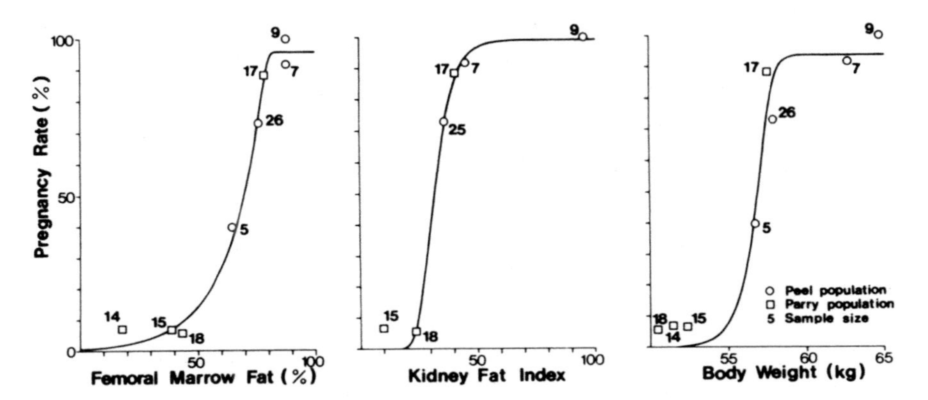

Figure 10.45 Relationships between pregnancy rates and condition indices in adult (more than 33 months) female Peary caribou collected in March and April (Thomas, 1982).

the product of some precarious balance between food intake and energy expenditure. Seasonal changes in body weight should be thought of, instead, primarily in terms of intrinsic cycles of growth and fattening mediated by appropriate changes in appetite rather than in terms of passive responses to seasonal changes in the quality, abundance and availability of food.

Starvation and hibernation

In spite of all the adaptations to a seasonally limited access to a changing energy density of the diet, some arctic forms are temporarily subjects to starvation, either as a result of acute lack of food, or as a lack of the ability to feed. We will address the latter first.

It is children's knowledge that bears go into dens in the late autumn and "sleep the winter away", to save energy when food is in short supply. In the case of polar bears, however, it is only the pregnant females that enter dens in winter, while subadults and males remain active throughout the year. Lønø (1970), who studied polar bears over a number of years in Svalbard, reported that the mature females accumulate an amazing 22 % of live body mass as fat, and Nelson *et al.* (1973), who primarily studied black bears revealed that bears, which neither eat, nor urinate or defecate during the denning period exhibit some rather amazing metabolic adaptations. Thus, Nelson *et al.* (1975) have shown that the female black bear is able to reabsorb both urea and water at the same rate as urine is produced, with the result that net protein catabolism is minimal and lean body mass is not reduced, while she is relying exclusively on stored fat to cover her energy expenditures. Unlike black bears and brown bears, the polar bear even seems to be able to pull this metabolic trick in response to temporary episodes of starvation, also outside the denning period (Ramsay *et al.*, 1989).

During the denning period these changes in the intermediary metabolism occur in concert with a reduction in metabolic rate of approximately 45 %. But, unlike small hibernators, like the arctic ground squirrel (*Spermophillus parryii*) that may cool even to sub-zero temperatures and become torpid during hibernation bouts (e.g. Barnes, 1989), the deep body temperature of bears is only reduced a few degrees, and the animals remain alert and may defend themselves and the cubs at any time. Thus, the beauty of the situation is that the bear can reduce its metabolic rate to almost the same extent as small hibernating animals and make their fat reserves last for several months without suffering the extensive cooling of the small hibernators, simply because of their prime insulation and, in particular, their large body mass with its low surface to mass ratio.

The arctic ground squirrel which is a relatively small (500-1,100 grams) hibernator, which is found throughout most of Alaska and mainland Arctic Canada, deserves further attention. This rodent regularly hibernates during winter in underground burrows, where the soil temperature may average -8 °C, with minima reaching -26 °C (Buck & Barnes, 1999a). This animal has typical 12-15 day hibernation periods separated by about 12 hours active periods throughout its 7 months hibernation season (e.g. Morrison & Galster, 1975).

Entry into hibernation in this species is usually preceded by a considerable fattening (e.g. Galster & Morrison, 1976; Buck & Barnes, 1999b), and usually, but not necessarily, triggered by low ambient temperature and lack of adequate food. Buck & Barnes (2000) have moreover shown that metabolic rate, body temperature and hence, the duration of the torpor bout is affected by ambient temperature in animals kept under experimental conditions, in such a way that each bout lasts for about 15 days at ambient temperatures around freezing, dropping to about 5 days at ambient temperatures of -20 and +20 °C. In a study of animals living in the wild, however, this relationship was confused , probably as a result of the use of food caches, different conductance of nests, or some other factor (Buck & Barnes, 1999a).

The yearly cycle of hibernation is also influenced by photoperiod and is controlled by hormonal cycles. It is therefore difficult, if not impossible, to induce hibernation during early summer, and especially during the reproductive season.

If body temperature decreases towards, or below, freezing, some hibernators die, while others respond either by arousal, or they avoid freezing by a regulated increase in metabolic rate. Thus, entry into hibernation, as well as the maintenance of the hypothermic state, is a well regulated process that results from a resetting of the "thermostat", demonstrating that the central nervous system is fully operational even at temperatures close to freezing in these species (Heller & Hammel, 1972).

Arousal is an active process that requires a considerable energy expenditure for a considerable time until the body temperature has reached normal. In the pocket mouse (*Perognathus californicus*), which has hibernation bouts of only a few hours, for instance, the cost of arousal amounts to 75 % of the total energy expenditure during the torpor cycle (Tucker, 1965). Thus, the arousing animal displays vigorous shivering, but before shivering starts, *brown fat*, which is found in all hibernating animals, is activated and produces heat at a very high rate (see pages 226–228 for further details). It is therefore characteristic of arousing hibernators that the temperature is not the same throughout the body (Lyman, 1948). In fact, measurements have shown that cardiac output initially is almost exclusively delivered to vital organs like CNS, heart and shivering muscles, but most of all to the brown fat (Johansen, 1961).

Another interesting example of more or less self imposed starvation in the Arctic is the fasting in both weaned and adult harp seals during moulting. We have already heard that the harp seal pups are abandoned by their mothers after weaning at the age of about 14 days, and that they subsequently moult their lanugo fur, while leading a "terrestrial" life on the ice for 2-3 weeks, without feeding or drinking. During the first weeks of this fasting period Nordøy, Ingebresen & Blix (1990) have shown in grey seal pups that their body mass decline exponentially, which implies that they lose a constant proportion of body mass each day; not a constant mass. We have also demonstrated that their metabolic rate is reduced by about 50 % (Fig.10.46), and that about 96 % of their energy expenditure is covered by combustion of their rich fat reserves, while only about 4 % is derived from catabolism of protein. This strategy is also reflected in high blood levels of free fatty acids and a marked increase in ketone bodies, such as beta-hydroxybutyrate, while blood glucose remains stable. Moreover, in spite of the fact that the animals don't drink during this period, they are only marginally dehydrated. This is due to the production and subsequent conservation of metabolic water, but in the end they still become hyperosmotic due to increased catabolism of protein, when sparing of the insulating blubber layer in preparation for life in cold water seems to gain preference.

Also the adult seals go through an annual moult each summer, during which they stay on the ice for most of the time and fast for periods up to 2-3 weeks. We have shown that, at least, hooded seals and harp seals fatten up after the termination of the breeding period for this occasion, and that they also reduce their met-

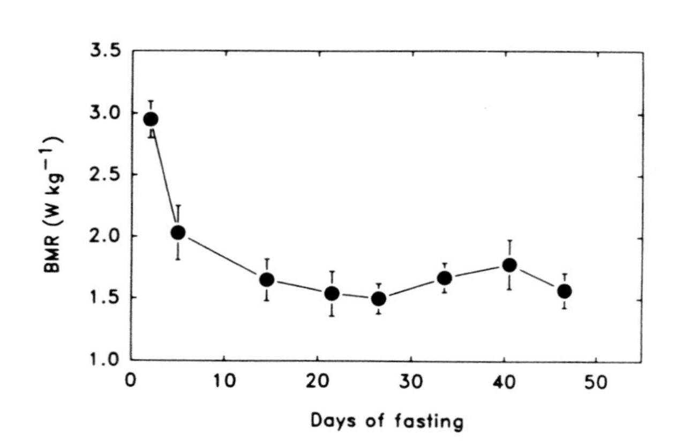

Figure 10.46 Basal metabolic rate from day 2 to day 47 of fasting in five grey seal (*Halichoerus grypus*) pups.Values are min ± SEM (Nordøy, Ingebretsen & Blix, 1990).

abolic rate substantially during fasting, presumably with much the same biochemical responses as in the young pups. There are probably several reasons for this annual "terrestrial" sojourn in the life of the seal, but most likely the fact that renewal of the hairs and, in seals, also the epidermis, require that the skin is extensively perfused with blood, which would have led to a catastrophic heat loss in ice water, is the most important one. But, evidence also suggests that the epidermal cells require rather high temperatures to be able to grow, and temporary haul-outs on the ice, where even solar radiation may aid in the warming, seems to be the obvious answer to these problems.

In the high Arctic, we have heard that the sun does not rise above the horizon for a period of up to 3 months during winter and that low temperatures and strong winds prevail to impose rather austere nutritional conditions on the animals that are stuck in the region. We have also heard, however, that most mammals and birds prepare themselves for the challenges of this long winter by deposition of large quantities of body fat during times of plenty in the autumn. Working at the time with Mortensen in our laboratory, we were able to show in the Svalbard ptarmigan that these fat reserves are not, as previously thought, used to supplement the daily energy budget during winter (Mortensen & Blix, 1985). It appears, instead, that they are carefully kept in store for episodes of acute starvation which might follow the winter storms. It is also most interesting that the fat reserves seem to be managed according to a seasonally *changing set-point* for body weight, in such a way that they are steadily reduced from their maximum at the time when the sun disappears in the autumn, down to zero at the time when the sun reappears in the spring. Thus, after the birds have been starved they subsequently overeat to replenish their fat reserves, not to the same level as they were before the period of starvation, but to the (lower) level birds which have not been starved would have kept their stores at the same time (Fig. 10.47). This is in keeping with the time-honoured principles of insurance policy, because here as everywhere else there is a premium to pay, which in this case is the considerable costs of carrying all this fat, and as the chances of death by starvation is steadily diminishing as the light season approaches, the insurance is reduced! It is also well worth noticing that we have found that the Svalbard ptarmigan, unlike most other birds, do not respond to starvation with increased activity, but do instead put on the fatalistic attitude of sitting quietly in wait for better days. This way, which seems to be the only adequate way, the fat reserves will, of course, last much longer. In fact, we have calculated that they would last for as much as 16 days during the most critical period from October till December.

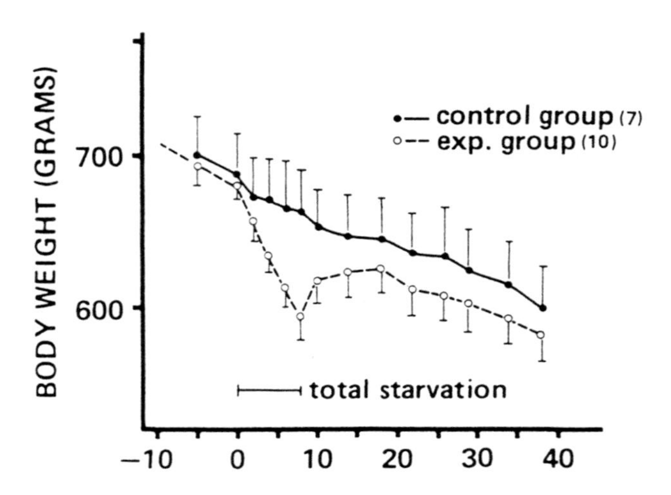

Figure 10.47 Food intake (mean ± SEM) of captive Svalbard ptarmigan before, during, and after a period of total starvation, starting at day 0. The control group was fed *ad libitum* throughout the experimental period (Mortensen & Blix, 1985).

Recently, we (Fuglesteg *et al.*, 2005) have also found evidence to the effect that the same principles for a changing set-point regulation of the fat reserves are in operation also in the arctic fox (Fig. 10.48), which even more than the ptarmigan is likely to be left without supper on occasions during the dark arctic winter.

The reindeer and muskoxen, being ruminants, face another grave problem, in addition to the purely energetic one, when met with episodes of starvation, which often occur due to overicing or heavy snowfall on their already marginal ranges. That is that, when left without food, the microorganisms of the rumen starts to die off at an alarming rate. This problem has been studied by Mathiesen and associates who found that the bacteria are reduced to only 0.3 % of the original population already after 3 days of starvation (Mathiesen, Rognmo & Blix, 1984). It goes without saying that the species of microorganisms that show the best tolerance to starvation, and hence, increase enormously in relative numbers in the rumen, are not necessarily those that are the most useful in effectively digesting what food eventually is provided to them. It therefore so happens that reindeer may develop grave gastro-intestinal disorders, and even die, upon refeeding after episodes of starvation. Luckily, however, it appears that lichen, which is usually prominent in the winter diet of continental reindeer, anyway, is the best "restarter-feed" for these animals, but still there is a limit to how long the animal can postpone to get its "engine" re-started.

Foxes are special in the Arctic in that they have two ways by which they store energy reserves in summer for subsequent use during winter. That is by the now

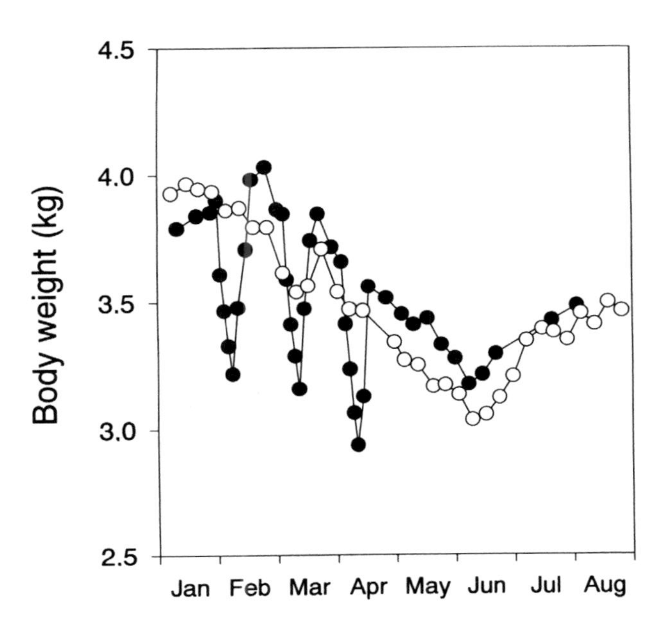

Figure 10.48 Mean body mass in four arctic foxes fed *ad libitum* from January to August (open circles), and in the same animals the year after, when they experienced three periods of acute starvation during winter (filled circles) (Fuglesteg *et al.*, 2005).

usual deposition of fat and by surplus killing and subsequent hoarding of food. Both so called scatter hoarding (hiding single or small numbers of prey at dispersed sites) and larder hoarding (hiding many prey items at or near den sites) have been documented (Sklepkovych & Montevecchi, 1996). Most such hoarding is associated with a superabundance of prey such as in birdcliffs, and the hoarding behaviour is likely to vary widely according to local conditions. The largest documented arctic fox larder contained no less than 136 sea-birds, but arctic foxes are also classical scavengers that feed opportunistically on dead reindeer and seal carcasses left by polar bears. Still, the availability of food to arctic foxes fluctuates both on a seasonal and on an annual basis, and starvation in winter is thought to be an important cause of mortality, particularly among juvenile foxes, which store less fat than the adults.

Underwood (1971) studied arctic foxes in Alaska, and found that the fat deposits, based on subjective fat indices, decreased from December to March, and concluded that fat is deposited in defence of food shortage during winter, while another study indicated that the fat deposits, instead, increased slightly over the same period. More recently, Prestrud & Nilssen (1992) have studied the arctic foxes in Svalbard, by determining body composition of minced carcasses of animals that had been trapped over a 7-year period. They found that fat was deposited both subcutaneously and viscerally in September-October,

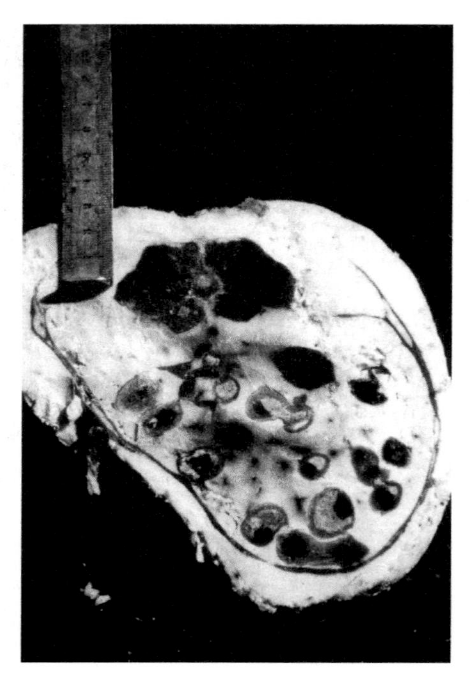

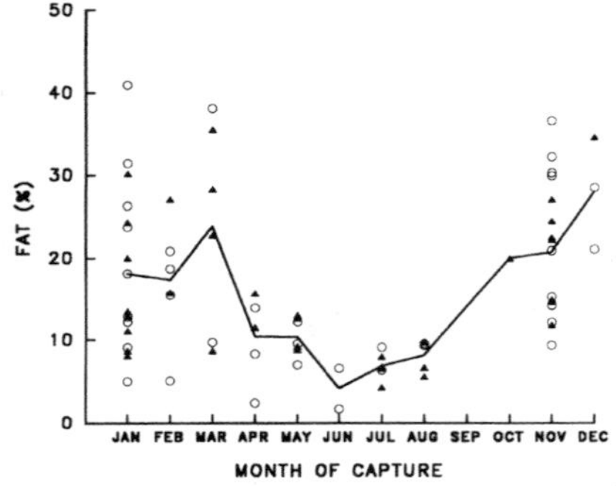

Figure 10.49 *Left*: Cross section through the frozen carcass of an arctic fox in winter, showing extreme fattening. (Photo: P. Prestrud). *Right*: Seasonal changes in fat content (% of skinned carcass mass) of arctic foxes from Svalbard, Norway. The line connects the means in different months. Males (open circles) and females (closed triangels) (Prestrud & Nilssen, 1992).

and reached, on average, a fairly stable maximum of about 20 % of the skinned carcass in November through March, whereafter it dropped to its all-time low of about 6 % in June (Fig. 10.49). However, while some of the foxes showed the record fatness of 40 % of carcass mass during winter, 15 % of the animals did not possess any visible fat deposits in winter. Thus, the fatness of the foxes differed in the extreme within any given month, probably just reflecting the dependence of the animals on these deposits in a situation with the sometimes rather wild variation in food supply which follows from the lack of lemmings in their habitat.

Seals also undergo marked seasonal changes in body fat, but interestingly, they show a high degree of fatness during winter, but lose weight during the moult in summer and stay lean until the autumn (Fig.4.1), when most arctic marine fish and invertebrates increase their fat content (Fig. 3.4), and hence, their energy density substantially. However, due to the dependence of marine mammals on the use of blubber for insulation against the chill of the arctic seas their blubber deposits are normally not depleted more than some 50 %, but high densities of animals and low abundance of food may sometimes result in lower than normal blubber thicknesses both summer and winter.

Adaptations to diving

Everybody know that seals spend most of their time, and whales all their time in water, but recent research has shown that several species of both orders spend as much as 80-90 % of the time *under* water. How is this achieved? Let us first look at what physiological problems life under water impose on air-breathing animals like seals and whales, and thereafter look more closely at the lifestyle of each individual species.

When mammals dive respiration has to stop immediately if drowning is to be avoided. But, since the tissues and cells continue to metabolize and blood circulation, in principle, is maintained this results in an ever-decreasing arterial oxygen tension and an ever increasing arterial carbon dioxide tension (Fig. 10.50), as first shown in a series of elegant experiments by P.F. Scholander (1940) in Oslo before the second world war.

Now, there is a straight-forward solution to this problem: when you cannot renew your oxygen pool, you bring as much oxygen as you can, you economize with it to the best of your ability, and you do so from the very moment of submersion. That is if you want to extend your diving capacity as much as possible. But, seals and whales do not always want to do that, and in the following we shall see how these animals have adapted to a variety of different ways of diving.

Common for all diving mammals (and birds) is that their oxygen stores have been increased. This is achieved in two ways. The blood volume of expert divers is more than twice that of terrestrial animals and the percentage of red blood cells, or the hematocrit value, in the blood is increased from about 45 % in terrestrial mammals to 65 and sometimes 70 % in diving mammals. This implies that the

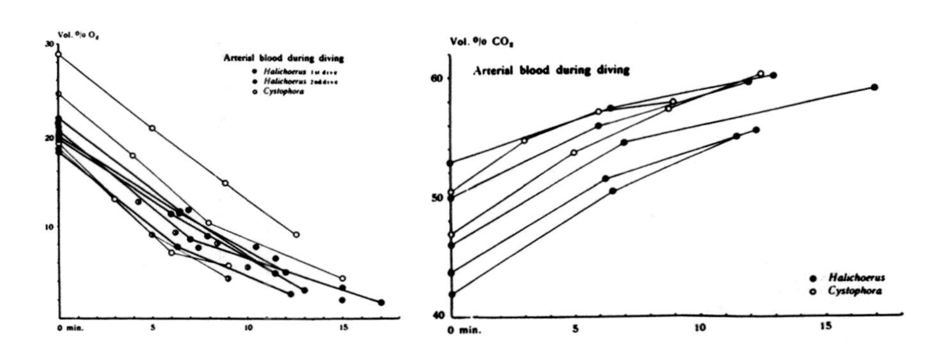

Figure 10.50 The basic problem facing air-breeding animals when submerged: Constantly decreasing arterial content of oxygen (*left*) and constantly increasing arterial content of carbon dioxide (*right*). Abcissa: Duration of dive (minutes) (Scholander, 1940).

amount of hemoglobin is sometimes 4 times greater in diving than in terrestrial animals, and the oxygen-carrying capacity of the blood equally improved. Diving mammals unlike their terrestrial counterparts also have a very high concentration of another oxygen carrying molecule, myoglobin, in their skeletal muscles and heart, and we shall se later how this second oxygen store is put to use during diving. The oxygen reserves of the lungs have very little use in diving mammals, partly because they usually expire prior to a dive to reduce buoyancy and to avoid divers disease (the "bends"), and partly because the lung is fully collapsed already at a depth of less than a hundred meters in most species.

Now, let us see what happens when a diving mammal, like the seal, is forced under water experimentally, since this is how we have learnt most of what we know of physiological adaptations to diving. It has been known for more than a hundred years that animals respond to forced submergence with a profound, and, in the case of seals, abrupt reduction in heart rate (Fig. 10.51), but again it was Scholander (1940) who first was able to put this dramatic, so called bradycardia, into perspective. In an elegant series of experiments, he demonstrated that the bradycardia is developed in concert with a widespread peripheral arterial constriction (Fig. 10.52). The result of this vasoconstriction is that the cardiac output almost exclusively is distributed to the brain (Blix, Elsner & Kjekshus, 1983), while the heart receives a much reduced supply (Blix *et al.*, 1976; Elsner *et al.*, 1985) and other organs, like the gut, kidneys and skeletal muscles are left without blood and have to rely on anaerobic metabolism, or in the case of muscle, the oxygen bound to the myoglobin. The end result of all this is that the oxygen store in the blood is deliv-

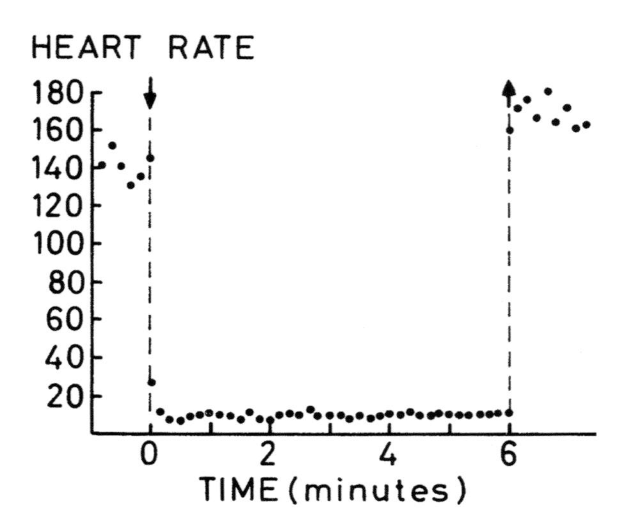

Figure 10.51 Heart rate response of a harbor seal during an experimental dive lasting 6 minutes (Elsner, 1965).

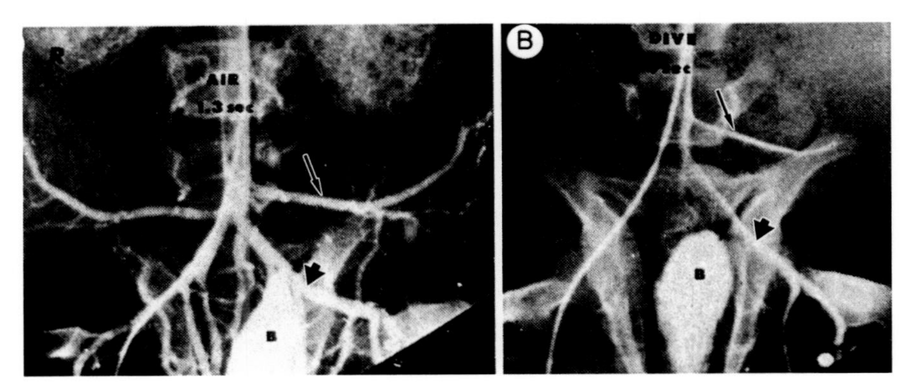

Figure 10.52 Angiograms of peripheral (abdominal) arteries of a harbour seal. **A**: During breathing in air at surface position, well-filled arteries of flanks *(thin long arrows)* and hind flippers *(short thick arrows)* are seen. **B**: during diving, same arteries in same animal are profoundly constricted and consequently poorly filled with contrast medium. Also shown is bladder (marked B) (Bron *et al.*, 1966).

ered selectively to the most oxygen sensitive organs, while arterial blood pressure is maintained due to reduced cardiac output as a result of the reduced heart rate. It follows that by such means the diving capacity of the animal can be much increased, but there is a price to pay! That is that the uncirculated, and hence, eventually rather ischaemic organs will produce and accumulate lactate, which is released into the blood when the animal returns to the surface and breathing is resumed. This lactate has to be eliminated by the liver to avoid undue pH problems before the animal can dive again, and it has been shown in the Weddell seal by Kooyman *et al.* (1980) that the recovery time at surface increases exponentially with the duration of the previous dive (Fig. 10.53). This implies that if the animal indulges in very long dives then, paradoxically, the total time the animal can spend submerged during a day will be reduced, when compared with a diving schedule of a series of short dives. We shall see now why this is so.

With the introduction of biotelemetry in the 1970ies it soon became clear that most seals and whales did not usually take advantage of their full diving capacity, but did instead perform series of short dives, during which there was no apparent cardiovascular adjustments. This caused quite some confusion at the time, and some even suggested that the time- honoured cardiovascular reactions to diving were, in fact, stress responses to the experimental insult (Kanwisher *et al.*, 1981), while others on the basis of brain-work and physiological insight argued that expert divers have voluntary control over their cardiovascular system and are able to respond in accordance with the challenge of each individual dive (Blix & Folkow, 1983; Blix, 1987).

So what happens during a series of short dives without apparent cardiovascular adjustments? The answer is: not much, except that the arterial oxygen content is now reduced at a much higher rate, a rate which is strongly influenced by the swimming activity of the animal. This is due to the fact that the working muscles are now circulated, and the muscular activity is therefore sustained from the blood oxygen stores, since myoglobin, which has a much higher affinity for oxygen than hemoglobin, will stay fully saturated almost as long as there is oxygen left in the blood (Fig. 10.54). In these cases the animal will therefore return to the surface to breathe with the blood oxygen stores pretty depleted, but without a lactate load, since the muscle (myoglobin) stores have been fully saturated throughout the dive. It follows, that the animal may continue to perform such short dives, only separated by short stops at surface, all day at end, and thereby manage to stay submerged for some 80 % of the time, but with little room for operations which call for continued attention each time. Such short dives are usually referred to as aerobic dives and several authors have recently wasted much time on the rather useless concept of aerobic dive limits (ADL). The ADL of an animal is determined as the fraction of the animals metabolic rate, or oxygen consumption, against estimated total body stores of oxygen. However, several investigators have found that ADL as calculated in this way very often falls short of the diving times which are actually recorded in many species of seals. The reasons for this may well, at least in part, be found in that we (Odden *et al.*, 1999) have found that seals, and even ducks (Caputa, Folkow & Blix, 1998) are able to selectively reduce their brain temperature during diving (Fig.10.55). This ability, which most likely is achieved by a controlled perfusion of the skin of the fore-flippers, which will bring cold blood into circulation, will reduce the metabolic rate of the brain, and hence its oxygen consumption, some-

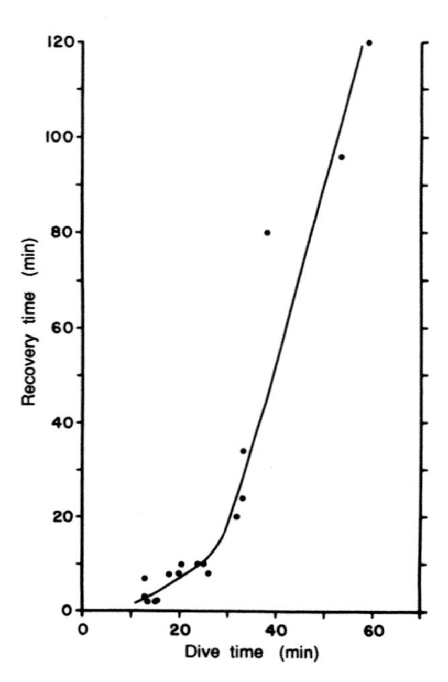

Figure 10.53 Recovery time required for various dive durations in Weddell seals. Recovery was considered complete when arterial lactic acid concentrations had returned to pre-dive levels (Kooyman *et al.*, 1980).

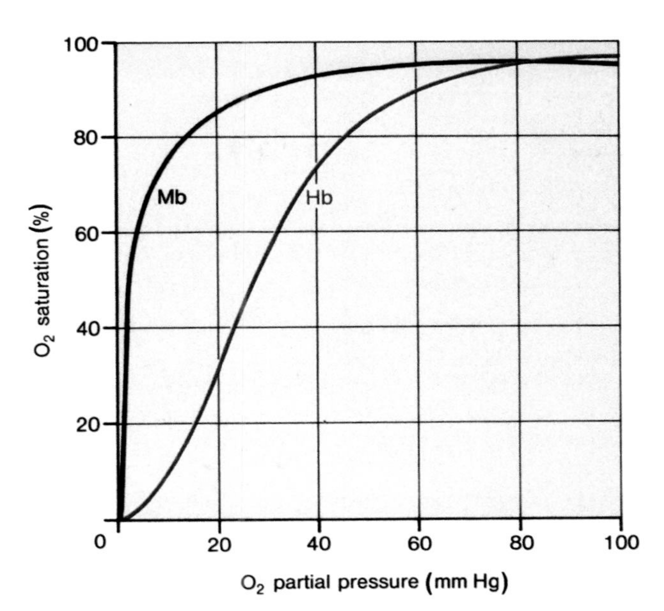

Figure 10.54 Hemoglobins (Hb) with multiple heme groups have sigmoid oxygen dissociation curves, whereas myoglobin (Mb) with only a single heme group has a hyperbolic dissociation curve. P_{50}, the partial pressure at which a respiratory pigment is 50 % saturated with oxygen, is a measure of oxygen affinity.

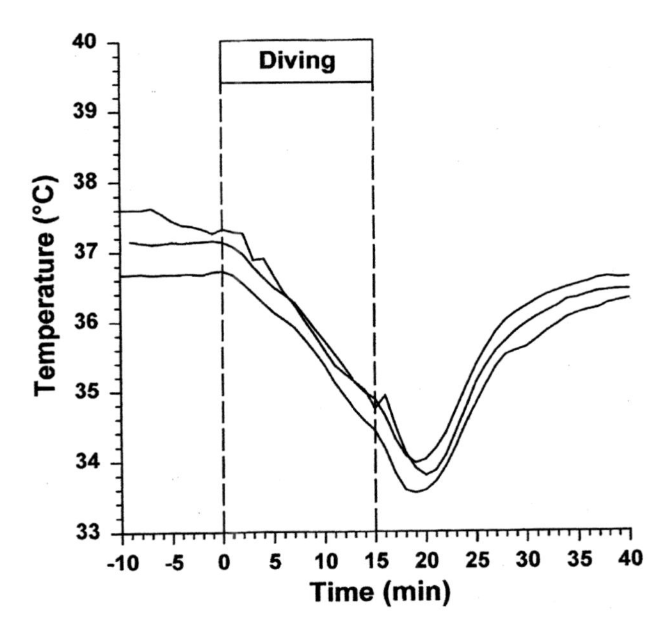

Figure 10.55 Changes in brain temperature during dives of 15 min duration in a juvenile (65 kg) female hooded seal (Odden *et al.*, 1999).

times as much as 20 %. This will of course increase the ADL in all circumstances, even during short aerobic dives, since it is primarily the brain which draws on the oxygen reserve in the blood. Recently, we (Kvadsheim, Folkow & Blix, 2005) have also shown that, even hypothermic, seals respond to (experimental) diving

Figure 10.56 Heart rate during a voluntary dive in grey seal (*Halichoerus grypus*) during diving at sea, obtained by radiotelemetry. Arrows mark the start and the end of the dive. During this dive heart rate averaged 6.5 beats min^{-1} and was below 4 beats min^{-1} for 90 % of the time (Thompson & Fedak, 1993).

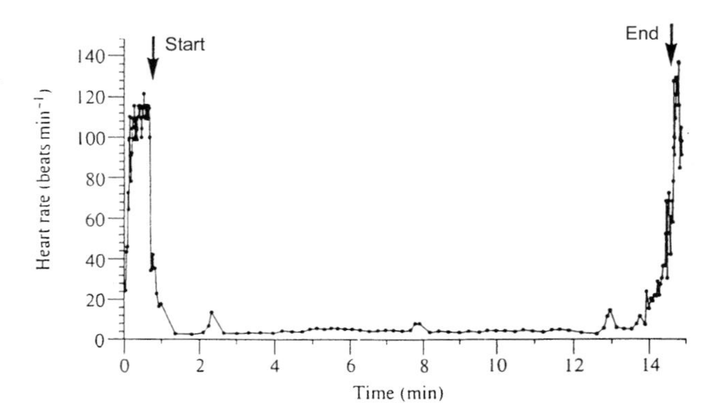

with an almost instantaneous inhibition of shivering. This thermogenic "blocade" allows cooling of the brain and other circulated tissues, whereby oxygen consumption is decreased and diving capacity extended.

If the seal should decide, for one reason or another, instead to embark on a long dive it will activate all its defensive mechanisms, including peripheral vaso-constriction and bradycardia, to the best of its ability, and do so from the very beginning of the dive. In so doing, the brain will be the main consumer of the blood oxygen, while the muscles, due to their isolation following the peripheral vaso-constriction, can now utilize the oxy-myoglobin and eventually turn to anaerobic metabolism. In such circumstances even a moderate brain cooling will extend the animals diving ability appreciably.

Time has shown that seals occasionally do, indeed, perform just like that, since Thompson & Fedak (1993) in Britain have shown that freely diving grey seals (*Halichoerus grypus*), some of which are in the habit of performing long duration dives, display a spectacular bradycardia (Fig. 10.56), while others which are in the habit of performing series of short dives do not.

The understanding of the finer details of the cardiovascular responses to diving and the historical development of this dramatic field of comparative physiology, which has attracted so many dramatic personalities over the times are sometimes beautifully outlined in several review articles (Andersen, 1966; Blix & Folkow, 1983; Butler & Jones, 1997).

Light

Arctic regions are more than anything else typified by long nights in winter and long days in summer. At the North Pole, the sun is continuously above the horizon for 6 months and continuously below it for the other half of the year (Fig. 1.1). In the arctic regions further south where animals live, periods of continuous light and darkness are gradually shorter down to the Arctic Circle, at which there is only one day of midnight sun and one day, when the sun stays below the horizon, each year. During the periods of alternating day and night between the periods of darkness and midnight sun the day length may change, say at Longyearbyen in Svalbard, by more than 30 min per day. *These rapidly changing photoperiods and, in particular, the long periods of continuous light and darkness are probably the only unique features of the Arctic region and the adaptations to it in fauna and flora the only thing that makes Arctic biology truly special.*

The most reliable source of information of the time of the day is the changes in the position of the sun as the earth turns around its axis, and at mid- and high latitudes the changes in photoperiod which follow from the rotation of the earth on its tilted axis around the sun, the best information about the seasons. The changes between day and night therefore provide information of both time and date.

Already in 1968, West kept willow ptarmigans at Fairbanks, Alaska, and found that they started their activity in the morning at the beginning of *civil twilight* and became inactive in the evening when civil twilight ended (Fig. 10.57). Civil twilight is defined as the indirect illumination that occurs when the sun is between 0 and 6° below the horizon, *at any place and elevation.* This is not to be confused by *nautical twilight* which is the illumination which follows when the sun is between 0 and 12° below the horizon *at sea level.* This implies that the ptarmigans are active during the day, but West also found that a major part of their activity took place within the periods of civil twilight throughout a major part of the year (West, 1968). Indoors, Stokkan has demonstrated a similar pattern in both captive willow and rock ptarmigan (Fig. 10.58). This kind of activity pattern with two maxima every day is typical for a number of both mammals and birds, and it is likely that herbivorous animals, like the ptarmigan, increase their feeding activity in the evening in preparation for a long quiet night, while the activity in the morning simply reflects that it is hungry after a night without eating. In addition, being less active during the lightest part of the day may also reduce the risk of being eaten by predators. It may appear from the figure that the ptarmigans responded directly to the dawn and sun-set with increased activity, but that is not the whole story.

It is by now well documented, both in animals, plants and microorganisms that most, if not all, of the 24-hour rhythms they display in their natural habitats do not simply reflect changes in their environment. In fact, in most cases the observed rhythms are endogenous and of physiological origin. Thus, while

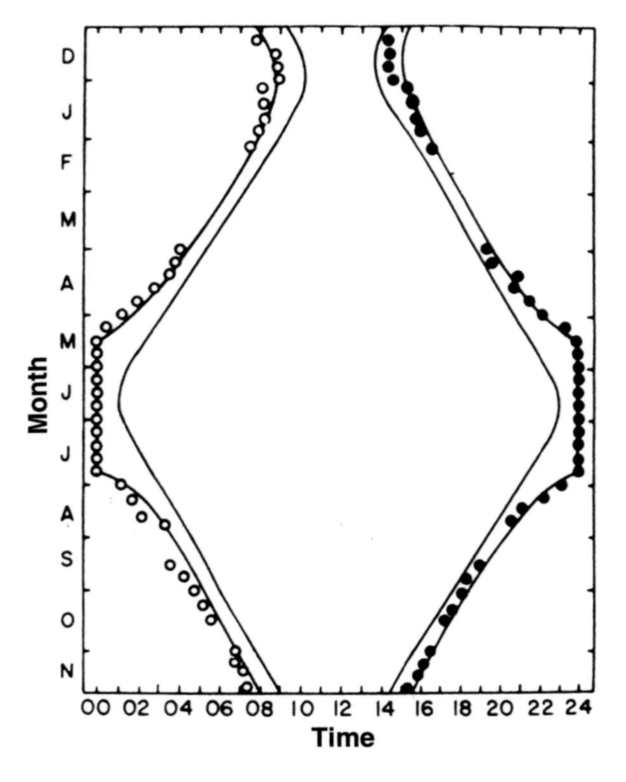

Figure 10.57 The beginning (open circles) and end (closed circles) of the daily activity period for willow ptarmigans kept outdoors in captivity at Fairbanks, Alaska, throughout the year. The outer lines mark the beginning and the end of civil twilight, and the inner lines the time for sunrise and sunset (West, 1968).

we can not completely rule out that the increased feeding activity in the morning is a response to the coming of the light, the increase in activity *before* sun-set (or light off in Fig. 10.58), must be caused by something from inside the animals themselves, and the only way the animals can know when the sun will set is from experience from the previous few days. And to make use of this experience the animal must be able to somehow measure time. Now, of course, the ptarmigans are not conscientiously concerned about time all day, but for the time being we can say that they depend on endogenous clocks. These clocks are oscillators with a period of approximately 24 hours that are located in the central nervous system of the animal. Our knowledge of how these clocks operate have traditionally been obtained in animals that are kept under absolutely controlled conditions.

A common denominator of rhythmic phenomena, like feeding activity, is that in the absence of any rhythmicity in the environment they never repeat themselves with a period of exactly 24 hours, and therefore they are known as "circadian" rhythms, from *circa dies*, which in latin means approximately one day. When these rhythms repeat themselves under natural conditions with exactly 24

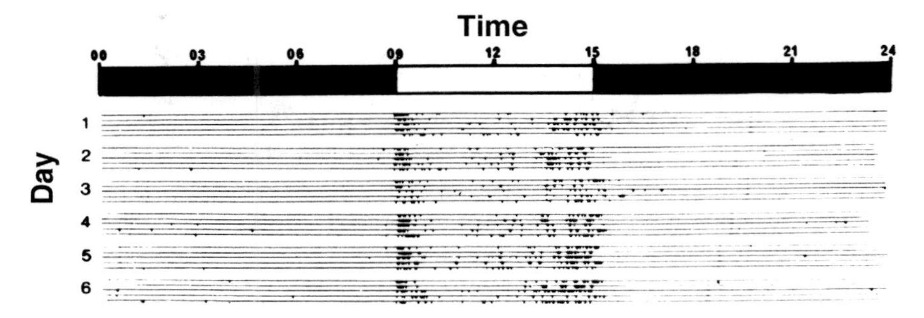

Figure 10.58 Feeding activity of five willow ptarmigans kept in captivity indoors at the University of Tromsø, recorded over a period of 6 days. The light was turned on abruptly at 0900 h and off at 1500 h every day (Stokkan, unpublished).

hour periodicities, they reveal another important feature, and that is that they are influenced by the environment. In that case we say that they have been syncronized, or *entrained*, by some external factor, of which the most important usually is the changes in light intensity between day and night. Such factors that are able to syncronize the rhythms are called *Zeitgebers* because this concept was developed by Aschoff, who was German, and these "time-keepers" are supposed to make sure that the clocks are on time so that things happen at the right time of the day. This is achieved by what is called a phase control, such that the endogenous rhythm attains a certain phase relationship with the exogenous rhythm. One such example is shown in Fig. 10.58, where the ptarmigans begin their afternoon feeding bout 4 hours before the light goes off.

In Fig. 10.59 we can see how the perch-jumping activity of a captive sparrow first is syncronized by the light-dark (LD) cycle indicated at the top of the diagram, while when kept in complete darkness (DD) it becomes free-running with a period, which was longer than 24 hours. The activity period is therefore displaced a bit every day. The results of another experiment which was made by Gabrielsen in our laboratory are illustrated in Fig. 10.60, showing the changes in deep body temperature in a willow ptarmigan, as measured by use of radio telemetry. The body temperature of these birds changes substantially throughout the day, being high during day-time and low at night. It is also quite clear that the birds are getting ready for the morning activity way ahead of time by increasing their body temperature several hours before the light goes on. More important, in the two days when the light was permanently turned off, body temperature still changed pretty much as before and at the same time as before, since two days is too short to reveal the circadian drift of the rhythm.

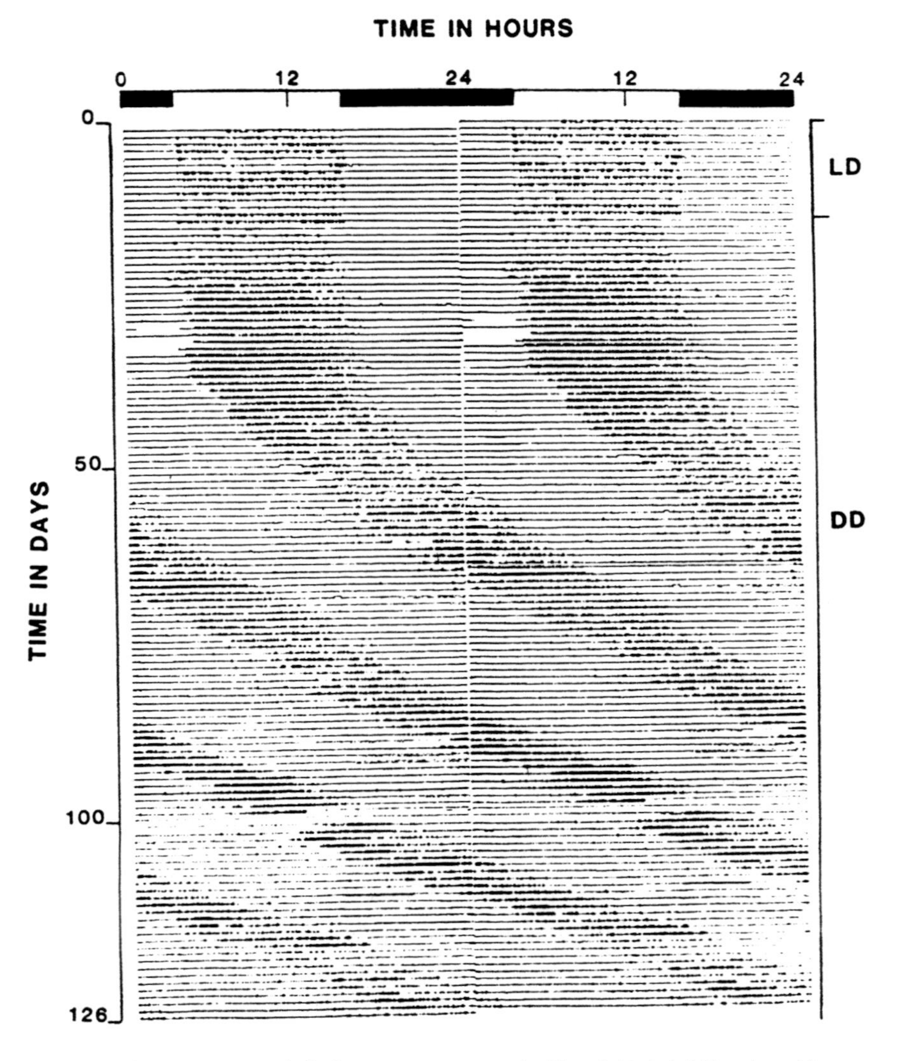

Figure 10.59 Locomotor record of a house sparrow syncronised to a light-dark (LD) cycle and free-running in darkness (DD), with the birds own characteristic period and phase (Cassone & Menaker, 1984).

So far, we have seen how the changes in day-length can act in concert with endogenous rhythms and tell the time of day, but how can the light regime provide the animals with a calendar? We know that several of the phenomena, like breeding, the autumnal deposition of fat and migration, that repeat themselves on an annual basis are also controlled by endogenous rhythms. Such rhythms

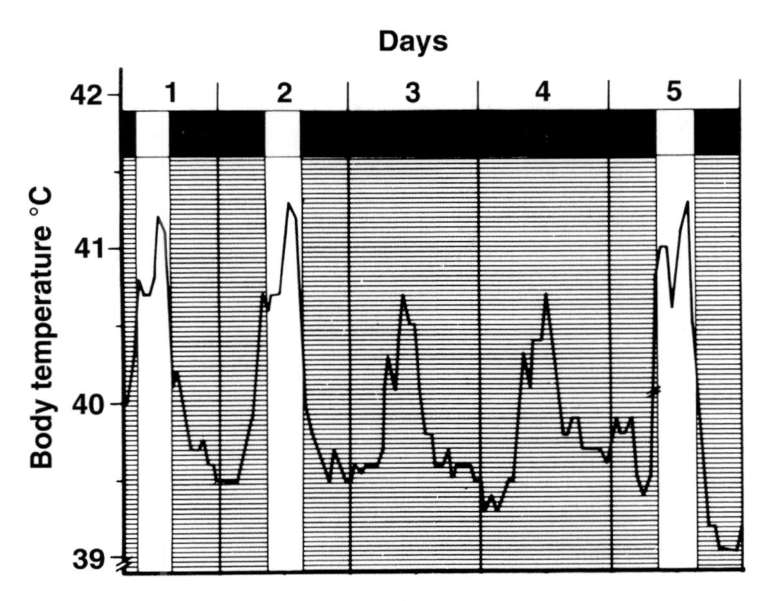

Figure 10.60 Changes in body temperature of a willow ptarmigan in captivity, recorded by use of radiotelemetry. The birds were kept indoors and the changes in light conditions are shown at the top of the figure. The light was turned on and off abruptly. Note that the animal maintained its daily body temperature rhythm even during two days without any light (Gabrielsen, unpublished).

have a periodicity of approximately one year, they are *circannual*, and the light conditions are again the most important Zeitgeber.

The most useful hypothesis for how daily photoperiodic information is converted into annual rhythms was already developed by the German biologist Bünning in the 1930ies. Bünning assumed that there exists a circadian rhythm for light (intensity) sensitivity. This implies that there is every day a period when the organism is sensitive to light, and that if light is perceived during this period, then a phenomenon, such as, for instance, the release of a certain hormone, will happen. It follows, that it is not the total amount of light that is measured, but *when* light occurs during the day. Thus, birds like ptarmigans have a circadian rhythm for light-sensitivity that is such that if they are exposed to light when they are light-sensitive, then they will produce and release, for instance, a hormone. The light-sensitive period is known to occur late in the day, and during winter, in the sub-Arctic, when the daily photoperiod is short, the light-sensitive period will occur during the night, and light then only serves one purpose; that of syncronizing the circadian rhythm. When the length of the day increases in the spring, the sensitive period will again, sooner or later, fall within the light period, and thereby cause certain hormones to be released, and, for instance, trigger gonadal growth and activation. As a consequence of these mechanisms, the extreme photoperiodic conditions in the Arctic stimulate gonadal activity in spring before environmental conditions, as snow cover, ambient temperature, and food availability, allow successful breeding to take place. At 70°N, for example, willow ptarmigan breed in late May after exposure to continuous light for one month while, at 80°N, Svalbard ptarmigan must control itself for more than two months of continuous light with maximal gonadal stimulation before breeding can commence in June. Thus, these birds, and presumably other resident animals at very high latitudes, possess physiological mechanisms which compensate for the fact that their biological clocks were phylogenetically established at much lower latitudes, where the photoperiodic and phenologic relationships are different.

Stokkan has studied these processes in willow ptarmigan in association with Sharp, and they (Stokkan *et al.*, 1986) found that reproductive activity is stimulated in the males when the daylength exceeds 12 hours in February-March. This was demonstrated by increased plasma levels of luteinizing hormone (LH) and testosterone and growth of supraorbital combs (Fig. 10.61). Maximum secretion of these hormones was reached in June, when the day has become continuous, and breeding occurs. The hormone levels then fell in July, signalling gonadal regression and that the birds had become what we call photorefractory, i.e. they no longer respond to the long days. This is important to ensure that chicks are not produced too close to the arctic winter. The neuroendocrine mechanisms involved in the development of such long-day refractoriness are not known.

The fact that birds, like ptarmigan, obviously are completely dependent upon being able to time such a crucial event as reproduction by responding to the rapid changes in day-length in the spring, begs the question of how they are able to respond. The physiological link between the organism and its photoperiodic environment is mediated by a rhythmic production and secretion of the hormone melatonin from the pineal gland, which serves as both clock and calendar. Daylight inhibits production of melatonin and the light-dark cycle syncronizes a biological clock which drives the pineal rhythmicity. Consequently, the concentration of melatonin in the blood is therefore high only during the dark phase of a 24-hour period and the duration of the nocturnal increase in melatonin closely matches the duration of the night.

Melatonin, which is a fat-soluble hormone, with the ability to cross any cellular boundry, was first isolated as late as in 1958. This hormone is synthesised from serotonin (the aminoacid tryptophane) through a cascade of enzymatic reactions. The control is somewhat different in birds and mammals in that mam-

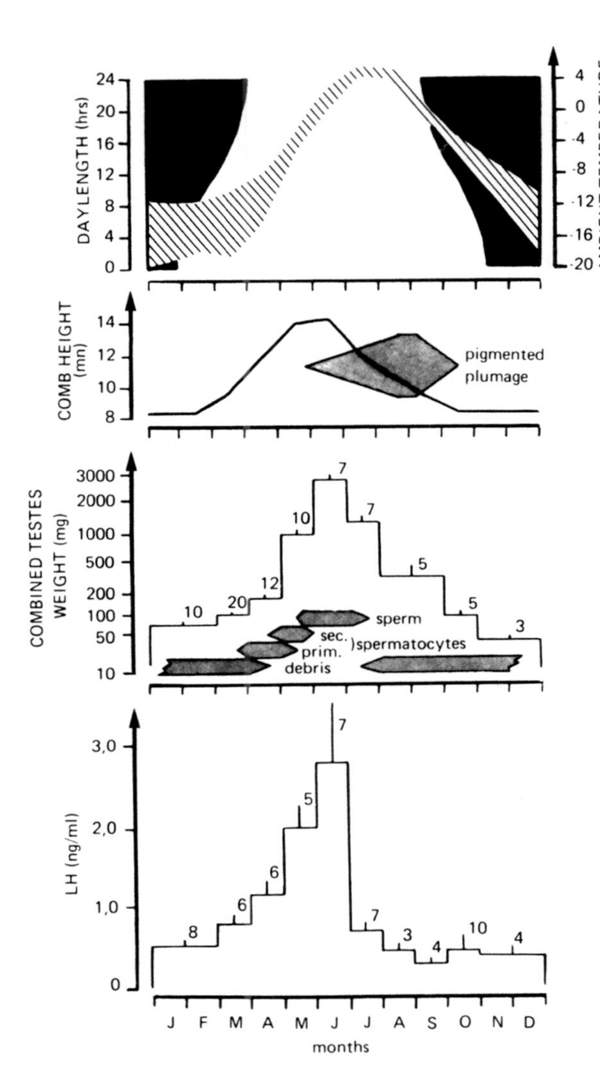

Figure 10.61 Annual changes in plasma luteinizing hormone (LH) levels, testicular weight, stages of spermatogenesis, height of the supraorbial combs (above the eyes), and plumage from Svalbard ptarmigan males shot at two locations (78 and 79°N) in Svalbard. In the top panel are shown the annual changes in daylight and ambient temperature (hatched area) in Ny Ålesund (79°N). Daylight includes the period of sunlight plus the morning and evening hours of civil twilight. Ambient temperature is represented by the area covering one SD on each side of the monthly mean (Stokkan, Sharp & Unander, 1986).

mals rely only on light syncronizing through the eyes, entraining a brain-clock which subsequently drives a passive pineal gland, while the avian pineal is light-sensitive in itself and have also built-in clock-mechanisms. In any case, this represents a neuro-endocrine transduction of photoperiodic information which is believed to be phylogeneticaly very old. A super simplified diagram of the control of the melatonin production is presented in Fig. 10.62.

These basic features, which cause a never-ending cycling of melatonin production in animals living at lower latitudes, begs an intriguing question: How do arctic animals gain sufficient photoperiodic information to syncronize their

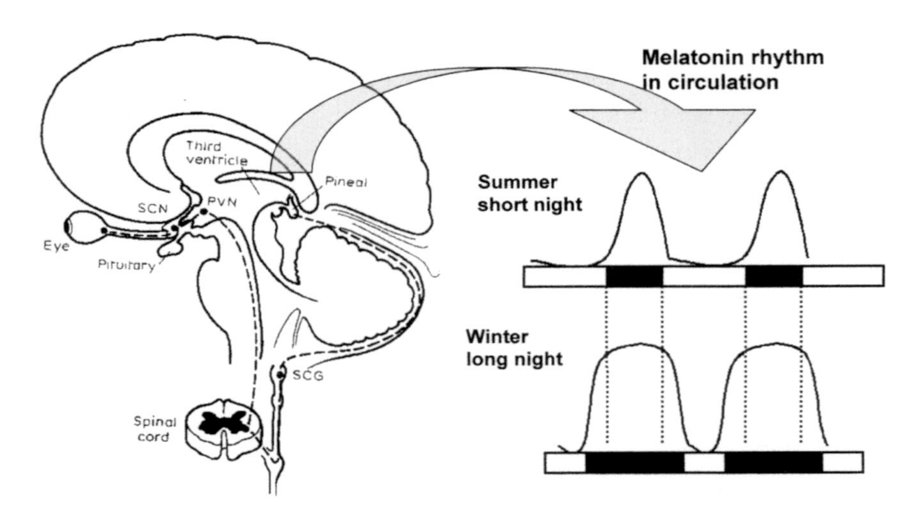

Figure 10.62 The production of the hormone, melatonin, and its release from the pineal gland in the brain is inhibited by light. Light which reaches the eye excites receptors in the retina to produce nerve impulses that reach the pineal by the long way of the supra-chiasmatic nuclei (SCN) at the base of the brain, via the paraventricular nuclei (PVN), down the spinal cord to the superior cervical ganglia (SCG) of the sympathetic nervous system, and from there up to the pineal gland. The resulting day – night rhythmicity in melatonin production, help synchronize the circadian, or daily, rythms of the body (Redrawn from Tamarkin *et al.*, 1985).

clocks when winter darkness prevails for months? Or, even worse, when the sun never sets for more than 4 months. This question may be of paramount importance, because, as I have already pointed out: it addresses the only adaptation which is truly unique to arctic animals!

Measurements of melatonin in the blood of ptarmigans and reindeer under natural light conditions at 70°N by Stokkan and associates (Stokkan, Tyler & Reiter, 1994; Reierth, Van't Hof & Stokkan, 1999) revealed unique patterns that are found in no other animals. They both appear to turn off the rhythmicity during summer and presumably also, to some extent, during mid-winter. Similar patterns also emerged when they monitored activity rhythms in these animals (see below). These results may reveal one fundamental adaptation to high-arctic photoperiodic conditions, whereby endogenous clock-mechanisms are turned down during those parts of the year when there is no ambient rhythmicity and when what I have called the "arctic resignation" (Blix, 1989) may be the optimal strategy of behaviour.

In an early attempt to reveal how the continuous polar day of the high-Arctic might convey temporal information Krüll (1976a) kept finches in dark cages and circled a light, reminiscent of the sun, around the cage with a period of 24 hours.

He found that this light source indeed acted as a weak Zeitgeber for the activity rhythm in these birds. So encouraged by this result he launched an expedition to Svalbard. There he studied the snow buntings during their breeding period and found that they were feeding actively during the "day" and resting during the "night" (Krüll, 1976b). To his elation, he moreover found that captive snow-buntings, as well as a greenfinch, which he had brought with him from Germany, also were active during the day and rested from about 2200 h to 0100 h every night. These birds were kept in cages in a valley near Longyearbyen, where they could *see* the sun only during mid-day and mid-night. Now he surmised that the very slight differences of light intensity which indeed exist between "day" and "night" at this high-arctic location in summer, cannot be effective as Zeitgebers, and, supported by other experimental work, he landed instead firmly on the conclusion that , "obviously, it is not the brightness but the colour temperature (of the light) that is effective as a Zeitgeber " (for the activity pattern in these birds in Svalbard).

These rather intriguing studies were followed by a study by Johnsson *et al.* (1979) in man, who is yet another migrant from the temperate zone to the Arctic. Unlike Krüll (1976b), they found that two graduate students that were studied over a period of 19 days under continuous light conditions showed free running rhythms with a period of about 26 hours and clearly were not affected by any Zeitgeber.

Now, where does this bring us? It brings us to the very obvious question of how do the *resident* species in Svalbard react to prolonged periods of continuous light, and even much more interesting; how do they react to the continuous darkness of the high arctic winter? At our laboratory in Tromsø we started to address these questions in the early 1980ies and we chose to study the control of the annual body mass cycle in Svalbard ptarmigan, being the most northern herbivorous resident of all birds. First, we studied the feeding activity rhythm in a group of captive birds that were offered high quality food *ad libitum* over a period of a whole year under natural light and temperature conditions at 78°N in Svalbard (Stokkan, Mortensen & Blix, 1986). It goes without saying that this was quite an undertaking, but we were abundantly awarded! We found that from the first week of February at about the end of the winter night, until the second week of April, at about the beginning of the long summer day, and from the third week of August, at about the end of the continuous day, until the third week of November, at about the beginning of a new winter night, all birds showed a clearly defined diel feeding rhythm with their feeding activity mainly confined to the light period and with increased feeding activity during the morning and evening hours of civil twilight, reminiscent of the activity pattern of the alaskan willow ptarmigans previously described. But, during the continuous polar night,

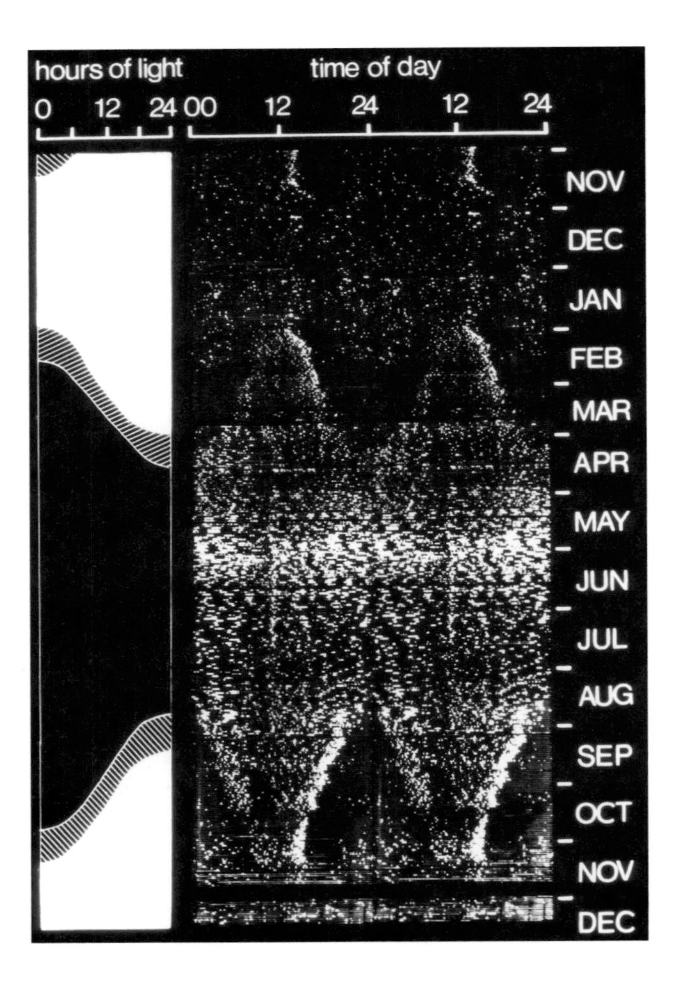

Figure 10.63 Typical feeding activity rhythm of one captive female Svalbard rock ptarmigan exposed to natural temperatures and light conditions for 13 months at Svalbard (79°N), and given high quality feed and water/snow *ad libitum*. Spots in actogram result from pen deflections caused by interruption of infrared (theft alarm) beam each time the bird puts its head into food box. Due to low paper speed events occasionally fuse into solid blocks when feeding is intensive. Resulting graph has been duplicated to facilitate visual inspection. Lack of record is due to instrumental malfunctions. Annual variations in daylight at Ny-Ålesund, Svalbard (79°N) are indicated to left (Stokkan, Mortensen & Blix, 1986).

from mid-November until the end of January, and during the continuous polar day, from mid-April until mid-August, the feeding activity of all birds was intermittently continuous around the clock (Fig. 10.63)! This implies that the Svalbard ptarmigan has no diel resting period for a total of about 8 months of the year, in spite of the 24-hour rhythms in, at least, the summer environment, which produce rhythmic behaviour in migratory passerines, as emphasized by Krüll (1976b). In this respect, it appears that the Svalbard ptarmigan is truly unique, at least in the northern hemisphere, with the emperor penguin *(Aptendocytes fosteri)*, as the only likely contender for global eminence, in the south. But, the emperor penguin does neither eat nor move during the winter, at least not the male who is left with the boring task of incubating the egg during the dark Antarctic winter, and therefore the Svalbard ptarmigans probably are unique!

References

Aagaard, K. & Carmack, E. C. (1994). The Arctic Ocean and climate: a perspective. In: *The Polar Oceans and Their Role in Shaping the Global Environment*, ed. O. M. Johannessen, R. D. Muench & J. E. Overland. Geophysical Monograph **85**, 5-20.

Andersen, H. T. (1966). Physiological adaptations in diving vertebrates. *Physiological Reviews*, **46**, 212-243.

Andreev, A. (1988). The ten year cycle of the willow grouse of Lower Kolyma. *Oecologia*, **76**, 261-267.

Anon. (1978). Polar regions atlas. *Central Intelligence Agency* (USA).

Anon. (1997). Arctic pollution issues: A state of the Arctic environment report. *Arctic Monitoring and Assessment Programme.*

Anon. (1998). Status of polar bear. IUCN/SSC polar bear specialist group. In: *Polar bears.* ed. A. E. Derocher, G. W. Garner, N. J. Lunn & Ø. Wiig. *IUCN.*

Anon. (2001). CAFF (Conservation of Arctic flora and fauna). *Arctic Flora and Fauna*: Status and Conservation. Helsinki: *Edita.*

Baker, M.A. (1979). A brain-cooling system in mammals. *Scientific American*, **240**, 114-122.

Bale, J.S., Hodkinson, I.D., Block, W., Webb, N.R., Coulson, S.C. & Strathdee, A. (1997). Life strategies of Arctic terrestrial arthopods. In: *Ecology of Arctic environments.* ed. S.J. Woodin & M. Marquiss. Pp. 137-165. Oxford : *Blackwell Science.*

Barnes, M. (1989). Freeze avoidance in a mammal: Body temperatures below 0 °C in an Arctic hibernator. *Science*, **244**, 1593-1595.

Barth, E. K. (1949). Kroppstemperatur hos fugler og pattedyr. *Fauna Flora*, 163-177.

Batzli, G.O. & Esseks, E. (1992). Body fat as indicator of nutritional condition for brown lemming. *Journal of Mammalogy*, **73(2)**, 431-439.

Best, R. C. (1988). Ecological aspects of polar bear nutrition. In: *Proceedings of 1975 Predator Symposium.* ed. R. L. Phillips & C. Jonkel. *University of Montana.*

Bliss, L. C., Matveyeva, N. V. (1992). Circumpolar arctic vegetation. In: *Arctic ecosystems in a changing climate: An ecophysiological perspective.* ed. F. S Chapin , R. L. Jefferies, J. F. Renyolds, G. R. Shaver & J. Svoboda, pp. 59-89. SanDiego: *Academic Press.*

Blix, A. S. (1987). Diving responses: Fact or fiction. *NIPS*, **2**, 64-66.

Blix, A. S. (1989). Arctic resignation: winter dormancy without hypothermia. In: *Living in the cold II.* pp. 117-119, ed. A. Malan & B. Canguilhem, Montrouge: *John Libbey Eurotext. Ltd.*.

Blix, A. S. & Folkow, B. (1983). Cardiovascular adjustments to diving in mammals and birds. In: *Handbook of Physiology. The cardiovascular System. Peripheral Circulation and Organ Blood Flow*, sect. 2, vol. III, chapt. **25**, 917-945. *American Physiological Society.*

Blix, A. S. & Folkow, L. P. (1995). Daily energy expenditure in free living minke whales. *Acta Physiologica Scandinavica*, **153**, 61-66.

Blix, A. S. & Johnsen, H. K. (1983). Aspects of nasal heat exchange in resting reindeer. *Journal of Physiology*, **340**, 445-454.

Blix, A. S. & Lentfer, J. W. (1979). Modes of thermal protection in polar bear cubs – at birth and on emergence from the den. *American Journal of Physiology*, **236**, R67-R74.

Blix, A. S., & Steen J. B. (1979). Temperature regulation in newborn polar homeotherms. *Physiological Reviews*, **59**, 285-304.

Blix, A.S., Elsner, R. & Kjekshus, J.K. (1983). Cardiac output and its distribution through A-V shunts and capillaries during and after diving in seals. *Acta Physiologica Scandinavica.* **118**. 109-116.

Blix, A. S., Fay, F. H. & Ronald, K. (1983). On testicular cooling in phocid seals. *Polar Research*, 1 n.s., 231-233.

Blix, A. S., Grav, H. J. & Ronald, K. (1975). Brown adipose tissue and the significance of the venous plexuses in pinnipeds. *Acta Physiologica Scandinavica*, **94**, 133-135.

Blix, A.S., Grav H. J. & Ronald, K. (1979). Some aspects of temperature regulation in newborn harp seal pups. *American Journal of Physiology*, **236(3)**, R188-R197.

Blix, A. S., Grav, H. J., Markussen, A. & White, R. G. (1984). Modes of thermal protection in newborn muskoxen (*Ovibos moschatus*). *Acta Physiologica Scandinavica*, **122**, 443-453.

Blix, A.S., Kjekshus, J.K., Enge, I. & Bergan, A. (1976). Myocardial blood flow in the diving seal. *Acta Physiologica Scandinavica.* **96**. 227-228.

Blix, A. S., Miller, L. K., Keyes, M. C., Grav, H. J. & Elsner, R. (1979). Newborn northern fur seals (*Callorhinus ursinus*) – do they suffer from cold? *American Journal of Physiology*, **236(5)**, R322-R327.

Bowen, W. D., Oftedal, O. T. & Boness, D. J. (1985). Birth to weaning in four days: Remarkable growth in the hooded seal, (*Cystophora cristata*). *Canadian Journal of Zoology*, **63**, 2841-2846.

Bron, K. M., Murdaugh, H. V., Jr., Millen, J. E., Lenthall, R., Raskin, P. & Robin, D. E. (1966). Arterial constrictor response in a diving mammal. *Science*, **152**, 540-543.

Buck, C. L. & Barnes, B.M. (1999a). Temperatures of hibernacula and changes in body composition of arctic ground squirrels over winter. *Journal of Mammalogy.* **80**. 1264-1276.

Buck, C. L. & Barnes, B. M. (1999b). Annual cycle of body composition and hibernation in free-living arctic ground squirrels. *Journal of Mammalogy.* **80**. 430-442.

Buck, C. L. & Barnes, B.M. (2000). Effects of ambient temperature on metabolic rate, respiratory quotient, and torpor in an arctic hibernator. *American Journal of Physiology. Regulatory, Integrative and Comparative Physiology.* **279**. R255-R262.

Butler, P. J. & Jones, D. R. (1997). Physiology of diving birds and mammals. *Physiological Reviews*, **77**, 837-899.

Caputa, M., Folkow, L.P. & Blix, A.S. (1998). Rapid brain cooling in diving ducks. *American Journal of Physiology.* **275**. R363-R371.

Casey, T.M. & Casey, K.K. (1979). Thermoregulation of Arctic weasels. *Physiological Zoology.* **52**. 153-164.

Cassone, V. M. & Menaker, M. (1984). Is the avian circadian system a neuroendocrine loop? *Journal of Experimental Zoology*, **232**, 539-549.

Cheng, K.-J., McAllister, T. A., Mathiesen, S. D., Blix, A. S., Orpin, C. G. & Costerton, J. W. (1993). Seasonal changes in the adherent microflora of the rumen in high-arctic Svalbard reindeer. *Canadian Journal of Microbiology*, **39**, 101-108.

Chernov. YU. I. (1985). *The living tundra.* Cambridge: *Cambridge University Press.*

Cuyler, C. & Øritsland, N. A. (2002). Effect of wind on Svalbard reindeer fur insulation. *Rangifer*, **22 (1)**, 93-99.

Danks, H. V. (1981). Arctic Arthropods: a review of systematics and ecology with particular reference to the North American fauna. Ottawa: *Entomological Society of Canada.*

Davenport, J. (1992). *Animal life at low temperature.* London: Chapman & Hall.

De Vries, A. L. (1980). Biological antifreezes and survival in freezing environments. In: *Animals and environmental fitness.* Volume I. ed. R. Gilles. Oxford: *Pergamon Press.*

Dietz, R. & Heide-Jørgensen, M.P. (1995). Movements and swimming speed of narwhals equipped with satellite transmitters in Melville Bay, northwest Greenland. *Canadian Journal of Zoology.* 73, 2106-2119.

Dingle, H. (1996). *Migration.* Oxford: *Oxford University Press.*

Eden, A. (1940). Coprophagy in the rabbit: Origin of "night" fæces. *Nature*, **145**, 628-629.

Elsner, R. (1965). Heart rate response in forced versus trained experimental dives in pinnipeds. *Hvalrådets Skrifter*, **48**, 24-29.

Elsner, R., Millard, R.W., Kjekshus, J.K., White, F., Blix, A.S. & Kemper, S. (1985). Coronary blood flow and myocardial segment dimensions during simulated dives in seals. *American Journal of Physiology.* **249**. H1119-H1126.

Elvebakk, A., Elven, R. & Razzhivin, V. Yu. (1999). Delimitation, zonal and sectorial subdivision of the Arctic for the panarctic flora project. *Skrifter.* Mat.-Naturv. Klasse. Ny Serie No. **38**, 375-386. Oslo: *The Norwegian Academy of Science and Letters.*

Eschricht, D. F. & Reinhardt, J. (1866). On the Greenland right-whale. (*Balaena mysticus*, Linn.), with especial reference to its geographical distribution and migrations in times past and present, and to its external and internal characteristics. In: *Recent Memoirs on the Cetacea by Professor Eschricht, Reinhardt and Lilljeborg*, ed. W. H. Flower. London: *The Ray Society.*

Fancy, S.G. & White, R.G. (1985a). Incremental cost of activity. In: *Bioenergetics of wild herbivores.* Ed. R.J. Hudson & R.G. White. Boca Raton, Florida : *CRC Press.*

Fancy, S.G. & White, R.G. (1985b). Energy expenditures by caribou while cratering in snow. *Journal of Wildlife Management.* **49**, 987-993.

Fay, F. H. (1982). Ecology and biology of the Pacific walrus. *North American Fauna, Number 74.* Washington, D.C., USA.

Finely, K. J. & Gibb, E. J. (1982). Summer diet of the narwhal (*Monodon monoceros*) in Pond Inlet, northern Baffin Island. *Canadian Journal of Zoology*, **60**, 3353-3363.

Fisher, H. I. & Dater, E. E. (1961). Esophageal diverticula in the redpoll, (*Acanthis flammea*), *Auk*, **78**, 528-531.

Flint, R. A. (1957). *Glacial pleistocene geology.* New York: *John Wiley & Sons.*

Folkow, L.P. & Blix, A.S. (1989). Thermoregulatory control of expired air temperature in diving harp seals. *American Journal of Physiology.* **257**. R306-R310.

Folkow, L. P. & Blix, A. S. (1992). Metabolic rates of minke whales (*Balaenoptera acutorostrata*) in cold water. *Acta Physiologica Scandinavica*, **146**, 141-150.

Folkow, L. P. & Blix, A. S. (1999). Diving behaviour of hooded seals (*Cystophora cristata*) in the Greenland and Norwegian seas. *Polar Biology*, **22**, 61-74.

Folkow, L. P., Blix, A. S. & Eide, T. J. (1988). Anatomical and functional aspects of the nasal mucosal and ophthalmic retia of phocid seals. *Journal of Zoology*, **216**, 417-436.

Folkow, L. P., Mårtensson, P.-E. & Blix, A. S. (1996). Annual distribution of hooded seals (*Cystophora cristata*) in the Greenland and Norwegian Seas. *Polar Biology*, **16**, 179-189.

Folkow, L.P., Nordøy, E.S. & Blix, A.S. (2004). Distribution and diving behaviour of harp seals from the Greenland Sea stock. *Polar Biology*. **27**. 281-298.

Framstad, E., Stenseth, N. C. & Østbye, E. (1993). Time series analysis of population fluctuations of *Lemmus lemmus* . In: *The biology of lemmings*. ed. Stenseth, N. C. & Ims, R. A. London: *Academic Press*.

Fuglesteg, B.N., Haga, Ø.E., Folkow, L.P., Fuglei, E. & Blix, A.S. (2005). Seasonal variations in basal metabolic rate, lower critical temperature and responses to temporary starvation in the arctic fox from Svalbard. *Polar Biology*.

Gabrielsen, G. W., Blix, A. S. & Ursin, H. (1985). Orienting and freezing responses in incubating ptarmigan hens. *Physiology & Behaviour*, **34**, 925-934.

Galster, W. & Morrison, P. (1976). Seasonal changes in body composition of the arctic ground squirrel. *Canadian Journal of Zoology*. **54**. 74-78.

Gasaway, W. C. (1976). Cellulose digestion and metabolism by captive rock ptarmigan. *Comparative Biochemistry and Physiology*, **54A**, 179-182.

Geist, V. (1987) Bergman's rule is invalid. *Canadian Journal of Zoology*. 65(4), 1035-1038.

Gessaman, J. A. (1972). Bioenergetics of the snowy owl (*Nyctea scandiaca*). *Arctic and Alpine Research*, **4 (3)**, 223-238.

Giertz, I., Lydersen, C. & Wiig, Ø. (2001). Distribution and diving of harbour seals (*Phoca vitulina*) in Svalbard. *Polar Biology*. 24, 209-214.

Gjertz, I., Kovacs, K. M., Lydersen, C. & Wiig, Ø. (2000) Movements and diving of adult ringed seals (*Phoca hispida*) in Svalbard. *Polar Biology*, **23**, 651-655.

Gjertz, I., Griffiths, D., Krafft, B. A., Lydersen, C. & Wiig, Ø. (2001). Diving and haul-out patterns of walruses (*Odobenus rosmarus*) on Svalbard. *Polar Biology*, **24**, 314-319.

Grav, H. J. & Blix, A. S. (1979). A source of nonshivering thermogenesis in fur seal skeletal muscle. *Science*, **204**, 87-89.

Grav, H. J., Blix, A. S. & Påsche, A. (1974). How do seal pups survive birth in Arctic winter? *Acta Physiologica Scandinavica*, **92**, 427-429.

Gray, D. R. (1987). *The muskoxen of Polar Bear Pass*. Markham: *Fitzhenry & Whiteside*.

Gulliksen, B. & Svensen, E. (2004). *Svalbard and life in polar oceans*. Kristiansund N: *Kom forlag*.

Hanssen, I. (1979 a). Micromorphological studies on the small intestine and caeca in wild and captive willow grouse (*Lagopus lagopus lagopus*). *Acta Veterinaria Scandinavica*, **20**, 351-364.

Hanssen, I. (1979 b). A comparison of the microbiological conditions in the small intestine and caeca of wild and captive willow grouse (*Lagopus lagopus lagopus*). *Acta Veterinaria Scandinavica*, **20**, 365-371.

Hansen, T. (1973). Variations in glycerol content in relation to cold-hardiness in larvae of *Petrova resinella* L. (Lepidoptera, Tortricidae). *Eesti NSV Tead. Akad. Toim. Biol.* **22**, 105-112.

Haug, T., Krøyer, A. B., Nilssen, K. T., Ugland, K .I. & Aspholm, P. E. (1991). Harp seal (*Phoca groenlandica)* invasions in Norwegian coastal waters: age composition and feeding habits. *IECS Journal of Marine Science*, **48**, 363-371.

Heide-Jørgensen, M. -P. & Dietz, R. (1995). Some characteristics of narwhal, (*Monodon monoceros*), diving behaviour in Baffin Bay. *Canadian Journal of Zoology*, **73**, 2120-2132.

Heide-Jørgensen, M.-P., Härkönen, T., Dietz, R. & Thompson, P.M. (1992). Retrospective of the 1988 European seal epizootic. *Diseases of Aquatic Organisms*, **13**, 37-62.

Heller, H. C. & Hammel, H.T. (1972). CNS control of body temperature during hibernation. *Comparative Biochemistry and Physiology. A Physiology*. **41**. 349-359.

Henshaw, R. E., Underwood, L. S. & Casey, T. M. (1972). Peripheral thermoregulation: Foot temperature in two Arctic canines. *Science*, **175**, 988-990.

Hertzberg, K., Leinaas, H.P. & Ims, R.A. (1994). Patterns in abundance and demography in patchy habitats : Collembola in a habitat patch gradient. *Ecography*. **17**. 349-359.

Hopkins, D.M. (1967). The cenozoic history of Beringia – a synthesis. In: *The Bering Land Bridge*, 451-484. ed. D.M. Hopkins. Stanford: *Stanford University Press*.

Horwitz, B. A. (1989). Biochemical mechanisms and control of cold-induced cellular thermogenesis in placental mammals. In: *Advances in Comparative and Environmental Physiology*.Vol. 4. ed. L.C.H. Wang. Berlin: *Springer-Verlag*.

Houghton et al. (1995). *Radiative forcing of climate change and an evaluation of the IPCC IS92 emission scenarios*. Cambridge: *Cambridge University Press*.

Hurst, R.J., Leonard, M. L., Watts, P.D., Beckerton, P. & Øritsland, N. A. (1982). Polar bear locomotion: body temperature and energetic cost. *Canadian Journal of Zoology*, **60**, 40-44.

Høst, P. (1942). Effect of light on the moults and sequences of plumage in the willow ptarmigan. *Auk*, **59**, 388-403.

Irving, L. (1972). *Arctic life of birds and mammals including man*. Berlin: *Springer-Verlag*.

Irving, L. & Krog, J. (1955). Temperature of skin in the Arctic as regulator of heat. *Journal of Applied Physiology*, **7**, 355-364.

Jackson, D. C. & Schmidt-Nielsen, K. (1964). Countercurrent heat exchange in the respiratory passages. *Proceedings National Academy of Science*, **51**, 1192-1197.

Jarrell, G. H. & Fredga, K. (1993). How many kinds of lemmings? A taxonomic overview. In: *The biology of lemmings*, ed. N. C. Stenseth & R. A. Ims. London: *Academic Press*.

Johansen, K. (1961). Distribution of blood in the arousing hibernator. *Acta Physiologica Scandinavica*. **52**. 379-386.

Johnsen, H. K. (1988). Nasal heat exchange. An experimental study of effector mechanisms associated with respiratory heat loss in Norwegian reindeer (*Rangifer tarandus tarandus*). *Dr. Philos. Thesis*, University of Tromsø, Norway.

Johnsen, H. K. & Folkow, L.P. (1988). Vascular control of brain cooling in reindeer. *American Journal of Physiology*, **254**, R730-R739.

Johnsen, H. K., Blix, A. S., Jørgensen, L. & Mercer, J. B. (1985). Vascular basis for regulation of nasal heat exchange in reindeer. *American Journal of Physiology*, **249**, R617-R623.

Johnsson, A., Engelmann, W., Klemke, W. & Ekse, A. T. (1979). Free-running human circadian rhythms in Svalbard. *Zeitschrift für Naturforschung*, **34**, 470-473.

Jørgensen, E. & Blix, A.S. (1985). Effects of climate and nutrition on growth and survival of willow ptarmigan chicks. *Ornis Scandinavica*. **16**. 99-107.

Jørgensen, E. & Blix, A. S. (1988). Energy conservation by restricted body cooling in cold-exposed willow ptarmigan chicks? *Ornis Scandinavica*, **19**, 17-20.

Kanwisher, J. (1959). Histology and metabolism of frozen intertidal animals. *Biological Bulletin*, **116**, 258-264.

Kanwisher, J., Gabrielsen, G. & Kanwisher, N. (1981). Free and forced diving in birds. *Science*, **211**, 717-719.

Keilhau, B. M. (1831). *Reise til Øst- og Vest-Finmarken samt til Beeren-Eiland og Spitzbergen, i aarene 1827 og 1828.* Christiania: *Johan Krohn.*

Kenagy, G. J. & Hoyt, D. F. (1980). Reingestion of feces in rodents and its daily rhythmicity. *Oecologia,* **44**, 403-409.

Kevan, P. G. & Danks, H. V. (1986). Adaptations of Arctic insects. In: *The Arctic and its wildlife,* ed. B. Sage. London: *Croom Helm,* pp. 55-57.

Klein, D. R. (1968). The introduction, increase, and crash of reindeer on St. Matthew Island. *Journal of Wildlife Management,* **32**, 350-367.

Kooyman, G. L., Wahrenbrock, E. A., Castellini, M. A, Davis, R. W. & Sinnett, E. E. (1980). Aerobic and anerobic metabolism during voluntary diving in Weddell seals: evidence of preferred pathways from blood chemistry and behavior. *Journal of Comparative Physiology,* **138**, 335-346.

Kost'yan, E. Y. (1954). New data on the reproduction of polar bears (in Russian). *Zoologiceskij Zurnal,* **33**, 207-215.

Krüll, F. (1976a). The position of the sun is a possible zeitgeber for Arctic animals. *Oecologia,* **24**, 141-148.

Krüll, F. (1976b). Zeitgebers for animals in the continuous daylight of high Arctic summer. *Oecologia,* **24**, 149-157.

Kvadsheim, P. H. & Aarseth, J. J. (2002). Thermal function of phocid seal fur. *Marine Mammal Science,* **18**, 952-962.

Kvadsheim, P.H., Folkow, L.P. & Blix, A.S. (2005). Inhibition of shivering in hypothermic seals during diving. *American Journal of Physiology.* (in the press).

Lange, R. & Staaland, H. (1970). Adaptations of the caecum-colon structure of rodents. *Comparative Biochemistry and Physiology,* **35**, 905-919.

Larsen, T. (1978). *The world of the polar bear.* London: *Hamlyn. Ltd.*.

Larsen, T. S., Nilsson, N. Ö & Blix, A.S. (1985). Seasonal changes in lipogenesis and lipolysis in isolated adipocytes from Svalbard and Norwegian reindeer. *Acta Physiologica Scandinavica,* **123**, 97-104.

Larter, N.C. (1999). Seasonal changes in Arctic hare diet composition and differential digestibility. *Canadian Field-Naturalist.* 113(3), 481-486.

Lavigne, D. M. & Øritsland, N. A. (1974). Black polar bears. *Nature,* **251**, 218-219.

Leader-Williams, N. (1988). *Reindeer on South Georgia.* Cambridge: *Cambridge University Press.*

Lentfer, J. W. (1975). Polar bear denning on drifting sea ice. Journal of Mammalogy. 56. 716-718.

Loeng, H. (1991). Features of the physical ocenanographic conditions of the Barents Sea. *Polar Research,* **10(1)**, 5-18.

Lundheim, R. & Zachariassen, K.E. (1993). Water balance of over-wintering beetles in relation to strategies for cold tolerance. *Journal of Comparative Physiology* B, **163**, 1-4.

Lydersen, C. & Hammill, M. O. (1993). Diving in ringed seal (*Phoca hispida*) pups during the nursing period. *Canadian Journal of Zoology,* **71**, 991-996.

Lydersen, C., Hammill, M.O. & Kovacs, K.M. (1994). Diving activity in nursing bearded seal pups. *Canadian Journal of Zoology.* **72**, 96-103.

Lyman, C. P. (1948). The oxygen consumption and temperature regulation of hibernating hamsters. *Journal of Experimental Zoology.* **109**. 55-78.

Lønø, O. (1970). The polar bear in the Svalbard area. *Norsk Polarinstitutts Skrifter,* No. **149**, 1-103.

MacQuarrie, B. (1996). *The Northern Circumpolar World.* Edmonton: *Reidmore Books.*

Markussen, K. A., Rognmo, A. & Blix, A. S. (1985). Some aspects of thermoregulation in newborn reindeer calves (*Rangifer tarandus tarandus*). *Acta Physiologica Scandinavica,* **123**, 215-220.

Martin, A.R., Kingsley, M.C.S. & Ramsay, M.A. (1994). Diving behaviour of narwhals on their summer grounds. *Canadian Journal of Zoology.* **72**, 118-125.

Martin, A. R., Smith, T. G. & Cox, O. P. (1998). Dive form and function in belugas of the eastern Canadian high Arctic. *Polar Biology,* **20**, 218-228.

Mathiesen, S. D., Rognmo, A. & Blix, A.S. (1984). A test of the usefulness of a commercially available mill "waste product" as feed for starving reindeer. *Rangifer,* **4(1)**, 28-34.

Mathiesen, S. D., Orpin, C. G., Greenwood, Y. & Blix, A. S. (1987). Seasonal changes in the cecal microflora of the High-Arctic Svalbard reindeer (*Rangifer tarandus platyrhynchus*). *Applied and Environmental Microbiology,* **53**, 114-118

Mech, L. D. (1970). *The wolf: The ecology and behavior of an endangered species.* New York: *The Natural History Press.*

Mech, L. D. (1988). *The Arctic wolf.* Stillwater: *Voyageur Press.*

Meng, M. S., West, G. C. & Irving, L. (1969). Fatty acid composition of caribou bone marrow. *Comparative Biochemistry and Physiology,* **30**, 187-191.

Miller, F.L., Gunn, A. & Broughton, E. (1985). Surplus killing as exemplified by wolf predation on newborn caribou. *Canadian Journal of Zoology.* **63**, 295-300.

Miller, L. K. (1965). Activity in mammalian peripheral nerves during supercooling. *Science,* **149**, 74-75.

Miller, L. K. (1969). Freezing tolerance in an adult insect. *Science,* **166**, 105-106.

Miller, L. K. (1982). Cold-hardiness strategies of some adult immature insects overwintering in interior Alaska. *Comparative Biochemistry and Physiology,* **73A**, 595-604.

Mirsky, S. D. (1988). Solar polar bears. *Scientific American,* March, 15-18.

Moore, S.E. & Reeves, R.R. (1993). Distribution and movement. In : *The Bowhead Whale.* ed. J.J. Burns, J.J. Montague & C.J. Cowles. *Society for Marine Mammalogy.* pp. 313-386.

Morrison, P. & Galster, W. (1975). Patterns of hibernation in the arctic ground squirrel. *Canadian Journal of Zoology.* **53**. 1345-1355.

Mortensen, A. (1984). Importance of microbial nitrogen metabolism in the ceca of birds. In: *Current perspectives in microbial ecology.* ed. M.J. Klug & C. A. Reddy, pp. 273-278. *American Society for Microbiology.*

Mortensen, A. & Blix, A. S. (1985). Seasonal changes in the effects of starvation on metabolic rate and regulation of body weight in Svalbard ptarmigan. *Ornis Scandinavica,* **16**, 20-24.

Mortensen, A. & Blix, A. S. (1986). Seasonal changes in resting metabolic rate and mass-specific conductance in Svalbard ptarmigan, Norwegian rock ptarmigan and Norwegian willow ptarmigan. *Ornis Scandinavica,* **17**, 8-13.

Mortensen, A. & Tindall, A. (1981). On caecal synthesis and absorption of amino acids and their importance for nitrogen recycling in willow ptarmigan (*Lagopus lagopus lagopus*). *Acta Physiologica Scandinavica,* **113**, 465-469.

Mortensen, A., Unander, S., Kolstad, M. & Blix, A.S. (1983). Seasonal changes in body composition and crop content of Spitzbergen Ptarmigan (*Lagopus mutus hyperboreus*). Ornis Scandinavica, **14**, 144-148.

Muus, B., Salomonsen, F. & Vibe, C. (1981). *Grønlands Fauna.* Copenhagen : *Gyldendal.*

Mårtensson, P.-E., Nordøy, E. S. & Blix, A. S. (1994). Digestibility of krill (*Euphausia superba* and *Thysanoessa* sp.) in minke whales (*Balaenoptera acutorostrata*) and crabeater seals (*Lobodon carcinophagus*). *British Journal of Nutrition,* **72**, 713-716.

Mårtensson, P.-E., Nordøy, E. S., Messelt, E. B. & Blix, A. S. (1998). Gut length, food transit time and diving habit in phocid seals. *Polar Biology,* **20**, 213-217.

Mårtensson, P.-E., Lager Gotaas, A. R., Nordøy, E. S. & Blix, A. S. (1996). Seasonal changes in energy density of prey of northeast Atlantic seals and whales. *Marine Mammal Science*, **12 (4)**, 635-640.

Nansen, F. (1897). *Farthest North.* 2 vols. London: *Constable.*

Nansen, F. (1925). *Hunting and adventure in the Arctic.* (Translated from "Blant sel og bjørn"). Gawler: *Duffield & Co.*

Nelson, R. A., Jones, J. D., Wahner, H. W., McGill, D. B. & Code, C. F. (1975). Nitrogen metabolism: Urea metabolism in summer starvation and winter sleep and role of urinary bladder in water and nitrogen concervation. *Mayo Clinic Proceedings*, **50**, 141-146.

Nelson, R. A., Wahner, H. W., Jones, J. D., Ellefson, R.D. & Zollman, P. E. (1973). Metabolism of bears before, during and after winter sleep. *American Journal of Physiology*, **224**, 491-496.

Nilssen, K. J., Mathiesen, S. D. & Blix, A. S. (1994). Metabolic rate and plasma T_3 in ad lib. fed and starved muskoxen. *Rangifer*, **14(2)**, 79-81.

Nilssen, K. J., Sundsfjord, A. & Blix, A. S. (1984). Regulation of metabolic rate in Svalbard and Norwegian reindeer. *American Journal of Physiology*, **247**, R837-R841.

Nilssen, K. J., Johnsen, H. K., Rognmo, A. & Blix, A. S. (1984). Heart rate and energy expenditure in resting and running Svalbard and Norwegian reindeer. *American Journal of Physiology*, **246**, 15, R963-R967.

Nordøy, E. S. (1995). Do minke whales (*Balaenoptera acutorostrata*) digest wax esters? *British Journal of Nutrition*, **74**, 717-722.

Nordøy, E. S., Ingebretsen, O. C. & Blix, A. S. (1990). Depressed metabolism and low protein catabolism in fasting grey seal pups. *Acta Physiologica Scandinavica*, **139**, 361-369.

Nordøy, E. S, Mårtensson, P.-E. & Blix, A.S. (1995). Food requirements of Northeast Atlantic minke whales. In: Whales, seals, fish and man, ed. A. S. Blix, L. Walløe and Ø. Ulltang. Amsterdam: *Elsevier Science B. V.*, pp. 307-317.

Norris, E., Norris, C. & Steen, J.B. (1975). Regulation and grinding ability of grit in the gizzard of Norwegian willow ptarmigan. *Poultry Science.* **54**. 1839-1843.

Odden, A., Folkow, L. P., Caputa, M., Hotvedt, R. & Blix, A. S. (1999). Brain cooling in diving seals. *Acta Physiologica Scandinavica*, **166**, 77-78.

Oftedal, O.T. (1984). Milk composition, milk yield and energy output at peak lactation: A comparative review. *Symposium of the Zoological Society, London.* **No. 51**. 33-85.

Olsen, M. A., Nordøy, E. S., Blix, A. S. & Mathiesen, S. D. (1994). Functional anatomy of the gastrointestinal system of Northeastern Atlantic minke whales (*Balaenoptera acutorostrata*). *Journal of Zoology*, **234**, 55-74.

Olsen, M. A., Blix, A. S. , Utsi, T. H. A., Sørmo, W. & Mathiesen, S. D. (2000). Chitinolytic bacteria in the minke whale forestomach. *Canadian Journal of Microbiology*, **46**, 85-94.

Orpin, C. G., Mathiesen, S. D., Greenwood, Y. & Blix, A. S. (1985). Seasonal changes in the ruminal microflora of the High-Arctic Svalbard reindeer (*Rangifer tarandus platyrhynchus*). *Applied and Environmental Microbiology*, **50**, 144-151.

Parker, K. L. (1983). Ecological energetics of mule deer and elk: Locomotion and thermoregulation, *PhD thesis.* Washington State University, USA.

Pedersen, A. (1934). *Polardyr.* Copenhagen: *Gyldendalske Boghandel Nordisk Forlag.*

Prestrud, P. & Nilssen, K. (1992). Fat deposition and seasonal variation in body composition of Arctic foxes in Svalbard. *Journal of Wildlife Management*, **56(2)**, 221-233.

Prestrud, P. & Stirling, I. (1994). The international polar bear agreement and the current status of polar bear conservation. *Aquatic Mammals*, **20**, 113-124.

Ramsay, M. A., Nelson, R. A. & Stirling, I. A. (1991). Seasonal changes in the ratio of serum to creatinine in feeding and fasting polar bears. *Canadian Journal of Zoology*, **69**, 298-302.

Rannie, W. F. (1986). Summer air temperature and number of vascular species in Arctic Canada. *Arctic*, **39**(2), 133-137.

Rasmussen, E.A. & Turner, J. (eds.). (2003). *Polar Lows*. Cambridge: *Cambridge University Press.*

Ray, C., Watkins, W.A. & Burns, J.J. (1969). The underwater song of *Erignatus* (bearded seal). Zoologica, **54**. 79-83.

Reed, M. & Balchen, J. G. (1982). A multidimensional continuum model of fish population dynamics and behaviour: Application to the Barents Sea capelin (*Mallotus villosus*). *Modelling Identification Control.* **3**(2), 65-109.

Reed, C. I. & Reed, B. P. (1928). The mechanism of pellet formation in the great horned owl (*Bubo virginianus*). *Science*, **68**, 359-360.

Reierth, E., Van't Hof, T. J. & Stokkan, K.-A. (1999). Seasonal and daily variations in plasma melatonin in the High-Arctic Svalbard ptarmigan (*Lagopus mutus hyperboreus*). *Journal of Biological Rhythms*, **14**, 314-319.

Reimers, E. & Ringberg, T. (1983). Seasonal changes in body weights of Svalbard reindeer from birth to maturity. *Acta Zoologica Fennica.* 175. 69-72.

Reynolds, J. E. & Odell, D.K. (1991). *Manatees and Dugongs*. New York : *Facts On File.*

Ringberg, T., Nilssen, K. & Strøm, E. (1980). Do Svalbard reindeer use their subcutaneous fat as insulation? In: *2nd Int. Reindeer/Caribou Symp.*, Røros, Norway. ed. E. Reimers, E. Gaare, & S. Skjenneberg, pp 392-395. Trondheim: *Direktoratet for vilt og ferskvannsfisk*, Norway.

Ryan, J. K. (1977). Synthesis of energy flows and population dynamics of Truelove Lowland invertebrates. In: *Truelove Lowland, Devon Island, Canada: A high Arctic ecosystem*, ed. L. C. Bliss. pp. 325-346. Edmonton: *University of Alberta Press.*

Sakshaug, E., Rey, F. & Slagstad, D. (1995). Wind forcing of marine primary production in the northern atmospheric low-pressure belt. In: *Ecology of Fjords and Coastal Waters*, ed. H.R. Skjoldal, C. Hopkins, K.E. Erikstad and H.P. Leinaas. Amsterdam: *Elsevier.*

Sakshaug, E., Bjørge, A., Gulliksen, B., Loeng, H. & Mehlum, F. (ed.) (1994). *Økosystem Barentshavet.* Oslo: *Universitetsforlaget.*

Salomonsen, F. (1950). *The birds of Greenland.* København: *Ejnar Munksgaard.*

Schmidt-Nielsen, K. (1990). *Animal Physiology*. 4th. Edit. Cambridge : *Cambridge University Press.*

Scholander, P. F. (1940). Experimental investigations on the respiratory function in diving mammals and birds. *Hvalrådets Skrifter*, **22**, 1-131.

Scholander, P.F. (1956). Climate rules. *Evolution.* **10**. 339-340.

Scholander, P. F., Walters, V., Hock, R., Irving, L. (1950 a). Body insulation of some Arctic and tropical mammals and birds. *Biological Bulletin*, **99**, 225-236.

Scholander, P. F., Hock, R., Walters, V., Johnson, F. & Irving, L. (1950 b). Heat regulation in some Arctic and tropical mammals and birds. *Biological Bulletin*, **99**, 237-258.

Scholander, P. F., van Dam, L., Kanwisher, J. W., Hammel, H. T. & Gordon, M. S. (1957). Supercooling and osmoregulation in arctic fish. *Journal of Cellular and Comparative Physiology*, **49**, 5-24.

Scoresby, W. (1820). *An account of the Arctic regions, with a history and description of the northern whale-fishery.* Edinburgh: *Constable.*

Shuert, P. G. & Walsh, J. J. (1992). A time-dependent depth-integrated barotropic physical model of the Bering/Chukchi seas for use in ecosystem analysis. *Journal of Marine Systems*, **3**, 141-161.

Silverman, H.B. & Dunbar, M.J. (1980). Aggressive tusk use by the narwhal. *Nature.* **284**, 57-58.

Skalstad, I. & Nordøy, E. S. (2000). Experimental evidence of seawater drinking in juvenile hooded (*Cystophora cristata*) and harp seals (*Phoca groenlandica*). *Journal of Comparative Physiology*, **170**, 395-401.

Sklepkovych, B. O. & Montevecchi, W.A. (1996). Food availability and food hoarding behaviour by red and arctic foxes. *Arctic*, **49**, 228-234.

Smith, M. & Rigby, B. (1981). Distribution of polynyas in the Canadian Arctic. In : *Polynyas in the Canadian Arctic.* ed. I. Stirling & H. Cleator. Canadian Wildlife Service Occasional Paper 45.

Smith, T.G. & Martin, A. R. (1994). Distribution and movements of belugas in the Canadian High Arctic. *Canadian Journal of Fisheries & Aquatic Science.* **51**, 1653-1663.

Sperber, I., Björnhaug, G. & Ridderstråle, Y. (1983). Function of proximal colon in lemming and rat. *Swedish Journal of Agricultural Research*, **13**, 243-256.

Springer, A. M., McRoy, C. P. & Flint, M. V. (1996). The Bering Sea green belt: shelf-edge processes and ecosystem production. *Fisheries Oceanography*, **5:3/4**, 205-223.

Staaland, H., Brattbakk, I., Ekern, K. & Kildemo, K. (1983). Chemical composition of reindeer forage plants in Svalbard and Norway. *Holarctic Ecology*, **6**, 109-122.

Stenseth, N.C. & Ims, R.A. (eds.) (1993). *The biology of lemmings.* London: *Academic Press.*

Stirling, I. (1988). *Polar Bears.* Ann Arbor : *The University of Michigan Press.*

Stirling, I. (1997). The importance of polynyas, ice edges, and leads to marine mammals and birds. *Journal of Marine Systems.* **10**. 9-21.

Stirling, I. & Oritsland, N.-A. (1995). Relationships between estimates of ringed seals (*Phoca hispida*) and polar bear (*Ursus maritimus*) populations in the Canadian Arctic. *Canadian Journal of Fisheries and Aquatic Sciences.* **52**. 2594-2612.

Stokkan, K.-A., Mortensen, A. & Blix, A S. (1986). Food intake, feeding rhythm, and body mass regulation in Svalbard rock ptarmigan. *American Journal of Physiology*, **251**, R264-R267.

Stokkan, K.- A., Sharp, P. J. & Unander, S. (1986). The annual breeding cycle of the high-Arctic Svalbard ptarmigan (*Lagopus mutus hyperboreus*). *General and Comparative Endocrinology*, **61**, 446-451.

Stokkan, K.-A., Tyler, N. J. C. & Reiter, R. J. (1994). The pineal gland signals autumn to reindeer (*Rangifer tarandus tarandus*) exposed to the continuous daylight of the Arctic summer. *Canadian Journal of Zoology*, **72**, 904-909.

Stonehouse, B. (1989). *Polar Ecology.* Glasgow: *Blackie.*

Strathdee, A.T. & Bale, J.S. (1998). Life on the edge : Insect ecology in Arctic environments. *Annual Review of Entomology.* 43. 85-106.

Syvertsen, E. E. (1991). Ice algae in the Barents Sea: types of assemblages, origin, fate and role in the ice-edge phytoplankton bloom. *Polar Research*, 10(1), 277-287.

Sømme, L. (1995). *Invertebrates in hot and cold environments.* Berlin: *Springer-Verlag.*

Tamarkin, K., Baird, C.J. & Almerida, O.F.X. (1985). Melatonin : a coordinating signal for mammalian reproduction. Science, **227**, 714-720.

Taylor, C. R., Schmidt-Nielsen, K. & Raab, J. (1970). Scaling of energetic cost of running to body size in mammals. *American Journal of Physiology*, **219**, 1104-1107.

Thomas, D. C. (1982). The relationship between fertility and fat reserves of Peary caribou. *Canadian Journal of Zoology*, **60**, 957-602.

Thomas, D. C. & Kroeger, P. (1980). Digestibilities of plants in rumen fluids of Peary caribou. *Arctic*, **33**, 757-767.

Thompson, D. & Fedak, M. A. (1993). Cardiac responses of grey seals during diving at sea. *Journal of Experimental Biology*, **174**, 139-164.

Tomczak, M. & Godfrey, J. S. (1994). *Regional Oceanography: An Introduction.* Oxford: *Pergamon.*

Timisjärvi, J., Nieminen M. & Sippola, A.-L. (1984). The structure and insulation properties of the reindeer fur. *Comparative Biochemistry and Physiology,* **79A**, 601-609.

Tullberg, T. (1899). Über das System der Nagethiere. *Nova Acta Reg. Soc. Scient. Upsal. Ser. tert.,* **18**, 1-511.

Tyler, N. J. C. (1987). Body composition and energy balance of pregnant and non-pregnant Svalbard reindeer during winter. *Symposium Zoological Society. London* , **57**, 203-229.

Unander, S., Mortensen, A. & Elvebakk, A. (1985). Seasonal changes in crop content of the Svalbard Ptarmigan (*Lagopus mutus hyperboreus*). *Polar Research,* 3. n.s., 239-245.

Underwood, L. S. (1971). The bioenergetics of the Arctic fox (*Alopex lagopus*). Ph.D. Thesis, *Pennsylvania State University.*

Vilhjalmsson, H. (1994). The Icelandic capelin stock. Reykjavik: *Rit Fiskideildar,* Vol. XIII No.1., 1-281.

Walker, M. D. (1995). Patterns and causes of Arctic plant community diversity. In: *Arctic and Alpine Biodiversity,* ed. F. S. Chapin & C. Körner. *Ecological Studies, 113.* Berlin: *Springer-Verlag.*

Wang, L.C.H., Jones, D.L., MacArthur, R.A. & Fuller, W.A. (1973). Adaptation to cold: energy metabolism in an atypical lagomorph, the arctic hare. *Canadian Journal of Zoology,* **51**, 841-846.

Welch, H.E., Bergmann, M.A., Siferd, T.D., Martin, K.A., Curtis, M.F., Crawford, R.E., Conover, R.J. & Hop, H. (1992). Energy flow through the marine ecosystem of the Lancaster Sound region, Arctic Canada. *Arctic.* **45**. 343-357.

Wheeler, P. A., Gosselin, M., Sherr, E., Thibault, D., Kirchman, D.L., Benner, R. & Whitledge, T. E. (1996). Active cycling of organic carbon in the central Arctic Ocean. *Nature,* **380**, 697-699.

White, R.G. & Yousef, M.K. (1978). Energy expenditure of reindeer walking on roads and on tundra. *Canadian Journal of Zoology.* **56**, 215-223.

Wiig, Ø. (1989). A description of common seals, *Phoca vitulina, L.* 1758, from Svalbard. *Marine Mammal Science,* **5(2)**, 149-158.

Wiig, Ø. (1995). Distribution of polar bears (*Ursus maritimus*) in the Svalbard area. *Journal of Zoology,* **237**, 515-529.

Wiig, Ø., Gjertz, I. & Griffiths, D. (1996). Migration of walruses (*Odobenus rosmarus*) in the Svalbard and Franz Josef Land area. *Journal of Zoology,* **238**, 769-784.

Würsig, B. & Clark, C. (1993). Behavior. In: *The Bowhead whale,* ed. J. J. Burns, J. J. Montague, C. J. Cowles. Special Publication Number 2, pp. 157 – 199. *The Society for Marine Mammalogy.*

Zachariassen, K. E. (1985). Physiology of cold tolerance in insects. *Physiological Reviews,* **65 (4)**, 799-832.

Zachariassen, K. E. & Hammel, H. T. (1976). Nucleating agents in the haemolymph of insects tolerant to freezing. *Nature,* **262**, 285-287.

Zachariassen, K.E. & Kristiansen, E. (2000). Ice nucleation and antinucleation in nature. *Cryobiology.* **41**. 257-279.

Zenkevitch, L. (1963). *Biology of the seas of the U.S.S.R.* London: *George Allen & Unwin Ltd..*

Zimov, S. A., Chuprynin, V. I., Oreshko, Chapin, F. S., Chapin, M. C. & Reynolds, J. F. (1995). Effects of mammals on ecosystem change at the pleistocene-holocene boundary. In: *Arctic and Alpine Biodiversity,* ed. F. S. Chapin & C. Körner. pp. 127-135. *Ecological Studies,* 113. Berlin: *Springer-Verlag.*

Øritsland, N. A. & Ronald, K. (1973). Effects of solar radiation and windchill on skin temperature of the harp seal. *Comparative Biochemistry and Physiology*, **44A**, 519-525.

Østbye, E. (1965). Development of thermoregulation in relation to age and growth in the Norwegian lemming (*Lemmus lemmus L.*). *Nytt Magasin for Zoology.*, **12**, 65-75, 1965.

Further Reading

Aleksandrova, V. D. (1980). *The Arctic and Antarctic: their division into geobotanical areas.* Cambridge: *Cambridge University Press.*

Anker-Nilsen, T., Bakken, V., Strøm, H., Golovkin, A.N., Bianki, V.V. & Tatarinkova, I.P. (ed.) (2000). *The status of marine birds breeding in the Barents Sea region.* Rapport nr. 113. Tromsø: *Norsk Polarinstitutt.*

Baker, R. R. (1978). *The evolutionary ecology of animal migration.* London: *William Clowes & Sons Ltd.*.

Berthelsen, C., Holbech Mortensen, I. & Mortensen, E. (ed.) (1990). *Greenland Atlas.* Copenhagen: *Pilersuiffik.*

Bliss, L.C. (ed.) (1977). *Truelove Lowland, Devon Island, Canada : a High-Arctic ecosystem.* Edmonton : *University of Alberta Press.*

Bliss, L. C., Courtin, G. M., Pattie, D. L., Riewe, R. R., Whitfield, D. W. A. & Widden, P. (1973). Arctic tundra ecosystems. *Annual Review of Ecology and Systematics,* **4**, 359-399.

Bobylev, L.P., Kondratyev, K.Y. & Johannessen, O.M. (2003). *Arctic environment variability in the context of global change.* Chichester: *Praxis Publishing Ltd.*.

Born, E.W. & Böcher, J. (2001) *The Ecology of Greenland.* Nuuk : *Ilinniusiorfik.*

Croxall, J. P., Evans, P. G. H. & Schreiber, R. W. (ed.) (1984). *Status and Conservation of the World's Seabirds.* ICBP Technical Publication NO. 2. Cambridge: *International Council for Bird Preservation.*

Dunbar, M. J. (1968). *Ecological development in polar regions.* Englewood Cliffs: *Prentice-Hall, Inc.*.

Elsner, R. & Gooden, B. (1983). *Diving and asphyxia.* Cambridge: *Cambridge University Press.*

Fogg, G. E. (1998). *The Biology of Polar Habitats.* Oxford: *Oxford University Press.*

French, H. M. & Slaymaker, O. (1993). *Canada's Cold Environments.* Montreal: *McGill-Queen's University Press.*

Frenzel, B., Pecsi, M. & Velichko, A.A. (1992). *Atlas of Paleoclimates and Paleoenvionments of the Northern Hemisphere. Late Pleistocene – Holocene.* Stuttgart: *Gustav Fischer Verlag.*

Heide-Jørgensen, M.P. & Lydersen, C. (ed.) (1998). *Ringed seals in the North Atlantic.* Tromsø: *NAMMCO.*

Horwood, J. (1990). *Biology and Exploitation of the Minke Whale.* Boca Raton: *CRC Press.*

Hudson, R.J. & White, R.G. (ed.) (1985). *Bioenergetics of wild herbivores.* Boca Raton: *CRC Press.*

Kassens, H., Bauch, H. A., Dmitrenko, I. A., Eicken, H., Hubberten, H.-W., Melles, M., Thiede, J. & Timokhov, L. A. (ed.) (1999). *Land-Ocean Systems in the Siberian Arctic.* Berlin: *Springer-Verlag.*

Kelsall, J. P. (1968). *The migratory barren-ground caribou of Canada.* Ottawa: *Queen's Printer.*

Kooyman, G. L. (1989). *Diverse Divers.* Berlin: *Springer-Verlag.*

Lavigne, D.M. & Kovacs, K.M. (1988). *Harps & Hoods.* Waterloo: *University of Waterloo Press.*

Longton, R. E. (1988). *The biology of polar bryophytes and lichens.* Cambridge: *Cambridge University Press.*

Mech, D. & Boitani, L. (2003). *Wolves: Behavior, ecology, and conservation.* Chicago: *University of Chicago Press.*

Nuttall, M. (2004). *Encyclopedia of the Arctic.* Fairbanks: *Arctic Info..*

Pickard, G. L. & Emery, W. J. (1982). *Descriptive physical oceanography.* Oxford: *Pergamon Press.*

Pielou, E. C. (1994). *A Naturalist's Guide to the Arctic.* Chicago: *The University of Chicago Press.*

Remmert, H. (1980). *Arctic Animal Ecology.* Berlin: *Springer-Verlag.*

Ridgway, S. H. & Harrison, R. J. (ed.) (1981). *Handbook of Marine Mammals.* Volumes 1, 2, 3 & 4. London: *Academic Press.*

Storey, K.B. & Storey, J.M. (1988). Freeze tolerance in animals. *Physiological Reviews.* **68**. 27-84.

Tener, J. S. (1965). *Muskoxen in Canada.* Ottawa: *Queen's Printer.*

Trayhurn, P. & Nicholls, D. G. (1986). *Brown Adipose Tissue.* London: *Edward Arnold.*

Vikingsson, G.A. & Kapel, F.O. (ed.) (2000). *Minke whales, harp and hooded seals: Major predators in the North Atlantic ecosystem.* Tromsø: *NAMMCO.*

Wadhams, P., Dowdeswell, J. A. & Schofield, A. N. (1996). *The Arctic and environmental change.* Amsterdam *: Gordon and Breach Publishers.*

Woodin, S. J. & Marquiss, M. (1997). *Ecology of Arctic Environments.* Oxford: *Blackwell Science Ltd..*

APPENDIX

An Outline Classification of Animals

Animals are classified into various taxonomic categories developed by the Swedish biologist Linnaeus in the eigthteenth century. What is provided below is a simplified outline of the animal classification that may differ from textbook to textbook.

PHYLUM PROTOZOA

CLASS FLAGELLATA. Flagellates.
CLASS SARCODINA. Amoeba, foraminifera, radiolaria.
CLASS CILIATA. Ciliates.
CLASS SPOROZOA. Coccidia.

PHYLUM PORIFERA. Sponges.

PHYLUM COELENTERATA

CLASS HYDROZOA. Hydrozoans.
CLASS SCYPHOZOA. Jellyfishes.
CLASS ANTHOZOA. Sea anemones and corals.

PHYLUM CTENOPHORA. Comb jellies.

PHYLUM PLATYHELMINTHES. Flatworms.

CLASS TURBELLARIA. Free-living flatworms.
CLASS TREMATODA. Flukes.
CLASS CESTODA. Tapeworms.

PHYLUM NEMATODA. Round worms.

PHYLUM ANNNELIDA. Segmented worms.
CLASS POLYCHAETA. Marine worms, tubeworms.
CLASS OLIGOCHAETA. Earthworms and many fresh-water annelids.
CLASS HIRUDINEA. Leeches.

PHYLUM TENTACULATA
CLASS BRYOZOA. Sea mats
CLASS BRACHIOPODA. Lamp shells.

PHYLUM MOLLUSCA. Molluscs.
CLASS AMPHINEURA. Chitons.
CLASS GASTROPODA. Snails, Sea slugs.
CLASS BIVALVIA. Bilvalve molluscs, clams, mussels.
CLASS CEPHALOPODA. Squids, octopuses, and *Nautilus.*

PHYLUM ARTHROPODA. Arthropods.
CLASS ARACHNIDA. Spiders, ticks, mites, scorpions.
CLASS CRUSTACEA. Crustaceans.
SUBCLASS ENTOMOSTRACA. Small crustaceans.
ORDER CLADOCERA. Water-fleas.
ORDER COPEPODA. Copepods.
ORDER CIRRIPEDIA. Barnacles.
SUBCLASS MALACOSTRACA. Large crustaceans.
ORDER AMPHIPODA. Amphipods.
ORDER ISOPODA. Isopods.
ORDER DECAPODA. Shrimps, lobster, crayfish, crabs.
CLASS CHILOPODA. Centipedes.
CLASS DIPLOPODA. Millipedes.
CLASS INSECTA. Insects.
ORDER COLLEMBOLA. Springtails.
ORDER ORTHOPTERA. Grasshoppers.
ORDER BLATTAEFORMIA. Termites.
ORDER LEPIDOPTERA. Butterflies.
ORDER COLEOPTERA. Beetles.
ORDER DIPTERA. Flies. Mosquitoes, midges, warble flies.
ORDER HEMIPTERA. Bugs.

ORDER APHANIPTERA. Fleas.
ORDER HYMENOPTERA. Wasps, ants, bees.

PHYLUM ECHINODERMATA. Echinoderms.
CLASS CRINOIDEA. Sea lilies.
CLASS ASTEROIDEA. Sea stars
CLASS OPHIUROIDEA. Brittle stars.
CLASS ECHINOIDEA. Sea urchins and sand dollars
CLASS HOLOTHUROIDEA. Sea cucumbers.

PHYLUM CHORDATA. Chordates.
SUBPHYLUM TUNICATA. Tunicates.
SUBPHYLUM CEPHALOCHORDATA. *Amphioxus.*
SUBPHYLUM VERTEBRATA. Vertebrates.
CLASS AGNATHA. Jawless fishers.
SUBCLASS CYCLOSTOMATA. Lampreys and hagfish.
CLASS CHONDRICHTHYES. Cartilaginous fishes (sharks, skates, rays).
CLASS OSTEICHTHYES. Bony fishes.
CLASS AMPHIBIA. Amphibians.
CLASS REPTILIA. Reptiles.
CLASS AVES. Birds.
CLASS MAMMALIA. Mammals.

Index

C

D

E

T

U

V

W

Z